Advances in Conservation Laws and Energy Release Rates

Advances in Conservation Laws and Energy Release Rates

Theoretical Treatments and Applications

by

YI-HENG CHEN

Dean and Professor,
School of Engineering and Mechanics,
Xi'an Jiaotong University,
Xi'an, P.R. China

KLUWER ACADEMIC PUBLISHERS
DORDRECHT / BOSTON / LONDON

A C.I.P. Catalogue record for this book is available from the Library of Congress.

ISBN 1-4020-0500-8

Published by Kluwer Academic Publishers,
P.O. Box 17, 3300 AA Dordrecht, The Netherlands.

Sold and distributed in North, Central and South America
by Kluwer Academic Publishers,
101 Philip Drive, Norwell, MA 02061, U.S.A.

In all other countries, sold and distributed
by Kluwer Academic Publishers,
P.O. Box 322, 3300 AH Dordrecht, The Netherlands.

Printed on acid-free paper

Printed in the Netherlands

To my father–Professor Shu-Tao Chen who died in 1968 at 55 years old.

To my tutor–Professor Zhao-Qian Tang who died in 1983 at 52 years old.

To thousands of Chinese former researchers who had sacrificed their lives for Natural Science.

Contents

List of Figures

List of Tables

Preface

This book deals with some recent advances in conservation laws and energy release rates. Unlike their conventional applications in Fracture Mechanics for single crack problems, the contents of this book are mainly concerned with the conservation laws and energy release rates in some relatively complex problems, *i.e.,* strongly interacting cracks, microcrack damage, and microcrack shielding problems in five kinds of materials, respectively. The five kinds of materials involve four traditional structural materials: brittle solids, metal/ceramic bimaterials, anisotropic elastic solids, and dissimilar anisotropic solids; and one functional material: piezoelectric ceramics. Although this book starts from the original concepts of the J_k vector, the M integral, and the L integral in single crack problems proposed by Eshelby (1956, 1970, 1975), Rice (1968a,b), Knowles and Sternberg (1972), Budiansky and Rice (1973), Berges (1974), Freund (1978), Cherepanov (1974, 1979), and Herrman and Herrman (1981), *etc.*, the extension from investigations of single crack problems to those of multiple crack interacting problems or microcrack damage problems is based on the author's (and/or co-workers) works published in recent years. From detailed manipulations and numerical examinations the author would like to show readers some new insights of the conservation laws and energy release rates. Readers will see that although these 'old' concepts were established more than 30 or 40 years ago, they play a quite important role in multiple crack interacting problems or microcrack damage problems of both structural and functional materials. These new insights are quite different from the traditional understandings of these concepts in Fracture Mechanics, which exhibit some new application possibilities in practice. The reason that their role and their new insights have not received much attention so far is just because almost all previous investigations for studying these 'old' concepts are limited to single crack problems. Therefore the purpose of this book can be considered as to supply the lack of the investigations in multiple crack interacting problems or microcrack damage problems. The major contents of this book may be briefly summarized as follows:

Chapter 1 attempts to circumscribe the essential concepts of the J_k integral vector, the M integral, and the L integral in plane elasticity briefly and the conventional conservation laws established by Eshelby (1956, 1970, 1975), Budiansky and Rice (1973), Freund (1978), Cherepanov (1974, 1979), and Herrman and Herrman (1981). In particular, the author reveals that there exist new conservation laws of the J_k vector for an array of arbitrarily oriented and randomly located cracks in strongly interacting situations. That is, the total contribution arising from the array of cracks and other discontinuities to the two components of the J_k vector vanishes, provided that there are no other discontinuities outside the closed contour chosen for calculating the vector. Detailed manipulations and numerical examinations are given. Readers will at once know that the extension of these four integrals mentioned above from single crack problems to multiple crack problems is not so straightforward. They will also see that this extension should associated with some illuminating works done by Herrmann and Herrmann (1981) and their serious contributions in understanding what role does the J_2 integral play. Of great interest is that these new laws are not covered by the conventional concepts of the conservation laws in Fracture Mechanics.

Chapter 2 attempts to present the applications of the new conservation laws of the J_k vector, from which the M integral and the L integral analyses in treating strongly interacting cracks in brittle solids and in the brittle phase of metal/ceramic bimaterials are presented. Readers may be interested in what role does the M integral play in describing damage level of a plane microcracking solid. Detailed manipulations and numerical examinations are given to clarify the role and some confusion on the dependence of the M integral from the global coordinate shifts. The author verifies that the M integral is equivalent to the decrease of the total potential energy owing to microcracking. Unlike those widely accepted, the M integral and the L integral are proved to be not two distinct integrals, quite contrary, they are closely related to each other by a very simple formulation. As a direct extension, how to treat a finite microcracking solid is discussed in some detail. It is found that the outside boundary of a finite solid should be considered as a special kind of discontinuity, *i.e.,* a special kind of interface between air and a solid or between a rigid body and a solid. Only in this way could the conservation laws of the J_k vector, the M integral, and the L integral be established in the finite microcrack solid, otherwise the M integral and the L integral will be dependent on the global coordinate shifts.

Chapter 3 deals with the microcrack shielding problem in brittle solids. Detailed manipulations and numerical examinations reveal that there exist a conservation law of the J integral for a discrete model in microcrack shielding problems. That is, there is a redistribution of the remote J integral when it transmits across the microcracking process-zone from infinity to the macrocrack tip. This redistribution is associated with the projected values of the two components of the J_k vector on the macrocrack line. So it could be called the projected conservation law of the J_k vector in shielding problems. An application of this law is presented, which directly leads to a consistency check for numerical examinations no matter how the results are derived by whatever kind of technique. However, it is emphasized that this projected conservation law falls in the conventional concept of the Energy Momentum Tensor (Eshelby, 1970, 1975).

Chapter 4 deals with an interaction problem between interface crack and near-tip sub-interface microcracks in metal/ceramic bimaterials. Detailed manipulations and numerical examinations reveal that the projected conservation law of the J_k vector are still valid, although the mismatch property between metal phase and ceramic phase yields the well known characteristic oscillating stress singularity near the interface crack tip. Numerical results are given for Cu/Al_2O_3 and Ni/MgO bimaterials. The role that the first Dundurs parameter plays is discussed.

Chapter 5 deals with an interaction problem between interface crack and near-tip sub-interface microcracks in two dissimilar anisotropic materials. The projected conservation law established in Chapters 3 and 4 is proved to be valid too, although the mismatch property between o ply and 90 ply of 0/90 fiber reinforced composites is quite different from that between metal and ceramic.

Chapters 6 and 7 present a direct extension of the projected conservation law from structural materials mentioned in above chapters to piezoelectric ceramics. Based on the pseudo-traction method or superimposing technique proposed by Chen (1984) and Horii and Nemat Nasser (1985,1987) in brittle solids, a pseudo-traction electric displacement method (Chen and Han, 1999a,b) is presented to treat multiple crack interacting problems or macrocrack microcrack interaction problems in piezoelectric materials. Based on the work done by Pak (1990) and Suo *et al.,* the second component of the J_k vector accounting for both mechanical and electric quantities is established and studied in some detail, although the mechanical quantities and electric quantities are coupled and the electric displacement and electric potential add complexities in the original formulation by Budiansy and Rice (1973). It is proved that the projected conservation law discussed in above chapters is still valid in mechanical electric coupled problems. The Mechanical Strain Energy Release Rates *MSERR* proposed by Park and Sun (1995a,b) are discussed in some detail for macrocrack microcrack interaction problems in piezoelectric materials. What role does the electric loading play is clarified and then its influence on crack tip stress intensity factors *SIF's*, electric displacement intensity factor *EDIF*, and *MSERR* is discussed in multiple crack interacting situations.

Chapter 8 presents some other developments. In order to establish the conservation laws in functional materials two essential ways are discussed in some detail. The first is based on the The Energy Momentum Tensor (Eshelby, 1970, 1975). As typical examples, the manipulations done by Pak (1990) in piezoelectric materials and Wang and Shen (1996) in linear electro magneto elastic media are discussed. The second is based on Betti's reciprocal theorem (Sokolnikoff, 1956) and the induced Work Conjugate Integral (Bueckner, 1973). As an example, the work done by Ma and Chen (2001a,b) is discussed. According to the work of Stroh (1958) and Ting (1982, 1986, 1990) the existence of the Pseudo-Orthogonal Property of the Williams eigenfunction expansion form (Chen, 1985) is proved for a semi-infinite crack in a plane piezoelectric material. Although mechanical quantities and electric quantities are coupled, this property yields such a conclusion that the J integral and the M integral are merely two special cases of the Bueckner integral. Some useful weight functions are proposed for distinctly calculating crack tip *SIF's* and *EDIF*. Finally, based on the work done by Gao (1994) an application of the J_k vector in nanomechanics with defects (singularities) is discussed. It is seen that there is an important property of surface evolution with a heteroepitaxial thin film concerned with the conservation laws about the J integral customarily adopted in macro-continuum mechanics. Whereas the J_2 plays an important role in roughness thin film problems.

It is found that, although the conservation laws and energy release rates are relatively 'old' concepts, they still play an important role in studying defect mechanics in functional materials and nanomaterials in the new century. Subsequent investigations remain adequately treated especially in applications of the so called 'Quasicontinuum method' (Ortiz, 1999). The author hopes that this book will be useful for researchers, engineers, college teachers, and graduate students, who are majoring in Fracture Mechanics and Damage Mechanics.

Dr. YI-HENG CHEN
Dean and Professor
School of Civil Engineering and Mechanics
Xi'an Jiao-Tong University, 710049
P.R. China

Acknowledgments

The author is grateful for the partial supports provided by the Kyushu University of JAPAN in 1999, the German DAAD foundation in 1999, and the K.C. Wang foundation of the Royal Society of UK in 2000.

The principal parts of this book were written during the author's research stay in Kyushu of JAPAN, in Magdeburg of Germany, in Cambridge of UK. The author is greatly indebted to Professor Y. Murakami of Kyushu University, Professor U. Gabbert of Otto-von-Geurick-university-Magdeburg, and Dr. T.J. Lu of University of Cambridge for their encouragement and constructive discussions.

The National Science Foundation of China is greatly appreciated too.

The preparation of the manuscript was skillfully and faithfully carried out when the author takes his research stay in the Max-Planck Institute for Metals Research at Stuttgart of Germany in 2001. Professor H. Gao and Dr. Yong Kong is gratefully acknowledged for assistance.

F inally, the author would like to express his thanks to the publisher, and particular to Drs. Arno Schouwenberg for excellent cooperation.

Chapter 1

HISTORICAL REVIEW AND FUNDAMENTALS OF CONSERVATION LAWS AND ENERGY RELEASE RATES

Within the framework of plane and linear elastic fracture mechanics (*LEFM*), many different kinds of approaches have been proposed for the determination of critical parameters associated with crack extension. Amongst these, path independent integrals (sometimes called invariant integrals) are extremely attractive. They need only the evaluation of contour integrals, which are available from numerical solutions for certain crack configurations and loading conditions without any special treatment for near-tip singular stress fields. This is the major reason why a number of investigators have focused their attention on analyses of path independent integrals in the thirty years since Rice reported his famous work (1968a,b).

As reviewed and discussed by Kanninen and Popelar (1985), although Eshelby (1956) was the first to derive this kind of integral, Rice (1968a,b) was apparently the first to recognize its potential use in fracture mechanics. Since then there have been a number of researchers who have focused their attention on this field. As widely accepted, the most famous invariant integrals are Rice's J integral, the M integral, and the L integral (Knowles and Stermberg, 1972; Budiansky and Rice, 1973). All these integrals are related to energy release rates (*ERR's*). For example, the J integral and the M integral are related to the *ERR* for crack extension (Eshelby, 1970, 1975) and uniform crack expansion (Budiansky and Rice, 1973), respectively, and the L integral corresponds to the *ERR* for crack rotation (Budiansky and Rice, 1973; Herrmann and Herrmann, 1981). However, the computations of the three integrals mentioned above cannot distinguish between the portion contributed by crack opening and that arising from shearing; this could be a disadvantage when studying mixed mode crack extension, *e.g.*, crack kinking or deflection. To overcome this shortcoming a number of other path independent integrals have been proposed. For example, Bergez (1974), Stern, *et al.* (1976), and Freund (1978) proposed some alternative methods which also lead to the computation of path independent contour integrals, but directly yield the separate values of the opening mode and shear mode stress intensity factors (*SIF*). Similarly, Bueckner (1973, 1989) proposed the famous weight function method, which is based on the Betti reciprocal theorem (Sokolnikoff, 1956) and the reduced Work Conjugate Integrals.

All the invariant integrals reported in the literature can be divided into two categories. Those in the first, including the J integral, the M integral, and the L integral, have a clear physical meaning as the *ERR*; those in the second, as proposed by Bergez (1974) *etc.*, are merely convenient computed quantities from which the crack tip parameters such as the *SIF's* and the T stress can be evaluated by using conventional finite element methods without the use of singular elements.

All the previous investigations are limited to single crack problems; the crack configurations and the formulations of the integrals are relatively simple. The goal of this book is to extend the analyses of conservation integrals and energy release rates to more complicated problems such as multiple crack interacting problems and microcrack damage problems. Owing to the strong interactions the analyses for multiple crack configurations are very complicated. It will be seen in the forthcoming chapters that these analyses are mainly based on the illuminating work of Herrmann and Herrmann (1981) as well as their series of investigations (Herrmann, 1981; 1982; King and Herrmann, 1981; Eischen and Herrmann, 1987). Some new features of the three kinds of path independent integrals , J, M, and L, and the relative *ERR's* in the crack interacting problems will be found. They will be discussed in detail for five kinds of materials: brittle materials, metal/ceramic bimaterials, anisotropic materials, dissimilar anisotropic materials, and piezoelectric materials. Before starting on these extensions we will address, in this chapter, the original definitions and basic concepts of the three kinds of path independent integrals, and provide an historical review of conservation laws and energy release rates presented by previous investigators over the past thirty years.

1. Definitions and basic formulations

1.1 Definitions of the J_k integral vector, M integral, and L integral

The J_k integral vector, the M integral, and the L integral proposed by Knowles and Sternberg (1972) and Budiansky and Rice (1973) are formulated in the following forms respectively

$$J_k = \oint_C (w n_k - u_{i,k} T_i) ds, \qquad (k = 1, 2), \tag{1.1}$$

$$M = \oint_C (w x_i n_i - T_k u_{k,i} x_i) ds, \tag{1.2}$$

$$L = e_{3ij} \oint_C (w x_j n_i - T_i u_j - T_l u_{l,i} x_j) ds, \tag{1.3}$$

where w is the strain energy density and T_k is the traction acting on the outside of a closed contour C, x_j with $j = 1, 2$ represents a rectangular plane coordinate system, n_i refers to the outside normal of the contour C, and u_i denotes the displacements, while e_{3ij} is the alternating tensor depending on the arrangement of the integer numbers i and j:

$$e_{3ij} = \begin{cases} -1 & \text{when} \quad i = 2 \quad \text{and} \quad j = 1, \\ 0 & \text{when} \quad i = j, \\ +1 & \text{when} \quad i = 1 \quad \text{and} \quad j = 2. \end{cases} \tag{1.4}$$

Equation (1.1) shows that J_1, J_2 are the two components of the vector J_k with $k = 1, 2$, respectively. This means that if the axes (x_1, x_2) are rotated through an angle $-\varphi *$, as in Figure 1.1, then

$$\begin{aligned} J_1 * &= J_1 \cos\varphi * - J_2 \sin\varphi *, \\ J_2 * &= J_1 \sin\varphi * + J_2 \cos\varphi *, \end{aligned} \tag{1.5}$$

where J_1*, J_2* are defined in the system (x_1*, x_2*) as shown in Figure 1.1.

Whereas the integrals J_1, J_2 are invariant under shift of the origin as shown in Figure 1.1:

$$\begin{aligned} x_{01} &= x_1 + \eta_1, \\ x_{02} &= x_2 + \eta_2, \end{aligned} \tag{1.6}$$

thus

$$\begin{aligned} J_{10} &= J_1, \\ J_{20} &= J_2, \end{aligned} \tag{1.7}$$

where J_{10}, J_{20} are defined in the system (x_{01}, x_{02}).

Since the M integral is an inner product of the J_k vector and the vector $\boldsymbol{x}_k = (x_1, x_2)$, it is a scalar and hence invariant under rotation:

$$M * (x_1*, x_2*) = M(x_1, x_2), \tag{1.8}$$

where the integrals $M * (x_1*, x_2*)$ and $M(x_1, x_2)$ are defined in (x_1*, x_2*) and (x_1, x_2), respectively.

It is not invariant under the shift of the origin; if relation (1.6) is used then

$$M_0(x_{01}, x_{02}) = M(x_1, x_2) + \eta_1 J_1 + \eta_2 J_2, \tag{1.9}$$

where the two additional terms in the right hand of equation (1.9) are an inner product of the J_k vector and the shift vector $\eta_k = (\eta_1, \eta_2)$.

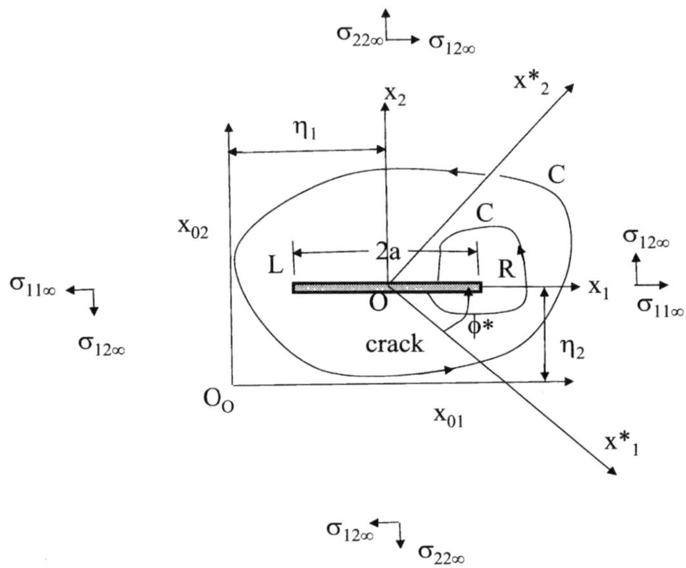

Figure 1.1. A plane crack in an infinite elastic body.

With the exception of the middle term $-e_{3ij}T_iu_j$ in equation (1.3), which is apparently independent from rotation, the L integral is also an inner product, and therefore a scalar:

$$L*(x_1*, x_2*) = L(x_1, x_2). \tag{1.10}$$

Under the shift of the origin with relation (1.6), as in Figure 1.1, it follows that

$$L_O(x_{01}, x_{02}) = L(x_1, x_2) + \eta_2 J_1 - \eta_1 J_2, \tag{1.11}$$

where the two additional terms in the right hand of equation (1.11) are the inner product of the J_k vector and the vector $(\eta_2, -\eta_1)$ orthogonal to the vector η_k with $k = 1, 2$.

Here we emphasize, as pointed out by Herrmann and Herrmann (1981), that the closed contour C introduced to define the J_k vector is, generally speaking, different from that used to define the M integral and the L integral. That for the J_k integral vector is a smooth closed contour surrounding either one tip of the crack or both tips of the crack (see Figure 1.1), but the latter, for the M and L integrals, should be chosen so that the whole crack with both tips is enclosed completely. In the forthcoming investigations, we always assume that the closed contour is defined in the latter way unless specially mentioned.

Here, for convenience in the forthcoming discussions, only cracks are taken into account. Other defects or discontinuities, such as voids or inclusions, do not alter the major features of the integrals; they can be treated in a similar way, although the details are different from those of cracks.

In Figure 1.1 the rectangular coordinate system has been chosen so that the x_1-axis and x_2-axis are parallel and perpendicular to the crack surfaces, respectively. In this case, it is well known that Rice's J integral (1968a,b) is the component, J_1, of the J_k integral vector parallel to the crack surfaces. Its relation to the crack tip *SIF's* is as follows (Freund, 1978)

$$J = J_1 = \frac{\kappa+1}{8\mu}[(K_{IR})^2 + (K_{IIR})^2] \tag{1.12}$$

(when the contour C encloses the right tip of the crack, see Figure 1.1),

$$J = J_1 = \frac{\kappa+1}{8\mu}[(K_{IR})^2 + (K_{IIR})^2 - (K_{IL})^2 - (K_{IIL})^2] \tag{1.13}$$

(when the closed contour C encloses the whole crack, see Figure 1.1).

Here μ is the shear modulus, $\kappa = (3-\nu)/(1+\nu)$ for generalized plane stress, and $\kappa = 3 - 4\nu$ for plane strain; the subscripts R and L denote the right tip and the left tip of the crack, respectively, and the subscripts I and II refer to the opening mode *SIF* and the sliding mode *SIF* in *LEFM*, respectively.

On the other hand, Freund (1978) has proved that the M integral is related to *SIF's* by:

$$\begin{aligned} M(x_1, x_2) &= M*(x_1*, x_2*) \\ &= \frac{\kappa+1}{8\mu}[(K_{IR})^2 + (K_{IIR})^2 + (K_{IL})^2 + (K_{IIL})^2]a, \end{aligned} \tag{1.14}$$

where a is the half length of the crack shown in Figure 1.1.

As regards the second component of the vector, J_2, that is perpendicular to the crack surfaces, Herrmann and Herrmann (1981) has shown that it has a quite different feature, *i.e.*, an unexpected path dependence owed to the contribution induced from the two traction-free surfaces. For a closed contour C surrounding the right tip of the crack only (see Figure 1.1), it is easy to note that $n_2 = -1$ and $n_2 = 1$ on the upper and lower surfaces of the crack, respectively. Therefore the formulation of J_2 calculated along C surrounding the right tip only, *i.e.*, equation (1.1) with $k = 2$, yields that (Herrmann and Herrmann, 1981)

$$J_2 = -\frac{\kappa+1}{4\mu}K_{IR}K_{IIR} + F_2(\text{surface}), \tag{1.15}$$

where F_2(surface) is contributed by the enclosed surfaces of the contour C, which is, of course, dependent on the selection of C, especially on the segments enclosed by C on the lower and upper surfaces, respectively, *i.e.*, the starting and ending points of C located on the lower and upper surfaces, respectively. This means that the second component of J_k, has apparent path dependence, which is mainly induced from the outside normal of the crack surfaces, *i.e.*, $n_2 = -1$ or $n_2 = 1$.

Herrmann and Herrmann (1981) was the first to recognize the different feature of J_2 from that of J_1. They added that only when taking the closed contour C enclosing the whole crack completely, is the second component path independent (see Figure 1.1). In this special case, the J_2 integral is related to the *SIF's* and the contribution from the traction-free surfaces by

$$J_2 = \frac{\kappa+1}{4\mu}[K_{IL}K_{IIL} - K_{IR}K_{IIR}] + F_{2a}, \tag{1.16}$$

where F_{2a} denotes the contribution arising from the whole traction-free surfaces, which can be put in the following form

$$F_{2a} = -\int_{-a}^{a}(w^+ - w^-)dx_1. \tag{1.17}$$

Here w^+ and w^- denote the boundary values of the strain energy density w on the upper and lower surfaces of the crack, respectively.

1.2 Path selections and conservation laws

According to Eshelby's Energy Momentum Tensor (1970, 1975), conservation laws in plane linear elasticity can be stated as follow:

If a closed contour C does not enclose any singularity, then the values of the four kinds of integral respectively defined by (1.1), (1.2), and (1.3) vanish.

Mathematically,

$$J_k = \oint_C (wn_k - u_{i,k}T_i)ds = 0, (k = 1, 2), \tag{1.18}$$

$$M = \oint_C (wx_in_i - T_ku_{k,i}x_i)ds = 0, \tag{1.19}$$

$$L = e_{3ij} \oint_C (wx_j n_i - T_i u_j - T_l u_{l,i} x_j) ds = 0. \tag{1.20}$$

However, when the closed contour C encloses a crack or several cracks, Eshelby's Energy Momentum Tensor (1970, 1975) has not indicated whether or not the values of the four integrals vanish. We need to discuss this problem in more detail.

Let us consider three selections of the closed contour enclosing a single crack completely as shown in Figure 1.2 (Herrmann and Herrmann, 1981). Here C_0 is infinitely large, C_D encloses the right crack tip only, and $C = C_R + C_+ + C_L + C_-$ consists of two infinite small circles surrounding the right and left tips, respectively, and two straight segments on the upper and lower surfaces of the crack, respectively. At infinity uniform loads $\sigma_{22}^{\infty}, \sigma_{12}^{\infty}$ are preferred. Since the plane region between C_0 and C contains no singularity, it is obvious that $J_k(C_0) = J_k(C)$, $M(C_0) = M(C)$, and $L(C_0) = L(C)$ owing to the conservation laws (1.18), (1.19),and (1.20).

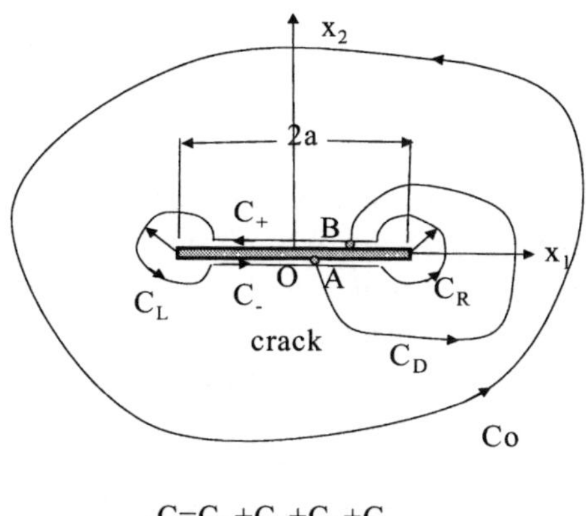

Figure 1.2. Three different path selections.

Recall the well known formulations of the stresses and displacements near the right crack tip, which can be put in the following forms:

$$\begin{aligned} \sigma_{ij}^m &= K_m (2\pi r)^{-1//2} f_{ij}^m(\theta), \\ u_i^m &= (K_m/2E)(r/2\pi)^{1/2} g_i^m(\theta), (i,j = 1,2; m = I, II), \end{aligned} \tag{1.21}$$

and

$$K_{IR} = K_{IL} = \sqrt{\pi a}\sigma_{22}^{\infty}, \tag{1.22}$$

$$K_{IIR} = K_{IIL} = \sqrt{\pi a}\sigma_{12}^{\infty}, \tag{1.23}$$

where $f_{22}^{I}, f_{11}^{I}, f_{21}^{II}, g_1^{I}, g_2^{II}$ are even functions of θ, whilst f_{21}^{I}, f_{11}^{II}, f_{22}^{II}, g_2^{I}, g_1^{II} are odd functions of θ, a is the half length of the crack.

Since the *SIF's* at the right tip are identical to those at the left tip and since the whole traction-free surfaces contribute nothing to J integral, equation (1.13) directly yields that the value of the J integral calculated over C or C_0 vanishes. However, it is not clear whether the J_2 integral vanishes owing to the existence of the second term in equation (1.16), although the first term in equation (1.16) does vanish.

As regards the closed contour C_D enclosing the right tip of the crack in Figure 1.2, we know that in plane stress condition:

$$J_2(C_D) = \int_{x_A}^{a} w^- dx_1 - \int_{x_B}^{a} w^+ dx_1 - 2K_{IR}K_{IIR}(1-\nu^2)/E \quad , \tag{1.24}$$

which is obviously dependent on the locations of the starting and ending points x_A and x_B. Here E and ν are Young's modulus and Poisson's ratio, respectively. For a plane strain condition E and ν should be replaced by $E' = E/(1-\nu^2)$ and $\nu' = \nu/(1-\nu)$, respectively

Indeed, as mentioned earlier and emphasized by Herrmann and Herrmann (1981), the second component J_2 over C_D does yield the path dependent nature because different starting and ending points A and B located respectively on the lower surface and the upper surface will lead to different contributions to J_2 in the first two terms of (1.24). In other words, a calculation of the contribution arising from the two segments enclosed by the path C_D to J_2 is needed from which whether the conservation law of J_2 holds can be clarified (see Figure 1.2).

Herrmann and Herrmann (1981) was the first to note this point. They found that in plane stress conditions, the strain energy density is related to the stress components σ_{ij} $(i, j = 1, 2)$ by

$$w = \frac{1}{2E}(\sigma_{11}^2 + \sigma_{22}^2 - 2\nu\sigma_{11}\sigma_{22}) + \frac{(1+\nu)}{E}\sigma_{12}^2. \tag{1.25}$$

For the right hand circle of the closed path C shown in Figure 1.2, the coordinate relations between (x_1, x_2) and (r, θ) yield that

$$\begin{aligned} x_1 &= a + r\cos\theta, \\ dx_1 &= -r\sin\theta d\theta, \\ x_2 &= r\sin\theta, \\ dx_2 &= r\cos\theta d\theta, \\ n_1 &= \cos\theta, \\ n_2 &= \sin\theta, \\ ds &= rd\theta, \end{aligned} \tag{1.26}$$

and

$$\begin{aligned} u_{i,1} = u_{i,r}\cos\theta - \tfrac{1}{r}u_{i,\theta}\sin\theta, \\ u_{i,2} = u_{i,r}\sin\theta - \tfrac{1}{r}u_{i,\theta}\cos\theta. \end{aligned} \tag{1.27}$$

The stress field and displacement field near the right tip can be considered to be two distinct parts induced from the Mode I fracture and Mode II fracture, respectively (Herrmann and Herrmann, 1981; Kanninen and popelar, 1985). Thus

$$\begin{aligned} \sigma_{ij} = \sigma_{ij}^{I} + \sigma_{ij}^{II}, \\ u_i = u_i^{I} + u_i^{II}. \end{aligned} \tag{1.28}$$

By noting the odd or even property of all the functions involved and performing some integrations from $-\pi$ to π over the small circle with radius r surrounding the right tip, we can easily derive equations (1.12), (1.13), (1.14), and (1.16).

Moreover, Herrmann and Herrmann (1981) found that the boundary values of the strain energy density w arising from the upper and lower traction-free surfaces of the crack are only related to the normal stress parallel to crack surfaces:

$$\begin{aligned} w^{+} = \tfrac{1-\nu^2}{2E}(\sigma_{11}^{+}(x_1))^2, \\ w^{-} = \tfrac{1-\nu^2}{2E}(\sigma_{11}^{-}(x_1))^2, \end{aligned} \tag{1.29}$$

in which $\sigma_{11}^{+}(x_1), \sigma_{11}^{-}(x_1)$ are the boundary values of the normal stresses parallel to the upper and lower crack surfaces, respectively. It is then concluded that

$$F_{2a} = \int_{C+} wn_2 ds - \int_{C-} wn_2 ds = \int_{-a}^{a} (w^- - w^+) dx_1 = -\int_{-a}^{a} [w] dx_1. \quad (1.30)$$

Here [ ] denotes the jump of the corresponding quantity across the crack. Whereas $[w] = w^+ - w^-$ is found to be, by means of complex representation (Herrmann and Herrmann, 1981) and equation (1.25),

$$[w] = -4\sigma_{12}^{\infty}(\sigma_{11}^{\infty} - \sigma_{22}^{\infty})x_1(a^2 - x_1^2)(1 - \nu^2)/E. \quad (1.31)$$

Since $[w]$ is an odd function of x_1, Herrmann and Herrmann (1981) concluded that

$$F_{2a} = -\int_{-a}^{a} [w] dx_1 = 0. \quad (1.32)$$

Consequently equations (1.13) and (1.16) together with (1.32) imply two new conservation laws when the closed contour encloses a single crack completely,

$$J_1(C_0) = J_1(C) = 0, \quad (1.33)$$

$$J_2(C_0) = J_2(C) = 0. \quad (1.34)$$

Here the only assumption is that the remote loads should be uniform.

What's more, by the definition of a plane vector (1.1) and (1.5), the two components of J_k should be automatically zero in any other coordinate systems, say, (x_1^*, x_2^*).

Keeping equations (1.33) and (1.34) in mind and turning back to consider equations (1.9) and (1.11), we conclude that the additional terms in the right hands of (1.9) and (1.11) arising from the origin shift vanish. This means that the values of the M integral and L integral over C or C_0 in Figure 1.1 are independent from the origin selection of coordinates.

Readers may raise an interesting and fascinating question as to whether the conservation laws (1.33) and (1.34) hold in multiple crack interacting problems. If so, then the original definitions of the four integrals (1.1), (1.2), and (1.3) established by Knowles and Sternberg (1972) and Budiansky and Rice (1973) can be extended to treat crack interacting problems or, more significantly, microcrack damage problems. In other words there should exist some new meanings and representations of energy release rates under strongly interacting crack situations. If not so, what happens about the four integrals under strongly interacting crack situations? In other words, unlike those in *LEFM*, do the four integrals under the situations lead to some useless results?

?

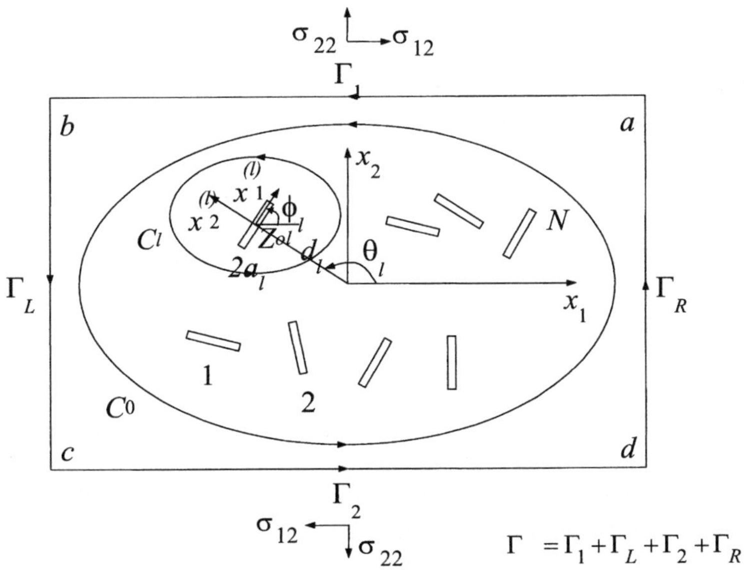

Figure 1.3. N interacting crack in an infinite plane elastic solid.

Chen and Hasebe (1998) and Chen (2001a) studied this question in some detail. They started their investigation from a typical problem in an infinite plane elastic solid with N strongly interacting cracks, as shown in Figure 1.3. At infinity uniform loads $\sigma_{12}^{\infty}, \sigma_{22}^{\infty}$ are preferred.

In order to verify the existence of the conservation laws of the J_k vector in the strongly interacting situation shown in Figure 1.3, we introduce a closed contour $\Gamma_{\infty} = \Gamma_1 + \Gamma_L + \Gamma_2 + \Gamma_R$ surrounding all the cracks and introduce a smaller closed contours C_l only surrounding the lth crack completely. According to the path independent nature of the J_k integral vector, the values of the vector calculated along Γ_{∞} in Figure 1.3 should be equal to the summation of the values induced from each crack:

$$J_1^{\infty} = J^{\infty} = \sum_{l=1}^{N} J_1^{(l)}, \tag{1.35}$$

$$J_2^{\infty} = \sum_{l=1}^{N} J_2^{(l)}, \tag{1.36}$$

where all the quantities are defined in the global coordinate system (x_1, x_2), the left terms are calculated over Γ_∞, whilst the right terms are calculated over C_l with *l=1, 2,...., N,* respectively.

Since every term in the summation on the right hand side of equation (1.35) or (1.36), generally speaking, is not equal to zero owing to interaction, it is unclear whether or not the summation of N terms on the right hand side of (1.35) or (1.36) vanishes. Indeed, as many people suspect, unlike the single crack situation the values of the stress intensity factors at the two tips of a typical crack l in Figure 1.3 are no longer the same in magnitude, owing to the interaction between the crack and its neighbors. We should focus our attention on this key problem to provide a detailed mathematical proof, from which our doubt can be clarified.

We can do this task by accounting the remote uniform stress strain field as well as the remote displacement field. Readers should remember that the closed rectangular contour Γ_∞ has been chosen as large as possible. The reason for introducing the large rectangular closed contour is to significantly simplify mathematical manipulations given below. If it is not so, for example, an arbitrary smooth closed contour C^O , whether it is large or not, see Figure 1.3, is chosen instead of the large rectangular contour, the results calculated overΓ_∞ and over C^O will be the same owing to the conservation law of the vector (1.18) applied in the plane region between Γ_∞ and C^O (This plane region encloses no singularity). It should also be mentioned that the second component of the vector is path independent, too, in the present case because the closed contours are always chosen either to surround all the cracks or surround a typical crack completely (Herrmann and Herrmann, 1981). Although the whole traction-free surface of each crack or the stress parallel to the surface leads to some nonzero contribution to the second component, the path independent nature has not been altered when the closed contour is chosen in the present way (Herrmann and Herrmann, 1981). As regards the array of the N cracks shown in Figure 1.3, generally speaking, the array is neither symmetrical with respect to the x-axis (x_1-axis), nor to the y-axis (x_2-axis).

Since *dy=0* in $\Gamma_1$and Γ_2, and *dx=0* in Γ_L and Γ_R, the left hand side of equation (1.35) becomes

$$J_1^{\infty} = \int_{\Gamma 1+\Gamma 2} (-\sigma_{i2}^{\infty} n_2 \frac{\partial u_i}{\partial x}) ds + \int_{\Gamma L+\Gamma R} (W dy)$$

$$+ \int_{\Gamma L+\Gamma R} (-\sigma_{i1}^{\infty} n_1 \frac{\partial u_i}{\partial x}) ds, \tag{1.37}$$

where $x = x_1$and $y = x_2$.

Noting that $n_2 = 1$ on Γ_1 and $n_2 = -1$ on Γ_2 and $n_1 = 1$ on Γ_R and $n_1 = -1$ on Γ_L, we find that

$$J_1^\infty = \sigma_{i2}^\infty \left[\int_a^b u_{i,1} dx - \int_d^c u_{i,1} dx \right] + \frac{1}{2} \sigma_{ij}^\infty \varepsilon_{ij}^\infty \left[\int_b^c dy - \int_a^d dy \right]]$$

$$+ \sigma_{i1}^\infty \left[\int_a^d u_{i,1} dy - \int_b^c u_{i,1} dy \right] , \quad (1.38)$$

Obviously the second term in equation (1.38) vanishes because $\left[\int_b^c dy - \int_a^d dy \right]$ = 0, whilst the first term depends on the value $\left[\int_a^b u_{i,1} dx - \int_d^c u_{i,1} dx \right]$ and the third term depends on the value $\left[\int_a^d u_{i,1} dy - \int_b^c u_{i,1} dy \right]$, which do confuse a number of people. They believe that equation (1.38) may not vanish for the unsymmetrical array of cracks in Figure 1.3. They may argue that, for example, in the first term of equation (1.38), the value of $\left[\int_a^b u_{i,1} dx - \int_d^c u_{i,1} dx \right]$ does not vanish since the array of the cracks shown in Figure 1.3 is, generally speaking, unsymmetrical to the x-axis and in turn the values of $u_{i,1}$ on the segment $\overline{ab}$ and those on $\overline{dc}$ will be different. Similarly, they may argue that in the third term of equation (1.38) the value of $\left[\int_a^d u_{i,1} dy - \int_b^c u_{i,1} dy \right]$ does not vanish either, since the array of the cracks is unsymmetrical to the y-axis. Indeed, attention should be focused on this major confusion, which is concerned with such a question as how to treat the remote stress strain field correctly when performing the integral (1.38). In fact, the reason that they are confused is the mistaken use of the remote conditions. As the rectangular closed contour shown in Figure 1.3 has been taken to be sufficiently large, this means that the asymptotic nature of the displacement u_i should be adopted when both segments $\overline{ab}$ and $\overline{dc}$ (or segments $\overline{ad}$ and $\overline{bc}$) are located far from the local microcrack damage zone near the coordinate origin. Because the remote stresses or strains are uniform, the partial derivations of the remote displacements with respect to x or y should be uniform too. Therefore, the asymptotic nature for very large values of $r = \sqrt{x^2 + y^2}$ shows that the partial derivation of the displacement with respect to the x-axis should be constant, *i.e.*, $u_{i,1}(\overline{ab}) = u_{i,1}(\overline{dc})$ =constant, such that the value of

$\left[\int_a^b u_{i,1}dx - \int_d^c u_{i,1}dx\right]$ should be zero without doubt. Similar discussions could be given for the third term $\left[\int_a^d u_{i,1}dy - \int_b^c u_{i,1}dy\right]$ on the right hand side of (1.38). It should be zero, too, without doubt owing to the remote asymptotic nature of $u_{i,1}$=constant. Consequently, all the three terms on the right hand side of equation (1.38) do vanish when taking the rectangular closed contour shown in Figure 1.3 as sufficiently large. Moreover, according to the conservation law of the J_1 integral (1.18) the right hand side of equation (1.35) should vanish, too, when taking another closed contour C^o surrounding all the cracks, whether it is large or not (see Figure 1.3).

In fact, the abovementioned confusion could be clarified in an alternative way. Assume that the original problem shown in Figure 1.3 is divided into two subproblems as shown in Figures 1.4(a) and 1.4(b), respectively. The first one shown in Figure 1.4(a) does not involve cracks, and the physical quantity $u_{i,1}$ should then be constant in the whole plane. Whereas the second subproblem shown in Figure 1.4(b) involves N cracks, each of which is loaded by the self-balance tractions $\sigma_l, \tau_l \quad (l = 1, 2, ..., N)$ acting on both upper and lower surfaces. However, the second subproblem yields no remote displacement field at infinity because the stresses, as well as the strains, induced from the self-balance tractions have an asymptotic nature of the order of r^{-2} (for large values of $r = \sqrt{x^2 + y^2}$). Thus the remote displacement field in Figure 1.3 is only governed by the first subproblem shown in Figure 1.4(a), which could certainly lead to the vanishing nature of J_1^∞ formulated by equation (1.35) owing to $u_{i,1}($ on $\Gamma_1) = u_{i,1}($ on $\Gamma_2)$ = constant and $u_{i,1}($ on $\Gamma_R) = u_{i,1}($ on $\Gamma_L)$ = constant for large values of $r = \sqrt{x^2 + y^2}$. Indeed, this does also mean that the first term and the third term on the right hand side of equation (1.38) vanish, no matter how many cracks are formed in the local finite region of the infinite solid and whatever the array of the cracks is symmetrical with respect to *x=0* (or *y=0*) or not. According to the conservation law of the J_1 integral (1.18) the value of (1.35) for the large rectangular closed contour should be the same as those for a smaller one C^o surrounding all the cracks although in the latter case the value of $u_{i,1}$ on C^o is no longer constant (Chen, 2001a).

As regards the J_2 integral, similar mathematical manipulations could be performed in a way similar to those for J_1. Readers who are interested in the topic can repeat the details without any difficulty.

Obviously the above discussions reveal that there really do exist new conservation laws about the J_k vector with strongly interacting cracks in an infinite

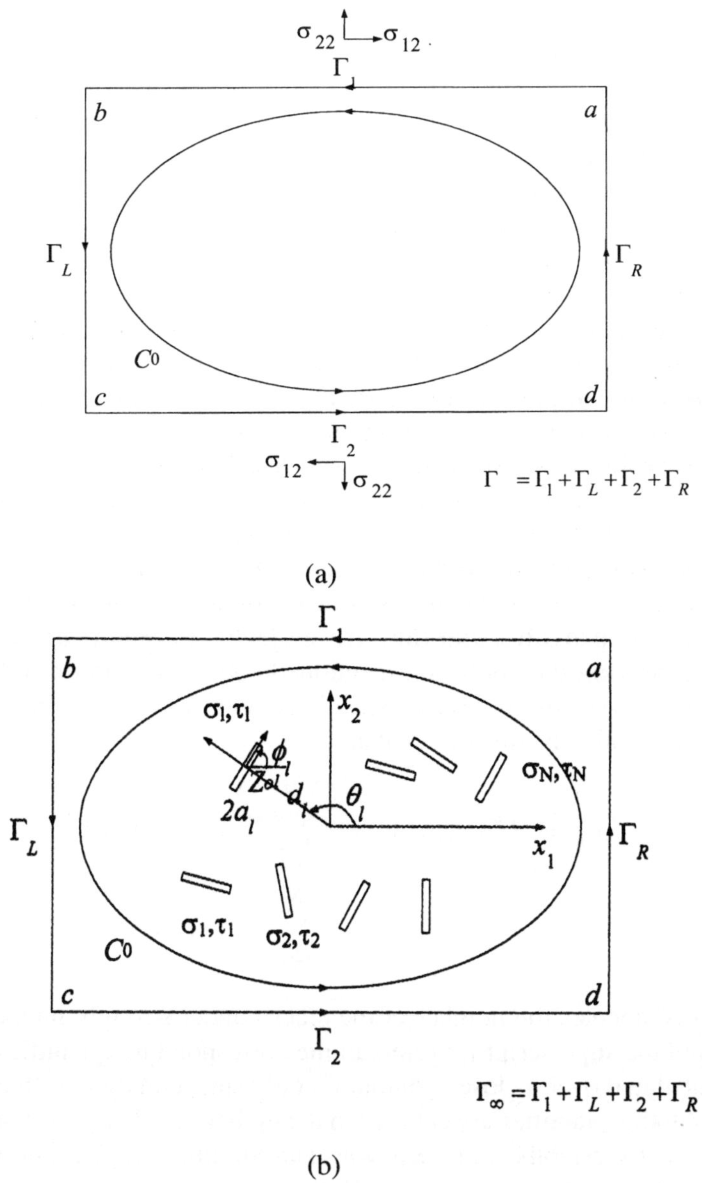

Figure 1.4. Subdivision of the original problem of Figure 1.3 to two subproblems.

plane elastic solid. These laws are not covered by the original understandings and applications of Rice's J integral (1968a,b) and Eshelby's Energy Momentum Tensor Theorem (1970, 1975), since the previous investigations were always limited to single crack problems and since the previous conservation laws were always concerned with the simplest cases that there are no singularities enclosed by the selected contours. Chen and Hasebe (1998) were apparently the first to recognize these new laws with interacting cracks and Chen (2001a,b) was the first to provide a detailed mathematical proof of the laws. Consequently, we conclude that the J_k vector can really be extended to treat strongly interacting crack problems in an infinite $2D$ brittle solid.

The new conservation laws of the J_k vector considering strongly interacting crack problems can be stated as follows:

Both components of the J_k vector defined in a global coordinate system vanish under remote uniform leads when the closed contour chosen to calculate the two-components of the vector encloses all the cracks completely (or voids, inclusions, *etc.*,) in an infinite plane brittle solid or when there are no other discontinuities outside the closed contour.

Alternatively, the new conservation laws of the J_k vector considering strongly interacting crack problems can also be stated as:

The total contributions induced from the formation of all the cracks in an infinite elastic solid to the two components defined in the global coordinate system vanish, provided that they are subjected to remote uniform loads.

After clarifying the confusion mentioned above, we can extend equations (1.33) and (1.34) from single crack situations to strongly interacting crack problems. Mathematically, this means that

$$J_1^\infty = J^\infty(x_1, x_2) = \sum_{l=1}^{N} J_1^{(l)}(x_1, x_2) = 0, \tag{1.39}$$

$$J_2^\infty(x_1, x_2) = \sum_{l=1}^{N} J_2^{(l)}(x_1, x_2) = 0, \tag{1.40}$$

in which N denotes the number of the cracks under strongly interacting situations, and the superscript (l) refers to the corresponding quantities of the lth crack calculated over a closed contour C_l only surrounding the lth crack completely. All the quantities are defined in the global coordinate system (x_1, x_2). Here the vector established by Knowles and Sternberg (1972) and Budiansky and Rice (1973) is given in equation (1.1).

Considering the global local coordinate transitions in Figure 1.3 and the relation (1.5), we conclude that

$$J_1^\infty = J^\infty(x_1, x_2) = \sum_{l=1}^{N} [J_1^{(l)}(x_1^{(l)}, x_2^{(l)}) \cos\phi_l - J_2^{(l)}(x_1^{(l)}, x_2^{(l)}) \sin\phi_l] = 0, \tag{1.41}$$

$$J_2^\infty(x_1,x_2) = \sum_{l=1}^{N}[J_1^{(l)}(x_1^{(l)},x_2^{(l)})\sin\phi_l + J_2^{(l)}(x_1^{(l)},x_2^{(l)})\cos\phi_l] = 0, \tag{1.42}$$

in which ϕ_l is measured from the global x_1 axis to the local $x_1^{(l)}$ axis.

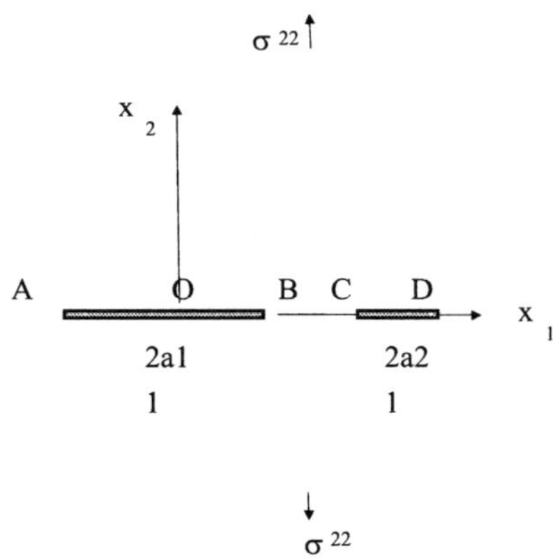

Figure 1.5. Two collinear interacting cracks.

As a direct application of the new laws (1.41) and (1.42) Chen and Hasebe (1998) applied them to be a consistency check for numerical results of multiple crack interacting problems. No matter how the results are derived by whatever kind of technique, any mistakes occurring in manipulations which may escape the traditional comparisons with previously known results in some special cases can then be recognized and avoid, since the mistakes will certainly lead to unsatisfactory conclusions contrary to the consistency check.

On the other hand, in order to confirm the present manipulations in more detail, a simple example is shown in Figure 1.5, in which two collinear cracks with different lengths (a_2=0.01a_1) are formed in an infinite plane elastic solid. The array of cracks is indeed not symmetrical with respect to $x = 0$ and the normalized *SIF's* at crack tips A, B, C, and D are quite different, which are equal to 1.002, 1.652, 1.469, and 0.655, respectively (Horii and Nemat Nasser, 1985, 1987). According to the doubt arising from some people, the value of J_1^∞ does not vanish owing to the nonzero value of the term $\left[\int_a^b u_{i,1}dx - \int_d^c u_{i,1}dx\right]$

in (1.38) induced from the unsymmetrical crack array. But, quite contrary, the normalized value of the integral $J_1^\infty = (K_{IB}^2 - K_{IA}^2) - (K_{ID}^2 - K_{IC}^2)$= $(1.652^2 - 1.002^2) + (0.655^2 - 1.469^2)$ =−0.004 is approximately equal to zero with a reasonable accuracy. This result does confirm the conservation law (1.39).

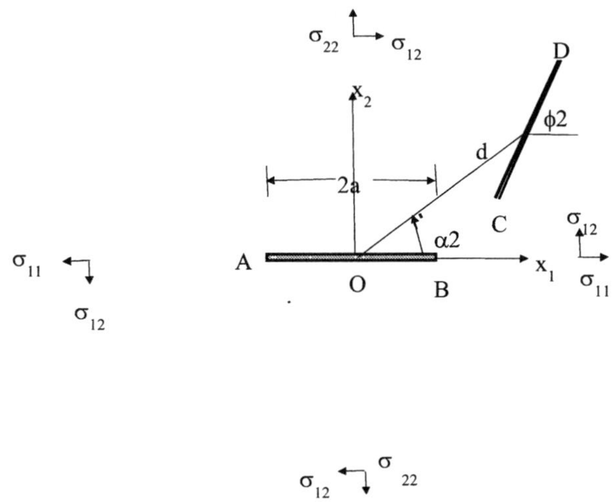

Figure 1.6. Two oriented cracks.

Moreover, a more complicated example is considered as shown in Figure 1.6, where two oriented cracks are made to interact with each other. One of them, crack 1, lies on the x_1-axis and its center coincides with the origin of the system (x_1, x_2). The other, crack 2, is located at a distance d with a located angle α_2 and it is oriented by an angle ϕ_2 with respect to the x_1-axis. The two cracks have the same length $2a$ and are subjected to remote uniform loads. This problem can be solved using the method proposed by Gross (1972), and Horii and Nemat Nasser (1987). Detailed process is no longer repeated here. Without going into the details, numerical results are listed in Table 1.1. It is seen that the contributions induced from crack 1 to both components of the vector do cancel with those induced from crack 2 although the array is neither symmetrical with respect to $x = 0$ nor to $y = 0$. It should be emphasized that the numerical results of the two components of the J_k vector in the global system (x_1, x_2) shown in Table 1.1 are related to those in the local system $(x_{1*}^{(l)}, x_{2*}^{(l)})$ $(l = 1, 2)$ by equation (1.5). Nevertheless, the second component in the local system is not only concerned with the *SIF's* at both tips of each crack, but also concerned with the non-uniformly distributed stress parallel to the traction free surfaces of the crack (Herrmann and Herrmann, 1981). Zhao and Chen (1997a, b) proposed a special numerical technique for calculating this kind of stress and its contribution to the second component. In Table 1.1 not only the *SIF's* at the tips of the two cracks but also the contribution of this kind of stress to the second component in the local system $(x_{1*}^{(l)}, x_{2*}^{(l)})$ are listed.

Table 1.1. Results of the J_k vector for two oriented cracks under remote uniform loading perpendicular to the x-axis (the first lies on the x-axis with tips A and B and the second with tips C and D is oriented by an angle φ_2 and a location angle α_2 with respect to the x-axis).

α_2(degrees)	0	30	60	90	15	45
φ_2(degrees)	30	60	45	75	75	60
K_I^A/K_I^∞	1.06644	1.02446	1.05011	1.00718	1.00392	1.04785
K_{II}^A/K_I^∞	−0.03632	0.01018	−0.02451	−0.01045	−0.01000	0.01474
K_I^B/K_I^∞	1.18405	1.09380	0.94884	0.97870	1.01272	1.08966
K_{II}^B/K_I^∞	−0.07413	0.03828	−0.07144	−0.00322	−0.01513	−0.03127
K_I^C/K_I^∞	0.96228	0.27555	0.57150	0.09196	0.15250	0.32666
K_{II}^C/K_I^∞	0.47719	0.52004	0.40760	0.12767	0.26399	0.61203
K_I^D/K_I^∞	0.82785	0.25963	0.56340	0.10117	0.06939	0.28885
K_I^D/K_I^∞	0.44085	0.50636	0.49304	0.20636	0.27046	0.51489
J_1^*(crack1)	0.26885	0.1482	−0.19793	−0.05664	0.01789	0.09012
J_2^*(crack1)	0.08233	−0.05953	0.10210	−0.01388	0.01072	0.10125
F_{2a}^*(crack1)	−0.01575	0.00335	0.01800	0.00087	0.00016	0.00222
J_1^*(crack2)	−0.27399	−0.02256	0.06776	0.02806	−0.01498	−0.13275
J_2^*(crack2)	0.06313	0.15814	−0.21216	−0.05112	0.01451	0.02742
F_{2a}^*(crack2)	−0.12532	0.13448	−0.12248	−0.03285	−0.02847	−0.07499
$\sum J_1$	0.00000	0.00000	0.00000	0.00000	0.00000	0.00000
$\sum J_2$	0.00000	0.00000	0.00000	0.00000	0.00000	0.00000

Here, for convenience, only cracks are taken into account. Readers can extend the main conclusions to other discontinuities, such as voids or inclusions, without any difficulty. Although the detailed contribution arising from each void or inclusion is different from a crack, the manipulations from (1.35) to (1.38) for other defects are treated as the same for the cracks when accounting for the remote uniform loads. Therefore the conservation laws (1.39) and (1.40) derived here are universal, which hold in linear elastic defect mechanics.

As known, equations (1.33) and (1.34) coincide well with a famous argument established by Budiansky and Rice (1973) for single defect problems. They argued that physically the two components of the J_k vector represent two special kinds of energy release rate (*ERR*) induced from the movements of the crack along the x_1-axis and the x_2-axis, respectively. Obviously, such movements do not yield any contribution to the *ERR* since the plane elastic body is infinitely large with no outside boundaries interacting with the single defect. Owing to these results, a number of investigators believe that the investigation of the J_k vector has come an end and that, from an engineering point of view, it has no significance and no longer needs further investigation.

However, we know that for problems of interacting multiple cracks, the situations for the two components of the J_k vector become quite different from those in single crack problems. The vector may be defined in a local system for a typical crack l or in a global system for all cracks. Generally, the contribution arising from each crack to the two components in the local or global system will no longer be zero owing to the non uniform loads arising from the interaction between its neighboring cracks. Of great interest is that the new conservation laws (1.39) and (1.40) still hold. In fact, these laws also coincide well with the argument established by Budiansky and Rice (1973) if we consider the multiple interacting cracks as **a single complicated defect**. Obviously, when the multiple cracks move simultaneously this movement does not yield any contribution to the *ERR*, since the plane elastic body is infinitely large with no outside boundaries interacting with the multiple cracks.

Therefore it seems important to reconsider the basic concepts of the J_k vector, the M integral, and the L integral in more detail. In particular, which roles they play in microcrack damage problems are needed to be studied. In the forthcoming chapters we will discuss the roles and derive some new features of the four integrals in microcrack damage problems.

2. Discussion of previous investigations for invariant integrals

Besides the work by Rice (1968a,b), Knowles and Sternberg (1972), and Budiansky and Rice (1973), there are a number of investigations about conservation laws or invariant integrals in the 1970s and the 1980s of the 20th century (see, *e.g.*, Begley and Landes, 1972; Blackburn, 1972; Rice, *et al.*, 1973; Hussain, *et al.*, 1973; Blackburn, *et al.*, 1977; Hutchinson, 1979; Elhaddad *et al.*, 1980; Castro, 1982; Ranaweera and Leckie, 1982; Batte, *et al.*, 1983). The goal of these attempts is to apply the J integral as a fracture criterion for predicting crack stability. This topic has been well summarized in the book written by Kanninen and Popelar (1985). It does not need to be repeated here. Other significant developments of invariant integrals, which will be associated with the forthcoming discussions in the following chapters, are discussed in this section.

Following the physical meaning of J, Hussain *et al.*, (1973) and Cherepanov (1974) made an attempt at supplying analogous physical interpretations for J_2 by postulating that the crack skews and the tip advances either in the direction of the vector J_k, or by claiming that if the crack advances at an angle ω with respect to its plane, the energy release should be

$$G(\omega) = J_1 con\omega + J_2 \sin\omega, \tag{1.43}$$

However, they did not note the contribution arising from the traction-free surfaces to the J_2 integral and the induced path dependence as Herrmann and Herrmann (1981) emphasized. For this reason Herrmann and Herrmann (1981) pointed out the correct conclusion that the L and M integrals provide a more natural description of energy release rates associated with plane cracks than the two components of the J_k vector since, if we consider a contour enclosing the whole crack, then both J_1 and J_2 vanish under the remote uniform loading, *i.e.*, (1.39) and (1.40); if we consider a contour enclosing only one crack tip, then J_2 is not path independent.

Indeed, during the 1980s of the 20th century Herrmann A.G. and his co-workers (Herrmann, 1981; Herrmann and Herrmann, 1981; King and Herrmann, 1981; Hermann, 1982; Eischen and Herrmann, 1987) made a significant advance and a series of investigations in understanding the physical meanings of the J_k vector, M and L integrals. Their results motivate a major development for investigation of conservation laws and energy release rates in multiple crack interacting problems or microcrack damage problems carried out by the present author and the co-workers (see, *e.g.*, Chen, 1996; Zhao and Chen, 1997a,b; Chen, 2001a, b; Tian and Chen, 2000a,b; Han and Chen, 2000a,b). What is more, as an experimental attempt King and Herrmann (1981) proposed a nondestructive evaluation of the J and M integrals without the requirement of crack tip advance. This topic will be discussed in the next section. The three-dimensional problem for the J integral analysis was also studied (see, *e.g.*, Bakker, 1984; Rice, 1989).

Hutchinson (1987) and Ortiz (1987, 1988) made another significant advance in invariant integrals. They applied the J integral in microcrack shielding problems. They provided the first physical example of adopting the conservation laws in microcrack damage problems. Their attempts directly lead to a quantitative evaluation of the shielding effect when a nearby microcracking zone surrounds a macrocrack tip. However, they either used the non-interacting crack assumption, under which the interactions between microcracks are neglected entirely, or used the direct continuum mechanics to model the microcrack saturation zone as a weaker material. No real microcrack interaction is dealt with in their works. More recently Chen (1996) applied the J integral in a discrete model of microcrack shielding problems. His detailed manipulations and numerical examinations reveal that there is a redistribution of the remote J integral when it transmits across the near-tip microcracks from infinity to the macrocrack tip. Hui and Riana (1995) and Chen and Hasebe (1997) studied the J integral formulation considering the higher order singular terms defined outside of the near-tip process zone in the Williams eigenfunction expansion form (EEF). They found that there is an additional contribution to the integral arising from the interaction between the higher order singular terms and the regular terms in the EEF.

Kirshmoto, *et al.*, (1980, 1982) proposed the $\hat{J}$ integral for dynamic fracture and Atlury (1982) established more general representations of path independent integrals, which account finite elasticity and inelasticity with body forces and inertia.

As known, the invariant integrals can be established from the famous Energy Momentum Tensor in Continuum Mechanics (Eshelby, 1970, 1975). Of great interest is that the invariant integrals can also be established in several different ways. For example, based on the well known Betti reciprocal theorem (Sokolnikoff, 1956), Bueckner (1973) proposed an alternative way to establish invariant integrals, the famous Work Conjugate Integrals, from which the weight function methods for calculating the crack tip *SIF's* can be reduced (Bueckner, 1989). Following his work, Chen (1985) restudied the Williams EEF and found that the Bueckner integral provides a more natural and universal description of path independent integrals since the *J* integral and the M integral can be deduced from the Bueckner integral by adopting two different complementary stress states. Sham and his co-workers (1988, 1989a,b) developed the theory of higher order weight functions to establish some invariant integrals for calculating crack tip parameters including the well known T stress in two-dimensional crack problems. Rice (1989) extended these investigations to the weight function theory for three dimensional elastic crack analysis and Gao (1991) and Chen and Hasebe (1994a,b) extended the weight functions to treat interface cracks in dissimilar materials. However, as mentioned earlier, these investigations seem falling in the second catalogue, that is, they only provide some powerful and effective computational tools with no special physical meanings concerning with the *ERR*.

Since conservation laws or invariant integrals not only provide some important physical meanings (Budiansky and Rice, 1973; Herrmann and Herrmann, 1981), but also provide powerful computational tools in linear elastic defect mechanics of solids (Bueckner, 1973; Rice, 1989), we can expect that these laws will play similar roles in modern materials with defects, *e.g.*, in functional materials. Recently, the investigations of conservation laws and invariant integrals in modern materials have become increasingly popular. Pak (1990) and Suo *et al.*, (1992) established the *J* integral formulation in piezoelectric materials. They found that the integral plays the same role as the total energy release rate per unit crack tip advance, which is similar to that in traditional structural materials, although both mechanical quantities and electric quantities are taken into account. However, whether the *J* integral is suitable for describing the stability of crack tips in such functional materials remains unclear (Park and Sun, 1995a,b). Following the Energy Momentum Tensor in Continuum Mechanics

(Eshelby, 1970, 1975), Wang and Shen (1996) provided some formulations of conservation laws and invariant integrals in magneto-strictive and electro-strictive materials, in which all the thermal quantities, the electric quantities, the magneto quantities, and the mechanical quantities are taken into account. However, they have not yet used their formulations in crack analyses for thermal electric magneto mechanical coupling problems to derive useful results in practice. Moreover, Gao (1994) adopted the J integral analysis in nano-mechanics with defects for calculating strain energy density of a very thin film (a heteroepitaxial thin film) when the film changes from a perfectly flat layer into an undulating configuration. Indeed, he did note that the projected value of the J_2 integral along a curved surface to the horizontal axis plays an important role in calculating the J integral being the change of the total potential energy . Suo (2000) provided a simple application of the M integral to the Zener crack. These attempts do show that conservation laws and energy release rates still play an important role in modern materials (*e.g.,* functional materials such as piezoelectric ceramics and nano-materials). Therefore, adequate investigations are absolutely necessary.

3. Physical meanings of the M integral and the L integral

This section addresses the physical meanings of the M integral and the L integral. For single crack problems Budiansky and Rice (1973) pointed out that the M integral represents the energy release rate (*ERR*) induced from the uniform expansion of a crack. Here the uniform expansion means that both tips of the crack advance a unit length in opposite directions, simultaneously. We can see this conclusion directly from Freund's formulation (1.14), in which the right hand terms actually refer to the energy release per unit crack tip advance arising from both left and right tips simultaneously. Whereas Herrmann and Herrmann (1981) pointed out that the L integral represents the energy release rate induced from the crack rotation when the remote uniform loading remains unaltered. This section deals with the physical significance of the L integral in more detail. We start from another physical quantity, the change of the total potential energy (*CTPE*) owed to the formation of a center crack in a plane elastic body such as shown in Figure 1.1.

The definition of the *CTPE* for a singe crack under remote uniform loading is

$$U = \frac{1}{2}\int_{-a}^{a} \sigma_{i2}^{\infty} \Delta u_i(x) dx, \qquad (i = 1, 2), \tag{1.44}$$

where Urefers to the *CTPE*, σ_{i2}^{∞} $(i = 1, 2)$ refers to the remote stress (see Figure 1.1), while $\Delta u_i(x)$ denotes the displacement jump across the crack.

In plane strain conditions, the *CTPE* induced from the single crack can be evaluated without any diff iculty by:

$$U = U_I + U_{II}, \tag{1.45}$$

where

$$U_I = \pi a^2 (\sigma_{22}^{\infty})^2 (1 - \nu^2)/E, \tag{1.46}$$

$$U_{II} = \pi a^2 (\sigma_{12}^{\infty})^2 (1 - \nu^2)/E. \tag{1.47}$$

Of great interest is the value of the M integral induced from the same crack (Freund, 1978, also see equation (1.14)):

$$M = \frac{\kappa + 1}{4\mu} \cdot \pi a^2 \{(\sigma_{12}^{\infty})^2 + \sigma_{22}^{\infty})^2\}, \tag{1.48}$$

where κ and μ have the same meanings as those in equations (1.14).

Thus, comparing (1.44), (1.45), and (1.46) with (1.48), we conclude that the M integral is equivalent to the *CTPE* by

$$M = 2U. \tag{1.49}$$

It is concluded that the M integral is twice the charge of the total potential energy owed to single cracking. In fact, this conclusion had already been implied in the previous work done by Budiansky and Rice (1973) and Herrmann and Herrmann (1981), but it has never been emphasized explicitly in the literature. Of great interest is that equation (1.49) will lead to a simple relation between the M integral and the L integral. Let us discuss this relation in more detail. This can be done by considering a rotation of a crack and by keeping the remote uniform load unaltered.

As shown in Figure 1.7, where a crack rotates through a small angle γ with respect to the x_1-axis and then the stresses in the new coordinate system (x_1^*, x_2^*) are related to those in the original system by

$$\begin{aligned}
\sigma_{11}^* &= \sigma_{11}(\cos\gamma)^2 + 2\sigma_{12} \cdot \sin\gamma \cdot \cos\gamma + \sigma_{22}(\sin\gamma)^2, \\
\sigma_{12}^* &= -\sigma_{11}\sin\gamma \cdot \cos\gamma + \sigma_{12} \cdot \ ((\cos\gamma)^2 - (\sin\gamma \cdot)^2) + \sigma_{22}\sin\gamma \cdot \cos\gamma, \\
\sigma_{22}^* &= \sigma_{11}(\sin\gamma)^2 - 2\sigma_{12} \cdot \sin\gamma \cdot \cos\gamma + \sigma_{22}(\cos\gamma)^2.
\end{aligned} \tag{1.50}$$

If the rotation angle γ is very small then equation (1.50) is approximately deduced as

$$\begin{aligned} \sigma^*_{22} &= \sigma_{22} - 2\gamma\sigma_{12}, \\ \sigma^*_{12} &= \sigma_{12} + \gamma(\sigma_{22} - \sigma_{11}). \end{aligned} \tag{1.51}$$

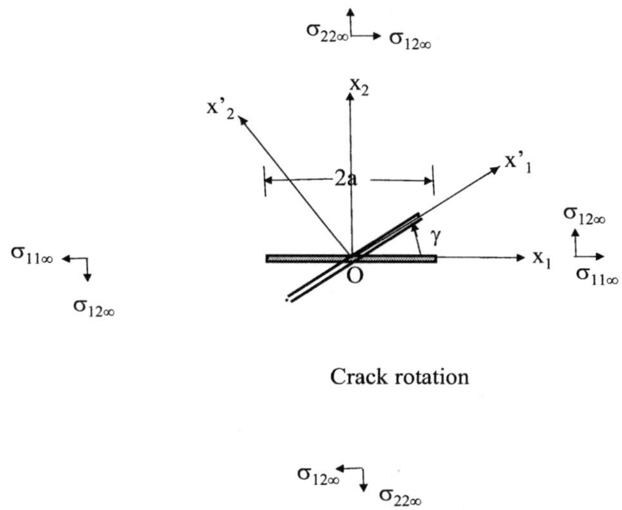

Figure 1.7. Crack rotation with an angle γ.

Introduce the definition of the rotation energy release rate G_R (for small γ):

$$G_R = \frac{dU}{d\gamma} \approx \frac{(U - U')}{\gamma}, \tag{1.52}$$

where the subscript R refers to the rotation. Now U is given in equation (1.44) and $U^*$is given by replacing $\sigma^{\infty}_{12}, \sigma^{\infty}_{22}$ by the starred quantities, using (1.51)

$$\begin{aligned} U^* &= \pi a^2 \{(\sigma^*_{22})^2 + (\sigma^*_{12})^2\}(1 - \nu^2)/E \\ &= \pi a^2 \{(\sigma^{\infty}_{22})^2 + (\sigma^{\infty}_{12})^2 - 2\gamma\sigma^{\infty}_{12}(\sigma^{\infty}_{22} + \sigma^{\infty}_{11})(1 - \nu^2)/E. \end{aligned} \tag{1.53}$$

Equations (1.44), (1.52), and (1.53) show that the rotation energy release rate G_R is

$$G_R = -2\sigma_{12}^{\infty}(\sigma_{22}^{\infty} + \sigma_{11}^{\infty})(1-\nu^2)/E. \tag{1.54}$$

As regards the value of the L integral for the single crack, we reconsider its original definition (1.3) and the special closed contour $C = C_R + C_- + C_L + C_+$ shown in Figure 1.2. It is seen that for the right tip along the circle path C_R the middle term $(-T_i u_j)$ is independent of the radius r, but ds is proportional to r, this term vanishes. Thus the only terms which contribute to L along the circles are the same as those in evaluating J_2 since $x_1 \to a$ and $x_2 \to 0$ as $r \to 0$. Consequently the contribution arising from the right and the left circles is given by:

$$L_R + L_L = -2aJ_{2R}, \tag{1.55}$$

where

$$J_{2R} = -\frac{2(1-\nu^2)}{E} K_{IR} K_{IIR} \tag{1.56}$$

and $K_{IR} = \sqrt{\pi a}\sigma_{22}^{\infty}$ and $K_{IIR} = \sqrt{\pi a}\sigma_{12}^{\infty}$.

Here it should be mentioned that Herrmann and Herrmann (1981) omitted the subscript R in (1.56), but the physical meaning is not altered. The contribution arising from the upper and lower traction-free surfaces of the crack is

$$L_{2a} = -\int_{-a}^{a} [w] x_1 dx_1 = 2\sigma_{12}^{\infty}(\sigma_{11}^{\infty} - \sigma_{22}^{\infty})\pi a^2 (1-\nu^2)/E, \tag{1.57}$$

or equivalently

$$L_{2a} = aJ_{2R} + 2\sigma_{12}^{\infty}\sigma_{11}^{\infty}\pi a(1-\nu^2)/E. \tag{1.58}$$

Therefore, the total contribution arising from the whole contour $C = C_R + C_- + C_L + C_+$ in Figure 1.2 is

$$L = L_R + L_L + L_{2a} = -2aJ_{2R} + 2\sigma_{12}^{\infty}(\sigma_{11}^{\infty} - \sigma_{22}^{\infty})\pi a^2(1-\nu^2)/E$$
$$= 2\sigma_{12}^{\infty}(\sigma_{11}^{\infty} + \sigma_{22}^{\infty})\pi a^2(1-\nu^2)/E. \tag{1.59}$$

By comparing equations (1.54) and (1.59) we conclude that

$$L = -G_R. \tag{1.60}$$

Indeed, as pointed out by Herrmann and Herrmann (1981), the rotation energy release rate (*RERR*) is identical to the L integral, except for a change in sign. Thus the value of the L integral does of course play a more natural description of the *ERR* associated with a plane crack than the integral J_2.

In order to obtain a new relation between the M integral and the L integral let us look at the above manipulations in more detail. From equation (1.49) we know that the M integral is twice the *CTPE*; but from equations (1.52) and (1.54) we know that the L integral (or $-G_R$) is just the derivative of the *CTPE* with respect to the crack rotation angle γ. Consequently we find a most significant conclusion about the inherent relation between the M integral and the L integral:

$$L = -\frac{1}{2}\frac{\partial M}{\partial \gamma}. \tag{1.61}$$

We conclude that the L integral is just the negative of the partial derivative of the M integral with respect to the rotation angle γ for a center crack in an infinite elastic body under remote uniform loading. This conclusion has never been reported by former researchers in the literature. Physically it means that the L integral and the M integral are not two independent scalars; rather, they are inherently related by equation (1.61). Other books have not covered this conclusion.

However, it is not clear whether or not the same conclusion could be given in multiple crack interacting problems. Since the stress displacement field near a typical crack is seriously disturbed by its neighboring cracks, the manipulations performed to obtain the *CTPE* or the G_R become very cumbersome. This topic will be discussed in the forthcoming chapters in some detail.

4. Nondestructive evaluation of the J and M integrals

Some readers may be interested in the practical applications of the J and M integrals. In this section we discuss some examples to satisfy their demands. As pointed out by Kanninen and Popelar (1985), with the exception of King and Herrmann (1981), there has been apparently little practical application of the M integral. Based on the above analyses we know that the M and L integrals provide a more natural description of the *ERR* associated with plane cracks than the integrals J_1 and J_2 emphasized by Herrmann and Herrmann (1981). If we consider a contour completely enclosing the whole crack then in the case under remote uniform loads both J_1 and J_2 vanish and we can derive no useful results. If we consider a contour enclosing only one crack tip then

J_2 is not path independent. It is noted, further, that the energy release rate for rotation, denoted by G_R, may be used experimentally in a capacity analogous to J. Instead of following the standard procedure of changing the crack length in compliance tests (see ASTM 1990), experiments could be performed by rotating the applied stress field with respect to the crack plane, keeping the crack length constant. King and Herrmann (1981) proposed a nondestructive evaluation of the J and M integrals. They started their investigation from the simple relation between the M and J integrals for a central crack under the remote uniform loading, $M=2aJ$. Thus, instead of working in the customary J integral evaluation, they focused attention on finding the M integral for a certain crack configuration under certain loading conditions, since knowing M they could find J. In other words, the M integral can be evaluated in cases where the J integral is inconvenient or inapplicable. For example, it is actually difficult to evaluate J if there is a wedge inserted into the crack surfaces (King and Herrmann, 1981).

4.1 Techniques for experimentally evaluating J and M

The customary technique for measuring the J integral is the well known Begley Landes (1972) compliance method, which takes advantage of the physical interpretation of the J integral as the *ERR*. Suppose the total mechanical energy of a specimen under load containing a crack is Π. If Π is measured twice for the same load, first for a specimen with crack length a, and secondly for an identical specimen but with crack length $a + \Delta a$, then the J integral is given approximately by

$$J \approx -[\Pi(a + \Delta a) - \Pi(a)]/\Delta a. \tag{1.62}$$

Here Δa should be much smaller than the original length a (for example, Δa has been customarily chosen as $\Delta a = 0.001a$).

Thus J is the area between the loading versus displacement curves for the two crack lengths. Equation (1.62) shows that if Π is measured for several different crack lengths J is obtained as a derivative: Π versus a. This technique has been well accepted (see ASTM 1990) and has been discussed in many books, for example, Kanninen and Popelar (1985).

The shortcoming of the compliance method (Begley and Landes, 1972) is that the accuracy of the J integral measurement depends on the accuracy of the measurement of the crack increment Δa. Rice *et al.* (1973) proposed an alternative method for determining the J integral from a load displacement curve for a single crack length for certain specimen configurations without any increment of the crack length. In contrast with the Begley Landes technique

(1972) this method could be called 'nondestructive' because it requires only a single specimen of fixed crack length.

In this section we discuss the nondestructive method proposed by King and Herrmann (1981). In order to avoid the use of a crack length increment we use the definition of J (or M) as a contour integral; this involves experimental evaluation of the integrand at various points along a contour (customarily taking the outside boundary of a certain cracked body) and then numerical integration.

Generally speaking, determination of the integrand of J or M involves knowledge of all the components in the (x_1, x_2) plane of the stress and the rotation tensor (or, equivalently, of the displacement gradient tensor). Determination of all these quantities experimentally would be quite difficult. However, with proper choice of the contour C in equations (1.1) with $k = 1$ and equation (1.2), the integrands of J and M can be simplified, so that they can be experimentally evaluated using gages and displacement transducers.

While the experimental results to be discussed below are limited to Linear Elastic Fracture Mechanics (*LEFM*), the method should be applicable for evaluating J and M in nonlinear elastic and elastic plastic cases (subject to the restrictions on validity of J and M in such situations, *i.e.,* the outside load should be monotonously increasing).

4.2 Edge crack

A panel with an edge crack is shown in Figure 1.8, where the cracked panel is subjected to remote uni-axial tension. The origin of the (x_1, x_2) coordinate system is placed at the center of the panel rather than at the crack tip. Strain gages are placed only on the upper half portions of both vertical edges owing to the apparent symmetry with respect to the x_1-axis.

The contour chosen for evaluation of the J integral follows the vertical edges of the specimen and then proceeds horizontally adjacent to the shoulders. Because of symmetry the value of the J integral for the entire contour is equal to twice the value of J along path *ABCD*. Along the traction-free vertical edges, the second term in the J integral formulation (1.1) with $k = 1$ vanishes and the only nonzero component of stress is σ_{22}.

The strain energy density w in plane stress is given by

$$w = \frac{1}{2}E\varepsilon_{22}. \tag{1.63}$$

And the contribution to J from the vertical edges is

$$J_v = \int_0^h \frac{E}{2}[\varepsilon_{22}(x_1 = b)^2]dx_2 - \int_0^h \frac{E}{2}[\varepsilon_{22}(x_1 = -b)^2]dx_2. \tag{1.64}$$

Assuming the value of σ_{yy} along the horizontal paths to be constant and equal to the far-field stress σ, and denoting by J_h the contribution of the upper horizontal portion of the contour, we find

$$J_h = -\int_b^{-b} (\sigma_{12}\varepsilon_{11} + \sigma u_{2,1})dx_1. \tag{1.65}$$

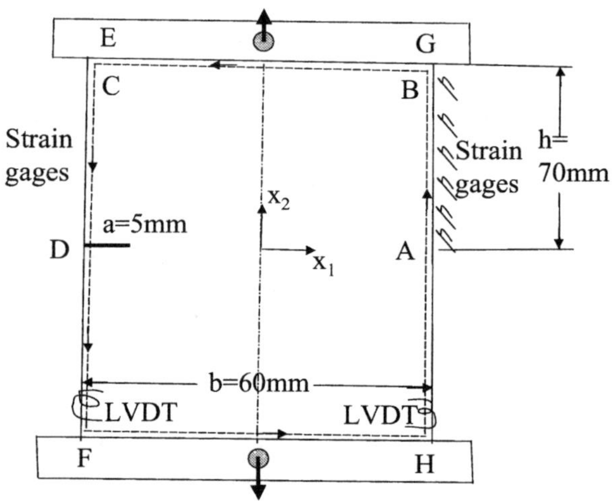

Figure 1.8. Schematic of an edge cracked panel under tension.

Comparing the two terms in equation (1.65) and assuming the shear stress near the shoulder to be sufficiently smaller than σ so that the first term may be neglected, we can write

$$J_h = -\sigma \int_b^{-b} (u_{2,1})dx_1 = \sigma(u_2(b) - u_2(-b)). \tag{1.66}$$

The J integral is then given by

$$J = 2(J_h + J_v). \tag{1.67}$$

King and Herrmann (1981) used an Aluminum 6061-T6 specimen for this kind of experiment. The dimensions of the specimen are shown in Figure

1.8. Five strain gages were placed along each of the vertical edges *AB* and *DC*, respectively, and the displacement between points E and F and G and H was measured using two linear variable differential transformers (*LVDT's*). The specimen was placed in a hydraulic testing machine using pin grips and a tensile load of 42000N was applied. The measured value of J was 1.54N/mm. Whilst J_h contributed 90 percent of this. The handbook value of *SIF* for this configuration (Tada, 1973), and relation (1.12) in plane stress conditions leads to the theoretical value J =1.49N/mm. The experimentally determined value of J agrees with this to within 3 percent.

4.3 Evaluation of the M integral

King and Herrmann (1981) performed an experiment involving evaluation of the M integral for a specimen containing an edge crack. A wedge of the crack was forced. A schematic of the specimen is shown in Figure 1.9. Freund (1978) evaluated the M integral for an edge crack in an infinite body and his reasoning is followed here. For the entire closed contour shown in Figure 1.9 the conservation law $M = 0$ applies. The portions of the contour along the crack's surfaces make no contribution to M: for a small crack opening these are very nearly radial segments, and the path *EF* is traction free, whilst on the path *DF* the tangential component of traction is zero and the normal component of displacement is zero too. The factor x_i in the integrand makes it nonsingular at the end of the wedge (point E), hence there is no contribution from a vanishingly small arc at the origin. A small arc at F contributes -*lJ*. Denoting the contribution of the portion of the edges of the specimen by M_0, and applying the conservation law $M = 0$ we find

$$M_0 - lJ = 0. \tag{1.68}$$

Experimental determination of M_0 will thus permit evaluation of the J integral. Since the edges are traction-free M_0 is given by

$$M_0 = \int_0^h \frac{E}{2}\varepsilon_{22}^2(b - a + l)dx_2 + \int_{l-a}^{b-a-l} \frac{E}{2}\varepsilon_{11}^2 h dx_1 + \int_0^b \frac{E}{2}\varepsilon_{22}^2(a - l)dx_2. \tag{1.69}$$

For an infinite body containing a semi-infinite crack, which is forced open a distance B, Freund (1978) determined the theoretical value of M_0 in plane stress to be

$$M_0 = \frac{EB^2}{8\pi}. \tag{1.70}$$

For a finite body M_0 should be approximated by adding a correction (King and Herrmann, 1981):

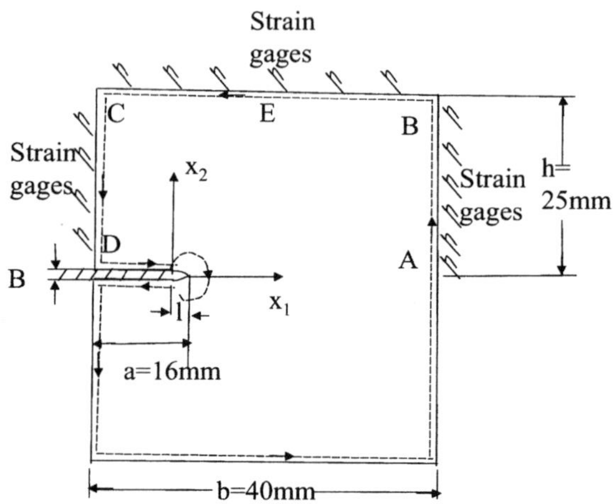

Figure 1.9. Schematic of specimen with wedged open crack used for the M integral experiment.

$$M_{AB} = \frac{-hb\sigma^2}{E} + \frac{4\sigma b}{E}[u_y(b,h) - u_y(0,h)], \tag{1.71}$$

where h and b are the vertical length and horizontal length of a finite body, respectively, $u_y(b,h)$ and $u_y(0,h)$ are vertical displacements at the right point and the middle point of the upper edge.

For the actual experiment King and Herrmann (1981) used a specimen of Aluminum 6061-T6. The specimen differed from the theoretical model in that the crack was machined so that it was initially slightly open, for example 15 degrees, to permit installation of the wedge and it was forced open further by the wedge. A steel wedge was rammed in place using a hydraulic hand press. The difference B between the crack opening distances before and after insertion of the wedge was measured using a travelling microscope. Ten strain gages were mounted along the edge *ABCD*. The value of M determined from the strain gage readings was 195.8N. Applying the measured value of B (B =0.00535in) to the theoretical solution for the infinite case and the finiteness correction, they found a theoretical value for M of 223.6N, which agrees with experiment to within approximately 12 percent. The discrepancy is probably owed to deviation of the specimen's geometry from the theoretical model used, and the approximate nature of the finiteness correction. It is unfortunate that they did not recognize the property that the arc at the notch root involved in the closed contour may yield some contribution to both J integral and M integral. Owing to this reason the discrepancy of 12 percent might also arise from neglecting the contribution of the arc at the notch root to the J integral (Chen and Lu, 2001).

4.4 Center crack

Another specimen configuration with a center crack is shown in Figure 1.10. The M integral is evaluated for the center-cracked panel under uniform tension.

The contour chosen follows vertically along the right and left edges of the specimen and horizontally a slight distance away from the upper and lower shoulders. By symmetry it is only necessary to evaluate one quadrant. Thus

$$M = 4(M_{AB} + M_{BC}), \tag{1.72}$$

where M_{AB} and M_{BC} are the contributions to M of paths *AB* and *BC*, respectively. On the traction-free vertical edge *BC*, we know that

$$W = (E/2)\varepsilon_{22}^2, x_i n_i = b, \tag{1.73}$$

so that

$$M_{BC} = \frac{bE}{2}\int_0^h \varepsilon_{22}^2 dx_2. \tag{1.74}$$

On the other hand, the contribution to M along the horizontal path *AB* is given by

$$M_{AB} = \int_0^b \{\frac{1}{2}[\sigma_{11}^2 - \sigma_{22}^2 + 2(1+\nu)\sigma_{12}^2]h dx_1$$

$$+ \int_0^b \{\frac{x_1}{E}(\sigma_{12}\sigma_{11} - \nu\sigma_{12}\sigma_{22}) - \sigma_{12}u_{1,2}h + x_1\sigma_{22}u_{1,2}\}dx_1. \tag{1.75}$$

Since the path *AB* is near the shoulder, σ_{22} may be assumed constant, but it is not obvious whether any further simplifying assumptions are justified. Consequently we investigate the importance of the various terms in M_{AB} by making use of the solution for an infinite cracked panel under remote uniaxial tension. The M integral is evaluated for a large rectangular path in the infinite plate, chosen to simulate the contour used in the present finite specimen. King and

Herrmann (1981) presented this investigation. They proved that the contributions of terms containing σ_{11} and σ_{12} are negligible, and if a quadratic-variation with x_i is assumed for u_2, the result for M_{AB} is

$$M_{AB} = \frac{-hb\sigma^2}{E} + \frac{4\sigma b}{3}[u_2(b,h) - u_2(0,h)]. \tag{1.76}$$

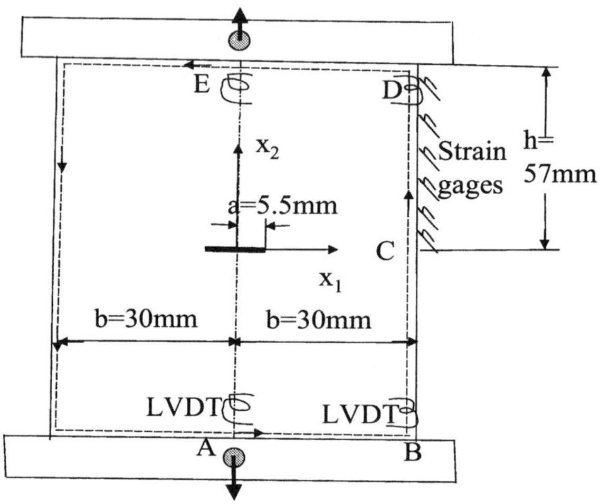

Figure 1.10. Schematic of center-cracked panel used for the M integral experiment.

In the experiment they used a specimen of Aluminum 6061-T6 with dimensions shown in Figure 1.10. Ten strain gages were mounted on the vertical edge BC and displacement transducers (*LVDT's*) were placed between points B, D and A, E, as shown in Figure 1.10. The specimen was placed in a hydraulic testing machine and loaded to a tension of 30000N. The experimental values for M were found to be 11.52N. The theoretical value for M is approximately

$$M = 2aJ_R = 2a\frac{K_I^2}{E} = 11.23N. \tag{1.77}$$

The measured value agrees with the theoretical value to within 3 percent.

From the three different kinds of experiments shown in Figures 1.8, 1.9, and 1.10, respectively, we see that for certain specimen geometries, by a convenient choice of contours the J and M integrals simplify sufficiently to permit experimental evaluation of the terms in their integrands, although care should be exercised in making simplifying assumptions regarding the terms in the integrand; this is especially true for the center-cracked panel. Experimental values of J and M thus determined agree well with corresponding theoretical values, so this appears to be a viable technique for non-destructive experimental evaluation of J and M integrals.

5. Brief Summary

Based on the discussions performed above, the following conclusions can be summarized:

(A) As pointed out by Herrmann and Herrmann (1981), the integrals M and L provide a more natural description of the *ERR* associated with plane cracks than the integrals J_1 and J_2, owing to different path selections and apparently different physical meanings. If we consider a contour completely enclosing a whole crack, then trivial results (1.33) and (1.34) are derived, *i.e.*, both J_1 and J_2 vanish. If we consider a contour enclosing only one crack tip, then J_2 is path dependent on the starting and ending points on the upper and lower surfaces of the crack (see, equations (1.15) or (1.24)).

(B) The M and L integrals are dependent on the coordinate shifts since they are coupled with the J_k vector when the shifts of system occur as formulated by equations (1.9) and (1.11). Only when the vector vanishes as represented by (1.33) and (1.34), will the M and L integrals be independent of the shifts.

(C) For multiple crack interaction in an infinite plane elastic solid, mathematical manipulations and numerical examinations reveal that both components of the J_k vector defined in a global coordinate system vanish under remote uniform leads when the closed contour chosen to calculate the two-components of the vector encloses all the cracks completely (or voids, inclusions, *etc.*,). These lead to new conservation laws (1.39) and (1.40).

(D) For single crack problems the M integral is equivalent to the change of the total potential energy (*CTPE*), *i.e.*, equation (1.49): $M = 2U$. Whereas the L integral is identical to the rotation energy release rate (G_R), *i.e.*, equation (1.60), except for a change in sign.

(E) Of the most significance is that there is a simple relation between M and L integrals, *i.e.*, equation (1.61): $L = -\frac{1}{2}\left\{\frac{\partial M}{\partial \gamma}\right\}$. It reveals that the L integral is just half the partial derivative of the M integral with respect to the rotation angle γ for a center crack under remote uniform loads, except for a change in sign. Thus, the M integral and the L integral proposed by Budiansky and Rice (1973) are not two distinct integrals; rather, they are inherently related by (1.61).

(F) Finally, whether the above mentioned conclusions (D) and (E) hold in multiple crack interacting problems or in microcrack damage problems is not clear at present. However, the above conclusions remind us to pay much attention to the path selections and to the roles that the J_k vector, the M and L integrals play when discussing more complicated problems in the forthcoming chapters.

References for Chapter 1

ASTM Standards, (1990), Vol. 03.01, 812 – 822.

Atluri,S.N.(1982)Path independent integrals in f inite elasticity and inelasticity, with body forces, inertia, and arbitrary crack face conditions, *Engineering Fracture Mechanics,* **18**, 341 – 364.

Bakker, A. (1984) The three-dimensional J integral, (Ph. D. Thesis), Technical University Delf.

Batte, A.D., Blackburn, W.S., Elsender, A, Hellen, T.K., and Jackson, A.D. (1983) A comparison of the J integral with other methods of post yield fracture mechanics, *International Journal of Fracture,* **21**, 46 – 66.

Begley, J.A., and Landes, J.D. (1972) The J integral as a fracture criterion, *Fracture Toughness, Proceedings of the 1971 National Symposium on Fracture Mechanics*, Part II, $ASTMSTP514$, 1 – 20.

Bergez, D. (1974) Determination of stress intensity factors by use of path independent integrals, *Mechanics Research Communications,* **1**, 179 – 180.

Blackburn,W.S. (1972) Path independent integral to predict onset of crack instability in an elastic-plastic material, *International Journal of Fracture,* **8**, 343 – 346.

Blackburn,W.S., Jack, A.D., and Hellen, T.K. (1977) An integral associated with state of a crack tip in a non-elastic material, *International Journal of Fracture,* **13**, 183 – 200.

Budiansky, B., and Rice, J.R.(1973) Conservation laws and energy release rates, *ASME Journal of Applied Mechanics,* **40**, 201 – 203.

Bueckner, H.F. (1973) *Field singularities and related integral representations,* in *Mechanics of Fracture*, Edited by G.C.Sih, Noordhoff, Leyden, Vol.1, 239 – 314.

Bueckner, H.F. (1989) Observations on weight functions, *Engineering Analysis with boundary Elements,* **6**, 3 – 18.

Castro, P.M.S.T, (1982) Comparison of J testing techniques and correlation J-COD using structural steel specimens, *International Journal of Fracture,* **17**, 83 – 96.

Chen Y.Z. (1985) New path independent integrals in linear elastic fracture mechanics, *Engineering Fracture Mechanics,* **22**, 673 – 686.

Chen Y.H. and Hasebe, N. (1994a) Further investigation of Comninou's EEF for an interface crack with completely closed faces, *International Journal of Engineering Science,* **32**, 1037 – 1046.

Chen Y.Z. and Hasebe, N., (1994b), Eigenfunction expansion and higher order weight functions of interface cracks, *ASME Journal of Applied Mechanics,* **61**, 843 – 849.

Chen Y.H. , (1996) On the contribution of discontinuities in a near-tip stress f ield to the J-integral, *International Journal of Engineering Science,* **34**, 819 – 829.

Chen Y.H. and Hasebe, N. (1997) Explicit formulations of J integral considering higher order singular terms in eigenfunction expansion form, Part I. Analytical treatment, *International Journal of Fracture,* **85**, 11 – 14.

Chen Y.H. and Hasebe, N. (1998) A consistency check for strongly interacting multiple crack problem in isotropic, anisotropic, and bimaterial solids, *International Journal of Fracture,* **89**, 333 – 353.

Chen Y.H. , (2001a) M-integral analysis for two-dimensional solids with strongly interacting cracks, Part I. In an inf inite brittle sold, *International Journal of Solids and Structures,* **38**, 3193 – 3212.

Chen Y.H. , (2001b)M-integral analysis for two-dimensional solids with strongly interacting cracks, Part II. In the brittle phase of an inf inite metal/ceramic bimaterial, *International Journal of Solids and Structures,* **38**, 3213 – 3232.

Chen Y.H. , and Lu, T.J. (2001) On the path dependence of the J-integral in notch problems, *International Journal of Solids and Structures,* in press.

Cherepanov, G.P. (1974) *Mechanics of Brittle Fracture Mechanics*, Moscow, Publishing House 'Nauka' , English translation published by McGraw-Hill (1979).

Eischen, J.W., and Herrmann, G. (1987) Energy release rates and related balance laws in linear elastic defect mechanics, *ASME Journal of Applied Mechanics,* **54**, 388 – 392.

Elhaddad, M.H, Dowling N.E, Topper T.H., and Smith K.N. (1980) J integral applications for short fatigue cracks at notches, *International Journal of Fracture,* **16**, 15 – 30.

Eshelby, J.D. (1956) *The continuum theory of lattice defects, Solid State Physics,* 3,F, Seitz and Turnbull, D. (edited), Academic, New York, 79 – 141.

Eshelby, J.D. (1970) *The Energy Momentum Tensor in Continuum Mechanics, Inelastic Behavior of Solids*, edited by Kanninen, M.F., *et al.,* McGraw-Hill, New York.

Eshelby, J. D. (1975) The elastic energy momentum tensor, *Journal of Elasticity,* **5**, 321 – 335.

Freund, L.B. (1978) Stress intensity factor calculations based on a conservation integral, *International Journal of Solids and Structures,* **14**, 241 – 250.

Gao, H. (1991) Weight function analysis of interface cracks: Mismatch versus oscillation, *ASME Journal of Applied Mechanics,* **58**, 931 – 938.

Gao, H. (1994) Some general properties of stress-driven surface evolution in a heteroepitaxial thin film structure, *Journal of Mechanics and Physics of Solids,* **42**, 741 – 772.

Gross, D., and Mendelson, A. (1972) *International Journal of Fracture Mechanics,* **8**, 267 – 276.

Han J.J, and Chen Y.H. ., (2000a) Interface crack interacting with micro-voids in the near-tip process zone, *International Journal of Fracture,* **102**, 223 – 245.

Han J.J., and Chen Y.H. , (2000b) T-effect for the interaction problem of an interface macrocrack with a near-tip micro-void, *International Journal of Fracture,* **102**, 205 – 222.

Herrmann, A. G. (1981) On conservation laws of continuous mechanics, *International Journal of Solids and Structures,* **17**, 1 – 8.

Herrmann, G.A., and Hermann, G. (1981) On Energy Release Rates for a Plane Cracks, *ASME Journal of Applied Mechanics,* **48**, 525 – 530.

Herrmann, A.G. (1982) Material momentum tensor and path independent integrals of fracture mechanics, *International Journal of Solids and Structures,* **18**, 319 – 326.

Horii, H., and Nemat Nasser, S. (1985) Elastic fields of interacting inhomogeneities, *International Journal of Solids and Structures,* **21**, 731 – 745.

Horii, M., and Nemat Nasser, S. (1987) Interacting microcracks near the tip in the process zone of a macrocrack, *Journal of the Mechanics and Physics of Solids,* **35**, 601 – 629.

Hui, C.Y., and Riana, A. (1995) Why K? Higher order singularities and small scale yielding, *International Journal of Fracture,* **72**, 97 – 120.

Hussain, M.C., Pu, S.L., and Underwood, J. (1973) Strain energy release rate for a crack under combined Mode I and Mode II, *Fracture Analysis*, Proceedings of the 1973 National Symposium on Fracture Mechanics, Part II, ASTM Special Technical Publication **560**, 2 – 28.

Hutchinson, J.W. (1979) *Nonlinear Fracture Mechanics,* Department of Solid Mechanics, The Technical University of Denmark.

Hutchinson, J. W. (1987) Crack Tip Shielding by Microcracking in Brittle Solids, *Acta Metall.,* **35**, 1605 – 1619.

Kanninen, M.F., and Popelar, C.F. (1985) *Advanced Fracture Mechanics,* Oxford University Press, New York.

King, R.B., and Herrmann, G. (1981) Nondestructive Evaluation of the J and M-Integrals, *ASME Journal of Applied Mechanics,* **48**, 83 – 87.

Kirshmoto, K., Aoki, S., and Sakata, M. (1980) On the path independent integral-$\hat{J}$., *Engineering Fracture Mechanics,* **13**, 841 – 850.

Kirshmoto, K., Aoki, S., and Sakata, M. (1982) Use of $\hat{J}$- integral in dynamic analysis of cracked linear viscoelastic solids by *FEM, ASME Journal of Applied Mechanics,* **13**, 841 – 850.

Knowles, J.K., and Sternberg, E. (1972) On a class of conservation laws in linearized and finite elastostatics, *Arch. Rat. Mech. Anal.,* **44**, 187 – 211.

Ortiz M. (1987) A continuum theory of crack shielding in ceramics, *ASME Journal of Applied Mechanics,* **54**, 54 – –58.

Ortiz M. (1988) Microcrack coalescence and macroscopic crack growth initiation in brittle solids, *International Journal of Solids and Structures,* **24**, 231 – –250.

Pak, Y. E. (1990) Crack extension force in a piezoelectric material, *ASME Journal of Applied Mechanics,* **57**, 647 – 653.

Park, S. B., and Sun, C. T. (1995a) Fracture criteria for piezoelectric ceramics, *Journal of the American Ceramic Society,* **78**, 1475 – 1480.

Park, S. B, and Sun, C. T. (1995b) Effect of electric fields on fracture of piezoelectric ceramics, *International Journal of Fracture,* **70**, 203 – 216.

Ranaweera, M.P. and Leckie, B. (1982) J integrals for some crack and notch geometries, *International Journal of Fracture,* **18**, 3 – 18.

Rice, J.R. (1968a) A path independent Integral and the Approximation Analysis of Strain Concentration by Notches and Cracks, *ASME Journal of Applied Mechanics,* **35**, 297 – 320.

Rice, J.R. (1968b) Mathematical analysis in the mechanics of fracture, *in Fracture*, edited by Liebowitz, H. Academic Press, New York ,Vol.2, 191–311.

Rice, J.R., Paris, P.C., and Merkie, J.G. (1973) Some Further Results of J–Integral Analysis and Estimates, *Progress in Flaw Growth and Fracture Toughness Testing, Proceedings of the 1972 National Symposium on Fracture Mechanics, ASTM STP 536*, 231 – 245.

Rice, J.R. (1989) *Weight function theory for three-dimensional elastic crack analysis, Fracture Mechanics, Perspectives and directions* (Twentieth Symposium), ASTM-STP-1020, Edited by Wei, R.P., Gangloff, R.P., America Society for Testing and Materials, Philadelphia, 29 – 57.

Sham, T.L. and Bueckner, H.F. (1988) The weight function theory for piecewise homogeneous isotropic notches in antiplane strain, *ASME Journal of Applied Mechanics,* **55**, 596 – 603.

Sham, T.L., and Zhou, Y. (1989a) Weight function in two-dimensional bodies with arbitrary anisotropy, *International Journal of Fracture,* **40**, 13 – 41.

Sham, T.L. (1989b) The theory of higher order weight functions for linear elastic plane problems, *International Journal of Solids and Structures,* **25**, 357 – 380.

Sokolnikoff, S. (1956) *Mathematical Theory of Elasticity,* McGraw-Hill, New York.

Stern, M., Becker, E.B., and Dunham, R.S. (1976) A contour integral computation of mixed-mode stress intensity factors, *International Journal of Fracture,* **12**, 359 – 368.

Suo, Z., Kuo, C. M., Barnett, D. M. and Willis, J. R. (1992) Fracture mechanics for piezoelectric ceramics, *Journal of the Mechanics and Physics of Solids,* **40**, 739 – 765.

Suo, Z. (2000) Zener's crack and the M-integral, *ASME Journal of Applied Mechanics,* **67**, 417 – 418.

Tada, H. (1973) *The Stress Analysis of Cracks Handbook,* Del Research Corp.

Tian W.Y., and Chen Y.H. , (2000a) A semi-infinite interface crack interacting with sub-interface matrix cracks in dissimilar anisotropic materials, Part I, Fundamental formulations and the J integral analysis, *International Journal of Solids and Structures,* **37**, 7717 – 7730.

Tian W.Y., and Chen Y.H. , (2000b) Interaction between an interface crack and sub-interface microcracks in metal/piezoelectric bimaterials, *International Journal of Solids and Structures,* **37**, 7743 – 7757.

Wang, X. M., and Shen, Y.P. (1996) The conservation laws and path independent integrals with an application for linear electro-magneto-elastic media, *International Journal of Solids and Structures,* **33**, 865 – 878.

Zhao L-G, and Chen Y.H. , (1997a) Further investigation of sub-interface cracks, *Archive of Applied Mechanics,* textbf67, 393 – 406.

Zhao L. G, and Chen Y.H. , (1997b) On the contribution of sub-interface microcracks near the tip of an interface crack to the J-integral in bimaterial solids, *International Journal of Engineering Science,* **35**, 387 – 407.

Chapter 2

CONSERVATION LAWS IN BRITTLE SOLIDS WITH STRONGLY INTERACTING CRACKS

Based on the discussions presented in Chapter 1, this chapter extends the concepts of conservation laws and energy release rates to treat multiple crack problems and to address an alternative description for damaged brittle solids with strongly interacting microcracks. The basic idea starts from the M integral analysis customarily used in single crack problems (Budiansky and Rice, 1973; Herrmann and Herrmann, 1981). The discussion is limited to the infinite 2D cases and the microcracks are assumed to be stationary (Chen, 2001a, b). Three kinds of bodies are considered corresponding to an infinite homogeneous brittle solid, an infinite metal/ceramic bimaterial, and a half plane brittle solid. Readers will see from the global local coordinate translations that in all the three kinds of solids the M integral is always divided into two distinct parts. The first of them is induced from the well known relation between the integral and the stress intensity factors (*SIF's*) at all the crack tips (Freund, 1978). The second is contributed by the two components of the J_k vector (Knowles and Sternberg, 1972; Budiansky and Rice, 1973) and the coordinates of each microcrack center. The latter is concerned not only with the crack tip *SIF's*, but also with the contribution arising from the traction free surfaces of each crack (Herrmann and Herrmann, 1981) as discussed for equations (1.16) and (1.17) in Chapter 1. As the detailed proof for the vanishing nature of the J_k vector along a closed contour surrounding all the microcracks has been given in Chapter 1 or Chen (2001a), the confusion about the dependence of the M-integral on the origin selection of global coordinates for a cloud of cracks is then clarified in this chapter. Several numerical examples will be shown in Tables and Figures to confirm these conclusions. It is shown that the M integral is equivalent to the decrease of the total potential energy of the solids Owing to microcracking, although the strongly interacting situations are taken into account. Therefore, from equations (1.49) and (1.61) in Chapter 1, readers will understand that there is still a simple relation between the M integral and the L integral in an infinite microcracking body under the stationary crack assumption. It will be concluded that the M integral analysis, from the physical point of view, does play an important role and provide an effective measure in evaluating the damage level of brittle solids with strongly interacting and randomly distributed microcracks. Although only the stationary microcracks are considered in this chapter, the derived conclusions could actually be extended to treat much more useful problems in which the multiple cracks may become critical and may grow during loading histories.

1. Historic reviews and engineering backgrounds of microcrack damage

Nonlinear mechanical responses of microcracking solids have received considerable attention in the past forty years (see, *e.g.,* Kachanov, 1958; Mori and Tanaka, 1973; Budiansky and O'connell, 1976; Gross, 1982; Chen, 1984; Horii and Nemat Nasser, 1983,1985; Benveniste, 1986; Chaboche, 1988a,b; Krajicinovic, 1989; Nemat Nasser and Hori, 1990; Christensen, 1990; Kachanov, 1992, 1993; Jun, 1991; Jun and Lee, 1991; Jun and Chen, 1994a,b; Chen and Hasebe, 1998). As is now generally accepted, the existence, the growth, and the nucleation of microcracks in brittle solids are of significant importance in both mechanical and civil engineering. The earliest work on this subject was given by Kachanov (1958) and several literature reviews were given by Chaboche (1988a,b), Krajcinovic (1989), and Kachanov (1992,1993). Now, it is a commonly recognized opinion that microcracks in brittle materials, *e.g.,* concrete, rocks and ceramics, often control overall deformation and failure mechanism because the distributed microcracks in such materials not only lead to macrocrack initiation and final fracture, but also induce progressive material damage (deterioration or evolutionary damage). Nevertheless, the progressive material deterioration could be measured through the decrease of strength, stiffness, toughness, stability and residual life of such materials. Chaboche (1988a,b) reviewed some general features to Continuum Damage Mechanics and summarized its main possibilities to present definitions and measures of damage, the description of the mechanical behavior of evolution of the corresponding damage variables. Moreover, there have been a number of investigators who focused their attention on the effective elastic moduli of two-dimensional brittle solids with strongly interacting, randomly distributed microcracks. For example, Jun and Chen (1994a,b) presented their statistical micro-mechanical formulations to investigate the effective elastic moduli of two-dimensional brittle solids with interacting slit microcracks. Besides these the nonlocal damage theory based on micromechanics of crack interactions was proposed (see *e.g.,* Bazant 1986; Bazant and Cedolin 1991; Zdenek and Bazant, 1992). Some 3D problems have also been treated (see *e.g.,* Budiansky and O'connell, 1976; Ju and Lee, 1991; Kachanov and Laures, 1989; Kachanov, 1993).

On the other hand, in Chapter 1 the historical review and fundamentals of conservation laws and energy release rates in single crack problems have been discussed in some detail. The basic concepts about the conservation integrals (path independent integrals or invariant integrals) such as Rice's J integral, the J_k vector, the L integral, and the M integral have been well addressed. Based on the work (Chen and Hasebe, 1998), the new conservation laws (1.39) and (1.40) of the J_k vector are presented in multiple crack interacting problems. However, to the present author's knowledge the path independent integrals mentioned above were always limited to single crack problems and a few researchers in the open literature had accounted for the roles that the conservation integrals play in multiple crack interacting problems or microcrack damage problems. Particularly, the implicit relation between the effective elastic moduli and the path independent integrals remains unclear even in the simplest cases, where all microcracks are assumed to be stationary (in other words, no microcracks are allowed to grow or nucleated during loading histories, the so called 'stationary' micro-mechanical models).

Recently Chen's (2001a,b) efforts supply the lack of the investigation of conservation laws in microcrack damage problems. Obviously his purpose is not to present some advances in the continuum damage mechanics, rather, based on the conservation laws, the purpose is to build a new framework and to provide a new way in understanding the microcrack damage in engineering materials. Only after doing so could an alternative description or a new evaluation for microcrack damage in brittle solids be established. The basic idea starts from the M integral analysis customarily used in single crack problems. The key is to select the integrand paths, which are always chosen to be paths either enclosing all microcracks or enclosing a typical microcrack completely whatever in a global system or in local systems. As an initial attempt the simplest cases only concerned with the stationary microcracks are considered and all the microcracks are assumed to be fully open and not intersecting with each other. In Section 2.2 some lengthy manipulations are performed to verify the independence of the M integral from the origin selection of global coordinates. It is found from the global local coordinate translations that the M integral is divided into two distinct parts for a cloud of microcracks in a two-dimensional brittle solid. The first of them corresponds to the contribution arising from *SIF's* of each crack formulated by Freund (1978), whilst the second corresponds to the contribution induced from the two components of the J_k vector and the center coordinates of each crack in the global coordinates. In Section 2.3 two numerical examples are presented. One of them is concerned with four regularly distributed microcracks and the other is concerned with twenty randomly distributed microcracks, respectively. Numerical results are shown in Tables and Figures, which not only confirm the independence of the M integral from the selection of the global system, but also reveal the equivalence between the M integral and the decrease of the total potential energy of the solid even under strongly interacting crack situations. Therefore, as in equation (1.61) in Chapter 1, the simple relation between the M integral and the L integral in the present problem is proved to be still valid under the assumption mentioned above.

Of the most interest is the implicit relation between the M integral and the reduction in the effective elastic moduli. It will be shown that the value of the integral against the remote loading angle increases as the moduli decrease and vice versa. The loading direction at which the maximum value of the integral occurs coincides well with the largest loss in effective moduli. In particular, numerical results for twenty randomly distributed microcracks show that the value of the M integral is not sensitive to the remote loading direction. This coincides well with the well known conclusion that the randomly distributed microcracks produce an isotropic overall elastic response (Jun and Chen, 1994a,b). A short summary is given to list the main conclusions derived in infinite brittle solids.

In section 2.4 an extension of investigation will be presented to treat multiple crack interacting problems in the brittle phase of an infinite metal/ceramic bimaterial. After considering the interaction between the interface and the cracks and making some cumbersome manipulations, it is concluded that the M integral does still play an important role in microcracking bimaterials. These conclusions do provide a possibility or evidence in establishing an alternative description or a new evaluation for microcracking damage. Although the M integral analysis has not been proved to be universal to account all aspects of microcrack damage, it provides, at least, an effective measure of the damage level for brittle solids or bimaterials with strongly interacting microcracks in the so called 'stationary' micro-mechanical models. It is concluded that an inherent relation does exist between the M integral and the effective elastic moduli for a microcracking brittle solid. In other words, the effective elastic properties could be reformulated by the M integral, although the detailed mathematical formulation between the moduli and the M integral remains to be adequately investigated.

In order to study microcrack damage problems in a finite solid, in section 2.5 the outside boundary of a half plane is treated as a special interface between air and brittle solid. It is verified from manipulations and numerical results that this treatment is quite fair and effective. Finally, in section 2.6 a brief summary is given.

2. Independence of the M integral from the origin selection of the global coordinates

This section addresses the new conservation laws of the J_k vector and the independence of the M integral from the origin selection of the global coordinates for multiple crack interacting problems in an infinite brittle body. The difficulties in extending the original definition of the M integral to treat multiple crack problems are overcome. Some confusion arising from misunderstanding the vector and the integral is clarified.

2.1 Application of the M integral in multiple crack interacting problems

The fundamental of the M integral customarily adopted in single crack problems is well discussed in Chapter 1 as given in equations (1.2), (1.8), (1.9), (1.14), (1.19) and (1.49). In this section we make an extension to more complicated cases considering a multiple crack interacting problem as shown in Figure 2.1. Assume that the closed contour C^o encloses all the cracks completely in the two-dimensional brittle solid. A global coordinate system, say (x_1, x_2), with the origin O, is introduced (see Figure 2.1). Here it should be emphasized that the original equation of the M integral (1.2) is defined in the global coordinate system (x_1, x_2). Of course, it can also defined in a local coordinate system $(x_1^{(k)}, x_2^{(k)})$ originated at the center of a typical microcrack (k) in Figure 2.1. However, the values of the M integral in the global and local systems are quite different because the M integral includes two additional terms when the coordinate shifts occur. Readers can understand this point from equation (1.9) in Chapter 1. In Figure 2.1 the local system $(x_1^{(k)}, x_2^{(k)})$ is parallel to the global system (x_1, x_2), but its origin $O(k)$ is different from the origin O of (x_1, x_2). Here $\xi_1^{(k)}$ and $\xi_2^{(k)}$ are coordinates of the origin $O(k)$ in the global system (x_1, x_2), and the following relations are valid:

$$x_1 = \xi_1^{(k)} + x_1^{(k)},$$
$$x_2 = \xi_2^{(k)} + x_2^{(k)}, \tag{2.1}$$

In order to define the M integral in the local system $(x_1^{(k)}, x_2^{(k)})$, a special closed contour $C^{(k)}$ is also introduced which encloses the microcrack (k) only (see Figure 2.1). Thus, in each local coordinate system we have

$$M^{(k)} = \oint_{C^{(k)}} (w x_i^{(k)} n_i^{(k)} - T_l^{(k)} \cdot u_{l,i}^{(k)} \cdot x_i^{(k)}) ds, \qquad (k = 1, 2, \cdots, N), \tag{2.2}$$

where the superscript (k) denotes the quantities defined in the local system $(x_1^{(k)}, x_2^{(k)})$.

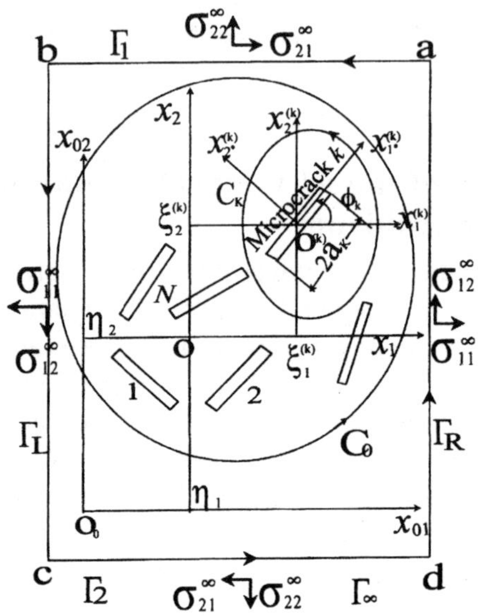

Figure 2.1. Strongly interacting microcracks and specially introduced closed contours in an infinite plane brittle solid.

Here it has been assumed that all the microcracks are not intersecting with each other although the microcrack concentrations are higher and microcrack spaces are closer. As pointed out by Jun and Chen (1994a,b), in this case the strong microcrack interactions occur and the effective medium theories are very cumbersome to be appropriated. The multiple crack interacting problem shown in Figure 2.1 could easily be solved as treated by Gross (1982) and Chen (1984). The detailed numerical technique is no longer discussed here. After doing so, the stress intensity factors at both tips of the kth microcrack denoted by $K_{IR}^{(k)}, K_{IIR}^{(k)}, K_{IL}^{(k)}, K_{IIL}^{(k)} \quad (k = 1, 2, \cdots, N)$ could be given and the $M^{(k)}$ integral in each local system $(x_1^{(k)}, x_2^{(k)})$ could then be evaluated by using Freund's formulation (1978):

$$M^{(k)} = \frac{\kappa + 1}{8\mu} \left[(K_{IR}^{(k)})^2 + (K_{IIR}^{(k)})^2 + (K_{IL}^{(k)})^2 + (K_{IIL}^{(k)})^2 \right] a_k, \tag{2.3}$$

where μ is the shear modulus of the brittle solid and $\kappa = 3 - 4\nu$ for plane strain, ν is the Poisson's ratio, subscripts R and L denote the stress intensity factors at the right tip and left tip of the microcrack (k), respectively, subscripts I and II denote the Mode I and Mode II fracture, respectively, and a_k refers to the half length of the kth microcrack.

It is mentioned in Chapter 1 that, as a scalar, the value of $M^{(k)}$ is independent of the rotation of a coordinate system when it is transferred from $(x_1^{(k)}, x_2^{(k)})$ to $(x_{1*}^{(k)}, x_{2*}^{(k)})$ with an oriented angle φ_k(see Figure 2.1). However, it can also be noted from equation (1.9) in Chapter 1 that the total contribution induced from the N microcracks to the M-integral could not be given by using a simple summation among $M^{(k)}(x_1^{(k)}, x_2^{(k)})$ with *k=1,2,...,N* owing to the additional terms contributed by the J_k vector (see equation (1.9)). This means that

$$M(x_1, x_2) \neq \sum_{k=1}^{N} M^{(k)}(x_1^{(k)}, x_2^{(k)}). \tag{2.4}$$

Readers recognize at once that equation (2.4) leads to significant trouble in evaluating the M integral in multiple crack interacting problems.

2.2 Independence of the M integral from the origin selection of the coordinates

Indeed, equation (2.4) implies that it is not clear whether the M integral depends on the origin shifts of the global coordinate system as many previous researchers suspected. Moreover, how to calculate the value of the M integral for multiple crack problems is worthy of further investigation, which has not been covered by other books.

For the forthcoming manipulations we should recall the new conservation laws of the J_k vector, as formulated by equations (1.39) and (1.40).

As discussed in Chapter 1, the values of the J_k vector are not dependent on the location of the coordinate system or the origin shift of the system. Therefore in the global system (x_1, x_2) the values of J_k can be evaluated by using a simple summation amongst the $J_k^{(k)}$ derived in the local system $(x_1^{(k)}, x_2^{(k)})$, that is, according to (1.39) and (1.40), we rewrite them as

$$J_1 = \sum_{k=1}^{N} J_1^{(k)} = \sum_{k=1}^{N} (J_{1*}^{(k)} \cos\varphi_k - J_{2*}^{(k)} \sin\varphi_k) = 0, \tag{2.5}$$

$$J_2 = \sum_{k=1}^{N} J_2^{(k)} = \sum_{k=1}^{N} (J_{1*}^{(k)} \sin\varphi_k + J_{2*}^{(k)} \cos\varphi_k) = 0, \tag{2.6}$$

and restate the key concepts of the new laws below:

Both components of the J_k vector defined in a global coordinate system vanish when the closed contour chosen to calculate the two components of the vector encloses all the cracks (or voids, inclusions, etc.) in an infinite brittle solid or there are no other discontinuities outside the closed contour.

Turning back to the M integral analysis and considering its original definition (1.2) and equations (2.1) and (2.2), it follows that

$$\begin{aligned}
M &= \sum_{k=1}^{N} M^{(k)}(x_1, x_2) = \sum_{k=1}^{N} \left\{ \oint_{C_k} (wx_i n_i - T_l u_{l,i} x_i) ds \right\} \\
&= \sum_{k=1}^{N} \left\{ \oint_{C_k} \left[w \left(x_i^{(k)} + \xi_i^{(k)} \right) n_i - T_l u_{l,i} \left(x_i^{(k)} + \xi_i^{(k)} \right) \right] ds \right\} \\
&= \sum_{k=1}^{N} \left\{ M^{(k)} \left(x_1^{(k)}, x_2^{(k)} \right) \right\} + \sum_{k=1}^{N} \left\{ \xi_1^{(k)} J_1^{(k)} + \xi_2^{(k)} J_2^{(k)} \right\} \\
&= M_N + M_A, \qquad (2.7)
\end{aligned}$$

$$M_N = \sum_{k=1}^{N} \left\{ M^{(k)} \left(x_1^{(k)}, x_2^{(k)} \right) \right\}, \qquad (2.8)$$

$$M_A = \sum_{k=1}^{N} \left\{ \xi_1^{(k)} J_1^{(k)} + \xi_2^{(k)} J_2^{(k)} \right\}, \qquad (2.9)$$

where $M^{(k)}(x_1^{(k)}, x_2^{(k)})$ is defined by equation (2.2) in the local system $(x_1^{(k)}, x_2^{(k)})$ and its value is calculated by equation (2.3), while $M^{(k)}(x_1, x_2)$ is considered in the global system (x_1, x_2). The subscripts N and A denote the net part and the additional part of the M integral in the present situation.

Intuitively, the first term in equation (2.7), *i.e.*, (2.8), denoted by M_N, which is called as the net contribution of the N cracks, does not influence the dependence of the M integral on the origin selection. Special attention should only be devoted to the second term in the right hand side of equation (2.7), *i.e.*, (2.9), since it involves the global coordinates of each crack center as well as the components of the J_k vector contributed from the formation of each crack. This term M_A represents the additional contribution of the N microcracks to the M integral in the global system. Owing to its existence, a considerable confusion arises. Many people believe that the value of the M integral is dependent on the origin selection of the global coordinate system (x_1, x_2) since different selections of the origin of the global coordinates lead to difference values of $\xi_1^{(k)}$ and $\xi_2^{(k)}$. This is the major reason why the present author should perform so lengthy and so cumbersome manipulations to clarify the confusion mentioned above. Only after doing so could readers understand what role is played by the M integral in microcrack damage.

Let us perform another coordinate transformation from (x_1, x_2) to (x_{01}, x_{02}) (see Figure 2.1). The latter is parallel to the former but originated at another point O_o as shown in Figure 2.1. The simple transformation relations between the two coordinate systems are:

$$\begin{aligned} x_1 &= x_{01} - \eta_1, \\ x_2 &= x_{02} - \eta_2, \end{aligned} \tag{2.10}$$

where η_1 and η_2 denote the coordinate shifts along the two original axes.

The value of the M integral in the new global coordinate system (x_{01}, x_{02}) denoted by M_0 should be given as

$$\begin{aligned} M_0 &= \oint_C (w x_{0i} n_i - T_l u_{l,i} x_{0i}) ds \\ &= \sum_{k=1}^{N} \left\{ M^{(K)}(x_1^{(k)}, x_2^{(k)}) + (\xi_1^{(k)} + \eta_1) J_1^{(k)} + (\xi_2^{(k)} + \eta_2) J_2^{(k)} \right\} \\ &= M + \eta_1 \sum_{k=1}^{N} J_1^{(k)} + \eta_2 \sum_{k=1}^{N} J_2^{(k)}. \end{aligned} \tag{2.11}$$

According to the new laws (1.39) and (1.40) or (2.5) and (2.6) readers can immediately see that the last two terms in the right hand side of (2.11) vanish although the multi-cracks' interaction seriously disturb the values of $M^{(k)}(x_1^{(k)}, x_2^{(k)})$ and $J_k^{(k)}$ in each local system. It is then concluded that

$$M_0 = M. \tag{2.12}$$

In other words, the value of the M integral does not depend on the origin selection of the global coordinates, either on the shifts of the origin, or on the coordinate rotation. The latter could be seen in the work done by Park and Earmme (1986). It should be emphasized that equation (2.12) is directly deduced from the new conservation laws (2.5) and (2.6) with no regards to the details of micro-defects. This means that other defects such as curve cracks, growing cracks, voids with any shapes, branched crack, and crack bifurcation during loading history do not alter the independence of the M integral from the origin selection if all of them are enclosed by the closed contour C in Figure 2.1. Nevertheless, different kinds of defects may contribute quite different values of the M integral and the growing microcracks enclosed by C may alter the values of the M integral.

3. Numerical examples

In order to support the above mentioned conclusions (2.7) and (2.12) and to provide necessary evidences of the role played by the M integral in microcrack damage problems, two numerical examples are given in this section. The first is concerned with four regularly distributed microcracks, while the second is concerned with twenty randomly distributed microcracks.

3.1 Four regularly distributed microcracks

Consider four strongly interacting microcracks regularly distributed in a plane elastic body as shown in Figure 2.2. The infinite body is loaded by the remote tensile stress σ_0 inclined by an angle ψ with respect to the x_{02}-axis. Here $2d$ and $2L$ refer to the distance between microcrack centers, $2a$ refers to the length of each microcrack, ϕ denotes the oriented angle of each crack with respect to the x_{01}-axis. Two coordinate systems, *i.e.*, (x_{01}, x_{02}) and (x_1, x_2), are introduced which are related by equation (2.10), arbitrarily taking, *e.g.*, $\eta_1/a = 0.8$, $\eta_2/a = 2.3$, $L/a = 1.2$, and $d/a = 1.2$.

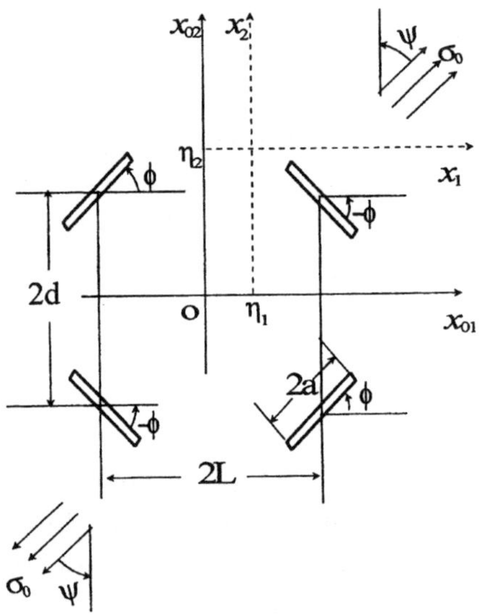

Figure 2.2. Four regularly distributed interacting cracks.

The four microcracks are assumed to be fully open. This assumption could be ensured when the values of the Mode I stress intensity factor at all tips are positive. It should be mentioned here that the purpose to present this example is not to study a real microcrack damage cell, rather, the purpose is to confirm the above derived independence of the M integral and to show the inherent relation between the value of the M integral and the reduction of the effective moduli arising from the formation of the four cracks.

Calculated values of the M integral are normalized by

$$M_R = \frac{\kappa + 1}{8\mu} \cdot \pi \cdot (\sigma_0 a_0)^2, \tag{2.13}$$

where a_0 refers to the average half length of all the microcracks. In this example, we know that $a_0 = a$. The subscript R refers to the remote loading level.

In order to ensure the assumption mentioned in section 2.1 that all the cracks are stationary, non-growing, the remote loading level is chosen in such a way that the numerical values of *SIF's* at every tip do not become critical. The critical condition is formulated as follows (Kanninen and Popelar, 1985):

$$(K_I^{/} K_{Ic})^2 + (K_{II}/K_{IIc})^2 = 1, \tag{2.14}$$

where the subscript c denotes the critical parameters of a certain brittle material. For example, for a certain kind of Al_2O_3 ceramic, K_{Ic}=0.259MN / $m^{3/2}$ and K_{IIc} = 0.518MN / $m^{3/2}$. The assumption of non-growing cracks is actually ensured when taking the half crack length a_0 to be 0.001m and the remote loading σ_O less than 2.737MN / m^2.

From equation (1.49) in Chapter 1 we have learned that the M integral is equal to twice the change of the total potential energy (*CTPE*) arising from the formation of a single crack in an infinite plane elastic body. However, as mentioned earlier, the M integral for multiple interacting cracks could not be given by using a simple summation (see equations (2.4) and (2.7). Therefore, it is not clear whether or not equation (1.49) holds in multiple cracks' situations since the displacement jump along the surfaces of each crack is not only induced from the remote loading, but also disturbed by the neighboring and interacting microcracks. In the present interacting situation, the change of the total potential energy (*CTPE*), denoted by U, should be (Chen, 2001a):

$$U = \sum_{k=1}^{N} \int_{-a}^{a} \sigma_{i2}^{\infty}(k) \cdot \Delta u_i^{(k)}(x_{1*}^{(k)}) dx_{1*}^{(k)}, \tag{2.15}$$

where the local coordinate $x_{1*}^{(k)}$ is chosen for the kth crack only, $\sigma_{i2}^{\infty}(k)$ refers to the released stresses when the kth crack is formed, N is the number of microcracks, and $\Delta u_i^{(k)}(x_{1*}^{(k)})$ is the displacement jump along the kth crack surfaces induced from the remote loading as well as the interaction with other cracks (see Figure 2.2).

Of great interest is the conclusion provided numerically by Chen (2001a), that reveals the validity of the relation (1.49) between the M integral (2.11) and the *CTPE* (2.15) even in the strongly interacting cracks situation. This means that, from the physical point of view, the M integral is identically equal to twice the *CTPE* irrespective of the shapes and numbers of the discontinuities involved in an infinite plane elastic body. Thus a definite value of the M integral exists for strongly interacting cracks. In turn, the value of the M integral does, of course, not depend on the selection of the global coordinates, either on the origin location, or on the coordinate rotation. Moreover, according to the conclusion for the L integral established by Budiansky and Rice (1973) and Herrmann and Herrmann (1981), this result leads to the simple relation between the M integral and the L integral in present interacting problems:

$$L = -\frac{1}{2}\frac{\partial M}{\partial \psi}, \tag{2.16}$$

since Herrmann and Herrmann (1981) have shown that the L integral represents the rotation energy release rate when the four cracks, as a single complex defect, rotate simultaneously. This rotation is equivalent to the applied loading rotation of an angle ψ.

Numerical results are listed in Table 2.1, in which the normalized values of the M integral and the change of the total potential energy are calculated by using equations (2.7) and (2.15), respectively.

Table 2.1. Normalized values of the M integral and the change of the total potential energy (*CTPE*) owing to the formation of the four microcracks shown in Figure 2.2 taking ϕ = 15 degrees.

ψ degrees	M_N	M_A	M	M_0	$2U$	M_T
0	3.5686	−0.0828	3.4858	3.4858	3.4858	3.7321
6	3.5442	−0.0875	3.4567	3.4567	3.4567	3.6942
10	3.5006	−0.0955	3.4051	3.4051	3.4051	3.6276
16	3.3931	−0.1128	3.2803	3.2803	3.2803	3.4689
30	2.9405	−0.1571	2.7834	2.7834	2.7834	2.8660
46	2.1080	−0.1596	1.9484	1.9484	1.9484	1.9396
60	1.2493	−0.0972	1.1521	1.1521	1.1521	1.1340
76	0.4479	−0.0006	0.4473	0.4473	0.4473	0.4707
90	0.1860	0.0371	0.2231	0.2231	0.2231	0.2680

It is shown that the values of the M integral calculated respectively in the two coordinate systems (x_{01}, x_{02}) and (x_1, x_2) (denoted by M_0 and M, respectively) coincide well with each other, which support the conclusion (2.12) given in

Section 2.2. Moreover, it is seen that the relation between the M integral and the *CTPE* derived in Chapter 1 is still valid in the four microcrack interacting case (Figure 2.2) as listed in the fifth and sixth columns of Table 2.1.

This reveals that the M integral for a microcracking solid always represents the change of the total potential energy owing to the formation of the microcracks irrespective of their shapes and numbers. Reader can imagine that the M integral for an evolutionary damage represents the progressive energy release owed to damage growth (*i.e.*, microcrack growth, microcrack coalescence, new microcrack nucleation (see, *e.g.*, Jun and Chen 1994b)).

In order to derive more aspects of the four microcrack array shown in Figure 2.2. Numerical results of the M integral and the *CTPE* for several fixed values of the location angle ($\phi = 0, 15, 30, 45, 60, 75, 90$ degrees) against the loading angle ψ from 0 to 90 degrees are plotted in Figures 2.3(a,b,c,d,e,f,g), respectively. Here the values of the net part M_N and the additional part M_A calculated respectively from (2.8) and (2.9) are plotted too for comparisons. Moreover, the values of M_T calculated from the so called Taylor's Model (Ju and Chen, 1994a,b) are also plotted, where the microcrack interactions are entirely neglected. Of great interest are the variable tendencies of the normalized values of the M integral against the loading angle for the seven fixed values of the oriented angle ϕ as shown in Figures 2.3a–2.3g. We will discuss them below in more detail.

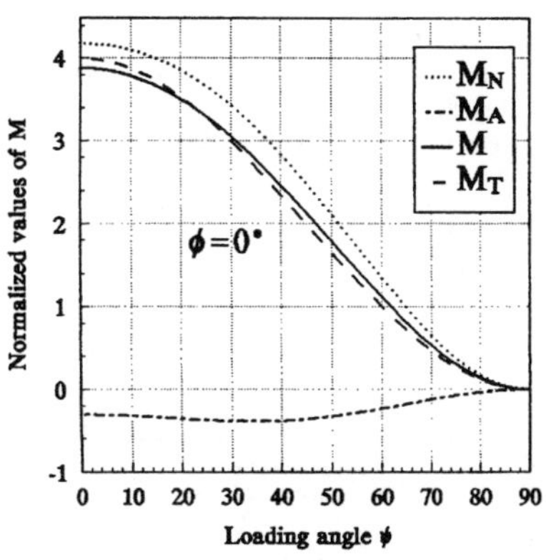

(a). Oriented angle $\phi = 0$.

It is seen from detailed comparisons between Figures 2.3a and 2.3g, or between Figure 2.3b and 2.3f, or between Figure 2.3c and 2.3e, that each pair shows the mirror effect with respect to the line of ψ=45 degrees when the inclined loading angle ψ increases from 0 to 90 degrees. These results confirm the independence of the values of the M integral from the rotation of the preferred coordinate system. A certain loading angle in Figures 2.3a, 2.3b and

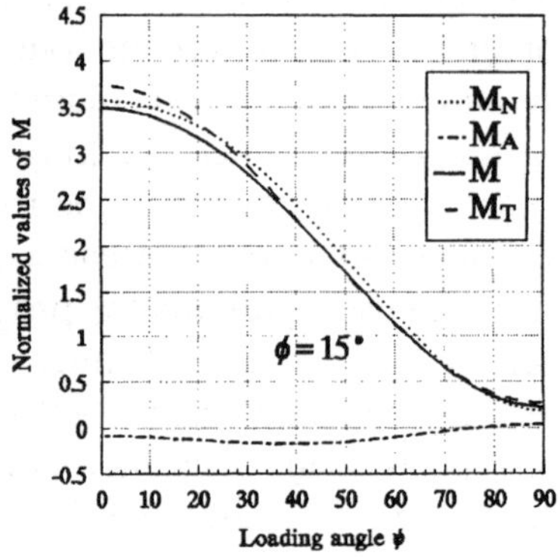

(b). Oriented angle $\phi = 15$ degrees.

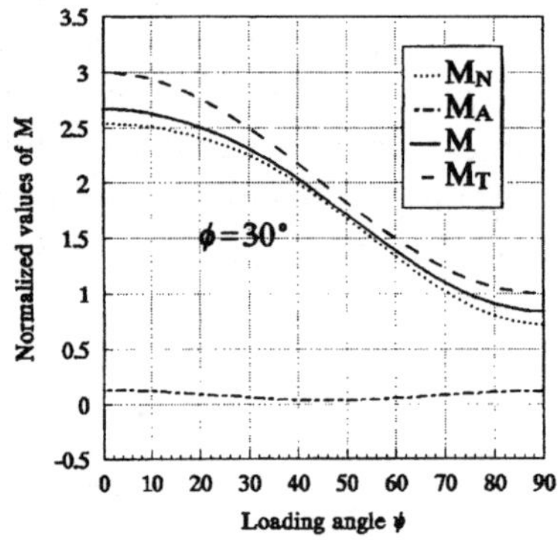

(c). Oriented angle $\phi = 30$ degrees.

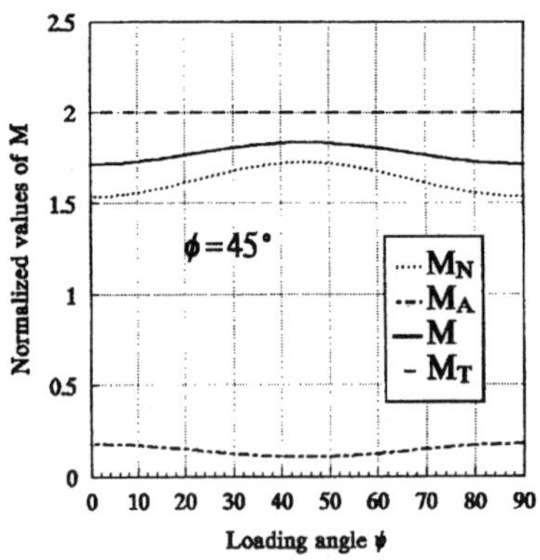

(d). Oriented angle $\phi = 45$ degrees.

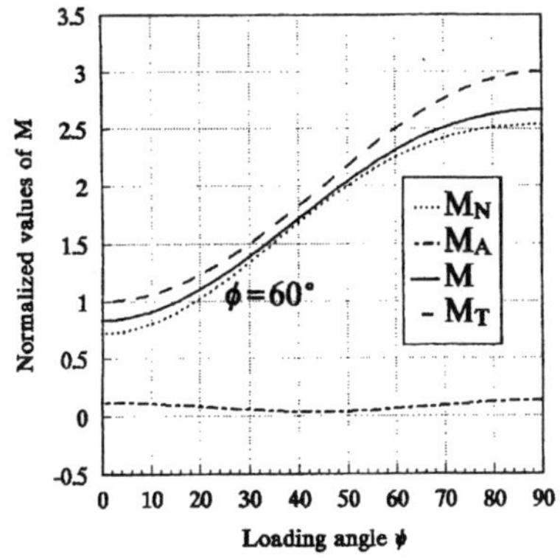

(e). Oriented angle $\phi = 60$ degrees.

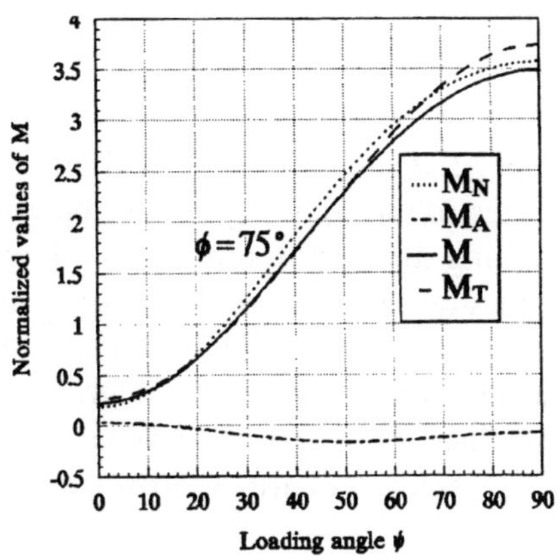

(f). Oriented angle $\phi = 75$ degrees.

2.3c, say ψ=16 degrees, is just corresponding to another certain loading angle in Figures 2.3g, 2.3f,and 2.3e, say, ψ=90−16=74 degrees, when the coordinate system rotates 90 degrees.

Of the most interest is the implicit relation between the values of M and the effective elastic moduli embedded in Figures 2.3a−2.3g for the four regularly microcracking solid. Readers can find that the maximum value of the M integral is just corresponding to the direction along which the largest reduction in the effective elastic moduli occurs. For example, ψ=0 degree in Figure 2.3a, 2.3b, and 2.3c, or instead, ψ=90 degrees in Figures 2.3e, 2.3f,and 2.3g. Correspondingly, the minimum value of the M integral is just corresponding to the direction along which the smallest reduction in the effective elastic moduli occurs. For example, ψ=90 degrees in Figures 2.3a, 2.3b and 2.3c, or instead,

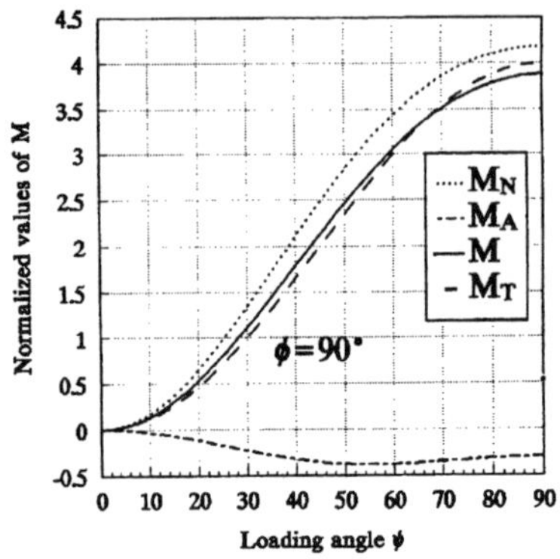

(g). Oriented angle $\phi = 90$ degrees.

Figure 2.3. M integral vs. loading ψ.

ψ=0 degree in Figures 2.3e, 2.3f and 2.3g. Indeed, there really exists an inherent relation between the values of the M integral and the effective elastic moduli induced from microcracking along different tensile loading directions. The larger value of the M integral along a tensile loading direction is, the larger reduction in the effective elastic moduli along the direction is. On contrary, Figure 2.3d shows that the values of the M integral are not sensitive to the tensile loading direction since the four microcracks configuration with ϕ=45 degrees in Figure 2.2 represents a nearly uniform reduction in the effective elastic moduli.

It is also seen from Table 2.1 and Figures 2.3a–2.3g that the divergence between the values of the M integral and those of M_T sometimes is remarkable owing to entirely neglecting the interacting effect amongst the four microcracks (the so called $Taylor$ model, Jun and Chen, 1994a,b). It is concluded that the divergence between Taylor's model and a more accurate damage model could be evaluated by using the M integral analysis. In most cases from Figure 2.3(a) to Figure 2.3(g), the values of Taylor's model overestimates the values of the M integral so that overestimates the reduction in the effective elastic moduli for microcrack weakened brittle solids.

3.2 Randomly distributed microcracks

In general, microcracks are randomly distributed in a practical structure. We need to consider a more complicated case, where twenty strongly interacting microcracks randomly are distributed in a plane elastic body as shown in Figures 2.4, 2.6, 2.8, and 2.10, respectively. Although the microcracks are randomly distributed, the length, the center coordinates, and the orientated angle of each crack are known beforehand for numerical calculations. Using the method proposed by Chen (1984), we obtain the calculated values of the M integral for each configuration of Figures 2.5, 2.7, 2.9, and 2.11, and these values are then normalized by equation (2.13). Moreover, the remote tensile loading is chosen in such a way that the stationary conditions of all the cracks should be satisfied. This could be done by substituting the calculated values of the *SIF's* at each tip of every crack one by one into equation (2.14) to ensure the tensile loading be less than the critical value.

Figures 2.5, 2.7, 2.9, and 2.11 show the variable tendencies of the M integral against the loading angle ψ for the crack configurations shown in Figures 2.4, 2.6, 2.8, and 2.10, respectively. It is seen that the values of the M integral are not sensitive to the tensile loading direction. This conclusion coincides well with the well known fact (Mori and Tanaka, 1973) that microcrack weakened brittle solids show an isotropic nature when microcracks are sufficiently randomly distributed although the effective elastic moduli are reduced significantly. It could be seen also that the additional part M_A always takes an inscrutable contribution to the M integral under strongly interacting situations. Generally speaking, such contribution cannot be neglected.

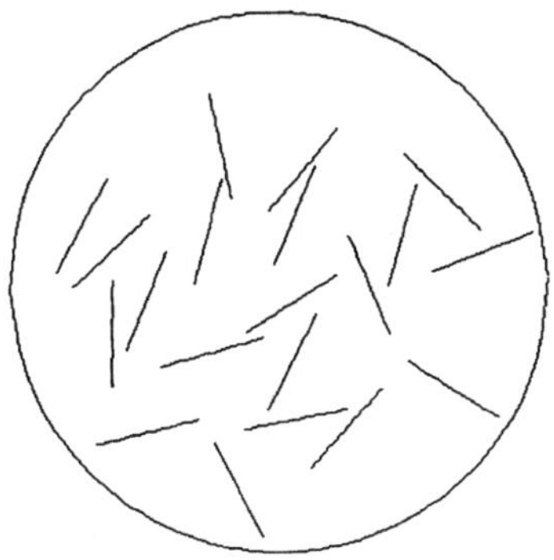

Figure 2.4. The first example of twenty randomly distributed cracks.

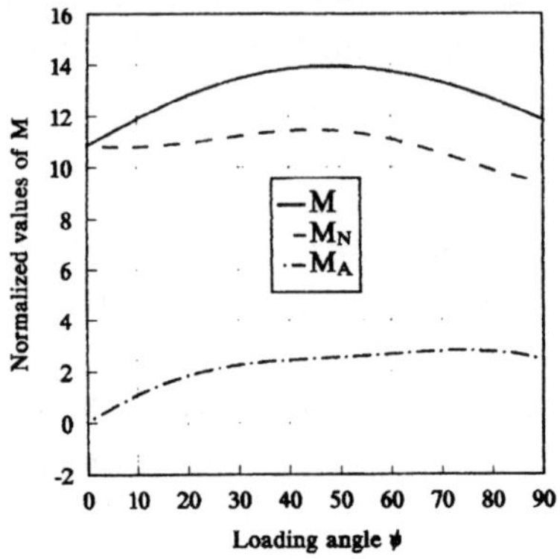

Figure 2.5. The computed values of the M integral for the first example.

Figure 2.6. The second example of twenty randomly distributed cracks.

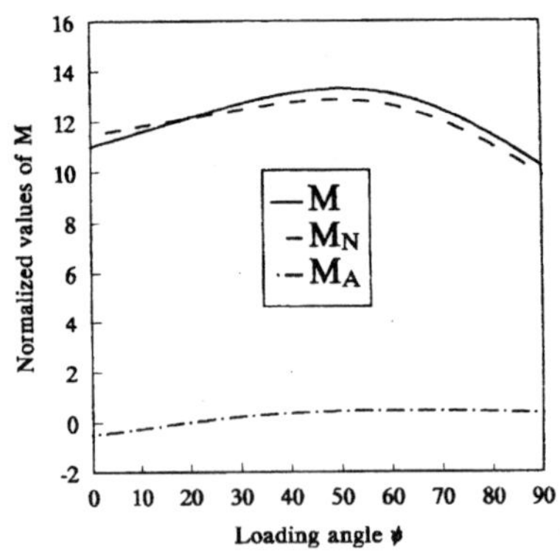

Figure 2.7. The computed values of the M integral for the second example.

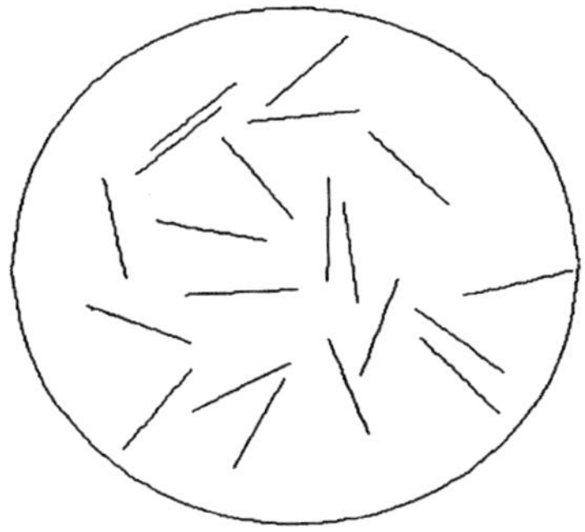

Figure 2.8. The third example of twenty randomly distributed cracks.

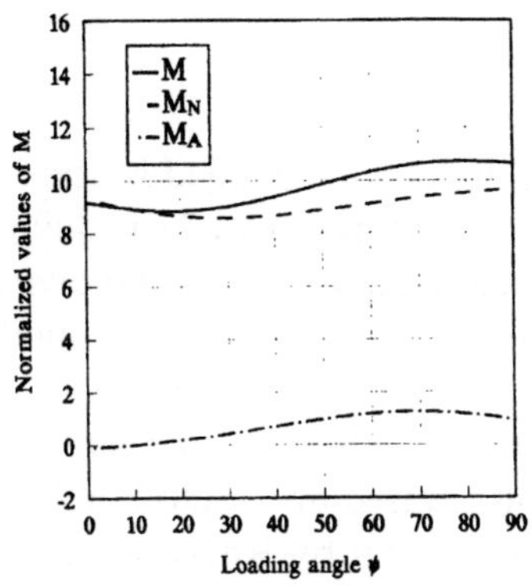

Figure 2.9. The computed values of the M integral for the third example.

Figure 2.10. The fourth example of twenty randomly distributed cracks.

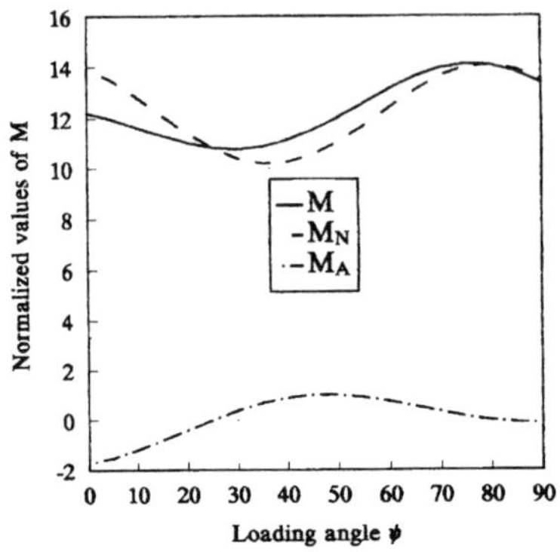

Figure 2.11. The computed values of the M integral for the fourth example.

Nevertheless, a real microcrack weakened brittle solid in practice with, for example, more than thousands of microcracks is quite difficult to be evaluated by using the traditional integral equation method owing to the intractably large amount of the computation. For this reason, a number of researchers focus their attention on the effective elastic moduli to account stiffness and stability of the solid with strongly interacting microcracks. For example, Jun and Chen (1994a,b) used the mixed fracture criteria in their effective elastic moduli methods. However, as pointed out by Kachanov (1992, 1993), what controls stiffness and what drives fracture in microcrack array settings is quite different. The first may shed little light on the second. From the discussions mentioned above, readers can understand that the M integral analysis does present a new way, from the phenomenological point of view, to describe the microcrack weakened problems in brittle solids.

Although the present M integral discussion only provides an initial attempt, which has not been proved to be universal to include all aspects of damage problems, it does provide some evidences or a possibility that there exists an alternative evaluation for microcracking brittle solids based on the conventional conservation laws such as M integral analysis. Therefore researchers may perform more efforts on such a new description to study the nonlinear mechanical behaviors of microcracking brittle solids. For example, the nondestructive technique customarily used for single crack problems, as proposed by King and Herrmann (1981) and discussed in Chapter 1, can be adopted to account the damage levels of microcracking brittle solid in experimental practices.

3.3 Short summary

From the above performed manipulations and numerical examinations, the following conclusions could be summarized:

(A) The M integral calculated along a closed contour surrounding all microcracks in an infinite two-dimensional microcracking solid is divided into two distinct parts as formulated by equation (2.7). One of them, called the net part and expressed by (2.8), is contributed by microcrack tip stress intensity factors as addressed by Freund (1978), while the other, called the additional part and expressed by (2.9), is contributed by the microcrack center coordinates in a global system and the J_k vector defined for each microcrack. Although the additional part involves the microcrack center coordinates, the M integral does not depend on the selection of the global coordinate system, either on the origin selection or on the coordinate rotation. This conclusion is directly deduced from the new conservation laws of the J_k vector, *i.e.,* (1.39) and (1.40) in Chapter 1.

(B) Although Kanninen and Popelar (1985) pointed out that there has been apparent little effort to apply the M integral in practical problems, the present discussion reveals that the M integral does play an important role in the description of damaged brittle solids owing to microcracking. Moreover, it does provide an effective measure for the evaluation of damage level in some examples shown in this section. In other words, the present investigation provides some important evidences that micro-structural statistical information is actually embedded in the formulation of a phenomenological parameter: the M integral.

(C) The dependence of the M integral values on the remote tensile loading direction reveals that the maximum value of the integral for a certain microcrack array coincides with the maximum loss direction in effective stiffness, whilst the minimum value coincides with the minimum loss direction in effective stiffness. This means that from the physical point of view the M integral (whose value has apparent directional nature with respect to different loading directions) is inherently related to the effective elastic moduli for stationary microcracking solids. Of course, the detailed relation between the effective elastic moduli and the values of the M integral remains to be adequately investigated.

(D) The M integral is proved numerically to be identically equal to twice the decrease of the change of the total potential energy owed to the formation of the pre-existing microcracks. Thus, there should be a simple relation between the M integral and the L integral, *i.e.,* equation (2.16), providing that the closed contour chosen to calculate the two integrals encloses all the microcracks or there are no other discontinuities outside of the contour. The discussion performed in this section supports the conclusion derived by Herrmann and Herrmann (1981) that the M and L integrals provide a more natural description of the *ERR* for plane cracks than the $J(= J_1)$ and J_2 even in strongly interacting situations.

(E) Potential applications of the new description based on the M integral analysis are remarkable. For example, the increasing values of the M integral during an evolutionary microcrack damage just represent the progressive energy release owed to the damage growth, *e.g.*, microcrack growth, microcrack coalescence, and new microcrack nucleation. This topic will inevitably require further investigation.

(F) The restriction of the present investigation is also considerable since the conservation laws (2.5) and (2.6) hold only in an infinite brittle solid as readers can see for equations (1.39) and (1.40) in Chapter 1. How to remove this restriction becomes a major task when we extend the M integral analysis to treat a finite microcracking solid. Chen (2001b) found that this could be done by evaluating the contribution induced from the interface between two dissimilar materials to the J_k vector, from which new formulations of the conservation laws accounting this contribution are derived instead of equations (2.5) and (2.6). Only in this way could the outside boundaries of a finite microcracking solid be considered as a special kind of interface between air and the solid or between a rigid body and the solid. This topic will be discussed in the next section in quite detail.

4. Conservation laws in bimaterials

In order to extend the discussion addressed in the above sections in infinite bodies to treat the multiple crack problem in a finite brittle solid, this section deals with the M integral analysis for microcrack damage in the brittle phase of a metal/ceramic bimaterial. The basic idea is based on such a concept that the outside boundaries of the finite solid could be considered as a special kind of interface between air and the solid or between a rigid body and the solid. Following the work did by Zhao and Chen (1997), who have found the contribution induced from the interface of the bimaterial to the J_2 integral, Chen (2001b) re-examined the conservation laws of the J_k vector in the interacting problem among the interface and multiple sub-interface microcracks. Unlike the conclusion given in the above sections, the vanishing nature of both components of the J_k integral vector at infinity should include the contribution arising from the interface (as another kind of discontinuity, see Smelser and Gurtin, 1977). Only after clarifying this point, could the new formulations of the conservation laws of the vector be deduced. This will lead to a correct way in which the extension of the investigation to treat a finite microcracking solid could be performed. Furthermore, another task of this section is aimed at providing a fundamental understanding of the interaction effect and the coalescence tendency for the microcracks based on the M integral analysis as influenced by the existence of the ductile phase of metal/ceramic bimaterials. Numerical results are given for the Cu/Al_2O_3 and the Ni/MgO bimaterial solids. It will be seen that the M integral actually plays an important role for microcrack damage in the brittle phase of the bimaterial solids, which could be considered as a measure of the damage level and the coalescence tendency among the microcracks. It will be seen also that for a certain microcrack array the value of the M integral in metal/ceramic bimaterial solids is always larger than that in homogeneous brittle solids. This means that the same microcrack array in the former cases shows lower stability and larger coalescence tendency than that in the latter cases owed to the existence of the ductile phase or owed to the interaction of the microcrack array with the interface. Of great interest is that the divergences of the M integral values in metal/ceramics bimaterials from those for the same microcrack array in homogeneous brittle solids are dominated by the first Dundurs parameter α with no regards to the second Dundurs parameter β. Extension of this investigation to treat microcrack damage in a half-plane brittle solid is addressed to show the potential applications of the M integral analysis when taking into account an outside boundary.

4.1 Conservation laws of the J_k vector in bimaterials

The microcrack damage problem considered here is shown in Figure 2.12, in which N microcracks are formed in the brittle phase (*i.e.*, the material 2 occupies the lower half plane) of a metal/ceramic bimaterial solid. The microcracks may be induced from thermal mismatch, impact or other sources, which may have different lengths (denoted by $2a_k$) and different oriented angles with respect to the interface (denoted by ϕ_k). Two global coordinate systems are introduced respectively, say, (x_1^0, x_2^0) and (x_1, x_2). The origin of (x_1^0, x_2^0) is taken to be located on the interface and the x_1^0-axis to be parallel to the interface, while the origin of (x_1, x_2) is taken to be not located on the interface and the x_1-axis to be still parallel to the interface (See Figure 2.12). Therefore, the coordinates of microcrack centers in (x_1^0, x_2^0) denoted by $(\xi_1^{0(k)}, \xi_2^{0(k)})$ and those in (x_1, x_2) denoted by $(\xi_1^{(k)}, \xi_2^{(k)})$ are related by

$$\begin{aligned} \xi_1^{(k)} &= \xi_1^{0(k)} + \eta_1, \\ \xi_2^{(k)} &= \xi_2^{0(k)} + \eta_2, \qquad\qquad k = 1, 2, ..., N \end{aligned} \tag{2.17}$$

where the superscript (k) refers to the kth microcrack, and η_1 and η_2 are constants which represent the translation from the system (x_1^0, x_2^0) to the system (x_1, x_2).

Besides these, two local coordinate systems are also introduced for a typical microcrack (k) below the interface. Here $(x_1^{(k)}, x_2^{(k)})$ is parallel to (x_1^0, x_2^0), whilst $(x_{1*}^{(k)}, x_{2*}^{(k)})$ is oriented with respect to $(x_1^{(k)}, x_2^{(k)})$ by ϕ_k. What is more, three kinds of closed contour, *i.e.*, Γ_0, Γ and Γ_k are introduced (See Figure 2.12). Here Γ_0 encloses all the microcracks and cuts the interface at the points A and$B$, Γencloses all the microcracks but does not cut the interface, and Γ_k encloses the kth microcrack only but does not cut the interface.

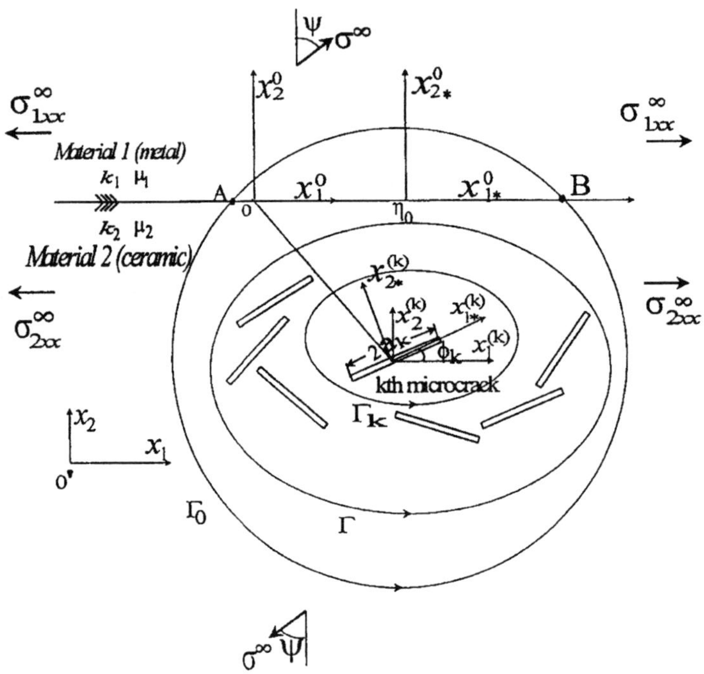

Figure 2.12. Multiple interacting cracks beneath an interface.

The bimaterial solid is loaded by the remote tensile stress $\sigma^\infty$with an inclined angle ψ and the remote horizontal stresses σ_{1xx}^∞ and σ_{2xx}^∞.

According to the path independent nature of the J_k integral vector, the following formulations are valid (Smelser and Gurtin, 1977; Zhao and Chen, 1997):

$$
\begin{aligned}
J_k(\Gamma o) &= \int_{\Gamma o} (W n_k - T_l u_{l,k}) ds \\
&= J_k(\Gamma) + J_k(\overline{AB}), \qquad (k = 1, 2),
\end{aligned}
\tag{2.18}
$$

where $J_k(\Gamma o)$,$J_k(\Gamma)$, and $J_k(\overline{AB})$ denote the values of the vector calculated along the three different contours Γo,Γ, and$\overline{AB}$, respectively (see Figure 2.12). The last two ones could be given as follows

$$J_k(\Gamma) = \sum_{K=1}^{N} J_k(\Gamma_K) \qquad (k=1,2), \tag{2.19}$$

$$J_k(\overline{AB}) = \int_{\overline{AB}} ([W]n_k - T_l[u_{l,k}])ds \qquad (k=1,2), \tag{2.20}$$

where [ ] denotes the jump of the corresponding quantities across the interface.
Obviously

$$J_1(\overline{AB}) = \int_{\overline{AB}} ([W]n_1 - T_l[u_{l,1}])ds = 0. \tag{2.21}$$

This is because the normal component n_1 vanishes and the displacement continuity conditions across the interface are met.

Therefore the conservation law for the first component of the vector is still valid (see equation (1.39) or (2.5). Attention should only be focused on the J_2 integral analysis in (2.6) to establish a new conservation law instead of equation (1.40) in Chapter 1 or (2.6) in section 2.2.

According to the work of Smelser and Gurtin (1977) the following equation is valid in a bimaterial without cracks

$$J_2 = \int_{\Gamma 0} (Wn_2 - \sigma_{ij}n_ju_{i,2})ds - \int_{l} ([w] - \sigma_{i2}[u_{i,2}])ds = 0, \tag{2.22}$$

where Γo refers to a closed contour cutting the interface at points A and B, while l refers to the segment of the interface from A to B.

For the multiple sub-interface cracks shown in Figure 2.12, it follows that

$$\int_{\Gamma 0} (Wn_2 - \sigma_{ij}n_ju_{i,2})ds = \int_{l} ([w] - \sigma_{i2}[u_{i,k}])ds - \int_{\Gamma} (Wn_2 - \sigma_{ij}n_ju_{i,2})ds, \tag{2.23}$$

where $\Gamma$is a closed contour surrounding all the cracks but does not cut the interface (see Figure 2.12).

Assume that the stress field and the displacement field consist of two distinct parts. One of them is induced from the uniform remote loading conditions and the other is induced from the existence of the multiple sub-interface cracks:

$$\sigma_{ij} = \sigma_{ij}^U + \sigma_{ij}^E,$$
$$u_i = u_i^U + u_i^E, \qquad (i, j = 1, 2), \tag{2.24}$$

where the superscripts U and E refer to the uniform remote loading and the existence of the cracks, respectively.

Thus, equation (2.23) could be reformulated as

$$\int_{\Gamma 0} ((W^U + W^E + W^{UE} + W^{EU})n_2 - (\sigma_{ij}^U + \sigma_{ij}^E)n_j(u_{i,2}^U + u_{i,2}^E))ds$$
$$= \int_l ([W^U + W^E + W^{UE} + W^{EU}]n_2 - (\sigma_{ij}^U + \sigma_{ij}^E)n_j[u_{i,2}^U + u_{i,2}^E])ds$$
$$+ \int_\Gamma ((W^U + W^E + W^{UE} + W^{EU})n_2 - (\sigma_{ij}^U + \sigma_{ij}^E)n_j(u_{i,2}^U + u_{i,2}^E))ds, \tag{2.25}$$

where the superscript UE and EU refer to the interleaving terms contributed by the stresses arising from the remote loading and the displacements arising from the existence of the cracks or vice versa.

Obviously, the summation among the interleaving terms vanishes as a result of the well known Betti reciprocal theory. Therefore equation (2.25) leads to

$$\int_{\Gamma 0} ((W^U + W^E)n_2 - (\sigma_{ij}^U n_j u_{i,2}^U + \sigma_{ij}^E n_j u_{i,2}^E)ds$$
$$= \int_l ([W^U + W^E]n_2 - (\sigma_{ij}^U n_j [u_{i,2}^U] + \sigma_{ij}^E n_j [u_{i,2}^E])ds \tag{2.26}$$
$$+ \int_\Gamma ((W^U + W^E)n_2 - (\sigma_{ij}^U + \sigma_{ij}^E)n_j(u_{i,k}^U + u_{i,2}^E))ds.$$

When setting the closed contour Γ_O to be infinite large so that the whole interface from $-\infty$ to ∞ is enclosed in it, the physical quantities σ_{ij}^E, u_i^E, W^E on Γ_O vanish owing to their asymptotic nature at infinity. But they may not vanish on the closed contour Γ or on the interface from $-\infty$ to ∞. By substituting equation (2.22) into (2.26), the latter becomes (Chen, 2001b)

$$\int_{-\infty}^{\infty} ([W^E]n_2 - \sigma_{ij}^E n_j [u_{i,2}^E])ds + \int_\Gamma (W^E n_2 - \sigma_{ij}^E n_j u_{i,2}^E)ds = 0, \tag{2.27}$$

which could be reformulated in a compact form as

$$\sum_{k=1}^{N} J_2^{(k)} + J_{2\infty} = 0, \qquad (2.28)$$

keeping in mind that the two terms in the left hand side of (2.28) denote the contributions of the N sub-interface cracks and the whole interface to the J_2 integral, respectively.

In other words, the contribution induced from the sub-interface multiple cracks cancels with those induced from the whole interface. Therefore equation (2.28) represents new conservation law in bimaterials with many cracks. Readers will find that this law is very useful in the forthcoming manipulations of the M integral analysis.

Now, we are ready to perform some numerical results. After using the technique proposed by Zhao and Chen (1997) and solving the interaction problem shown in Figure 2.12, we can calculate first term in equation (2.28) by making a summation of the contributions induced from each sub-interface crack as we did equation (1.36) presented in Chapter 1. Of course, the local global coordinate translation of the J_k vector should be adopted for each oriented microcrack (see Figure 2.12) as we discussed in section 2.2 and the contribution induced from the traction-free surfaces of each crack should also be taken into account (Herrmann and Herrmann, 1981). However, the second term in equation (2.28) provides a new problem, which should be solved by developing a new technique for treating the infinite integral intervals involved. Such a technique is similar to that cited in Appendix of this chapter by using the Chebyshev polynomial and the Chebyshev integration together with some variable translations (Fleck, *et al.*, 1991, Zhao and Chen, 1997; Chen 2001b).

In order to confirm the new conservation law (2.28) a numerical example is presented here. Without loss of generality, a single sub-interface crack is considered, which is parallel to the interface in the brittle phase of a bimaterial solid (Figure 2.13). The distance between the crack and the interface is denoted by h, while a refers to the half length of the crack. The mathematical technique for solving this simple problem is based on the work did by Zhao and Chen (1997), from which all values of stresses and displacements at any point in the metal/ceramic bimaterial could then be given without any difficulty. Equations (2.27) and (2.28) can then be calculated by using a similar technique as cited in Appendix.

Thus, the contributions of the sub-interface crack and the whole interface $(-\infty, +\infty)$ to the J_2 integral could be calculated numerically. Numerical results

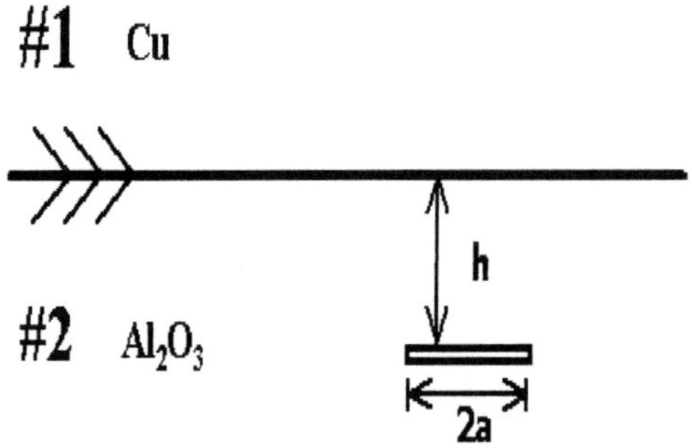

Figure 2.13. A sub-interface crack beneath the interface

for Cu/Al_2O_3 bimaterial solid and under the remote loading $\sigma_{yy}^{\infty} = \sigma_{x2}^{\infty} = \sigma_{\infty}$=constant are shown below:

$$J_2 = \int_{-\infty}^{+\infty} ([W] - \sigma_{k2}[u_{k,2}])dx = -0.1387576 \frac{\sigma_{\infty}^2 (1 - \nu_2)\pi a^2}{2\mu_2}, \qquad (2.29)$$

for the whole interface, and

$$J_2 = \int_{-a}^{a} ([W])dx = 0.1387576 \frac{\sigma_{\infty}^2 (1 - \nu_2)\pi a^2}{2\mu_2}, \qquad (2.30)$$

for the sub-interface crack, where [] denotes the jump of the corresponding quantities across the interface or across the sub-interface crack..

Equations (2.29) and (2.30) are well agreed with each other, with a change of sign. They do confirm the new conservation law (2.28) for the J_2 integral. In other words, the total contribution of the sub-interface crack and the whole interface to the J_2 integral vanishes, providing that the closed contour is taken to be infinite large, which encloses the whole interface as well as the crack. Indeed, this supports such a concept that the whole interface represents a different kind of discontinuity from cracks.

In fact, by comparing equation (2.6) in section 2.2 of this chapter with equation (2.28) in this section, it is seen that the latter could be considered as a special case of the former. If we use N to denote the number of all the discontinuities in the bimaterial rather than the number of the cracks only, the former (2.6) will become universal. Of course, other discontinuities, such as voids, kinking cracks, curve cracks, or inclusions, will not alter the conservation laws of the J_k vector if the closed contour encloses all kinds of discontinuities completely.

It should be emphasized also that the conservation laws of (2.28) derived in this section and (2.5) derived in section 2.2 are only suitable to treat the special case, where the interface is perpendicular to the x_2-axis. If it is not so, for example, the interface is perpendicular to the x_1-axis rather than the x_2-axis, the above manipulations should be repeated for evaluating the J_1 integral rather than the J_2 integral. The new conservation laws should then be replaced by

$$\sum_{k=1}^{N} J_1^{(k)} + J_{1\infty} = 0, \tag{2.31}$$

$$\sum_{k=1}^{N} J_2^{(k)} = 0, \tag{2.32}$$

keeping in mind that $J_{1\infty}$ refers to the contribution induced from the whole vertical interface to the first component of the vector.

4.2 Independence of the M integral from the coordinate selection in bimaterials

After clarifying the role played by the J_k vector in bimaterials with multiple sub-interface cracks, the independence of the M integral from the coordinate selection shown in Figure 2.12 could be re-examined. In this section, analytical treatments are divided into two sides. First of them is associated with the global coordinates originated on the interface, while the second with the global coordinates not originated on the interface.

Special attention is always focused on the contribution arising from the segment $\overline{AB}$. Indeed, confusion, if it exists, is always induced from the misunderstanding or oversight when treating the segment. At present, we have not made it clear as what happens to the M integral in the present problem shown in Figure 2.12 when choosing the different closed contours Γ (below the interface) or Γ_0 (across the interface), respectively. In fact, many people believe that the value of the M integral should depend on the location of the coordinates. This confusion will be clarified below.

(A) The origin of the coordinate system located on the interface

The global coordinate system (x_1^0, x_2^0) is assumed to be originated on the interface (see Figure 2.8). From the original definition of the M integral as cited by (1.2) and (1.19) in chapter 1 (Budiansky and Rice, 1973), the M integral calculated along Γo in Figure 2.12 should be equal to the summation of the contributions induced from the segment $\overline{AB}$ and Γ, respectively:

$$M_{\Gamma_0}(x_1^0, x_2^0) = M_{\overline{AB}}(x_1^0, x_2^0) + M_{\Gamma}(x_1^0, x_2^0), \tag{2.33}$$

where the subscripts Γo, $\overline{AB}$, and Γ refer to the contributions calculated over the three contours, respectively (see Figure 2.12). Since the origin of the system (x_1^0, x_2^0) lies on the interface (Park and Earmme, 1986), the first term in the right hand side is deduced to

$$M_{\overline{AB}} = \oint_{\overline{AB}} ([W]x_i^0 n_i - T_i[u_{i,l}]x_l^0)ds, \; (l, i = 1, 2). \tag{2.34}$$

Obviously, it vanishes in the present case because $x_2^0 \equiv 0$, $n_1 \equiv 0$, and $[u_{i,1}] \equiv 0$. Therefore the following manipulation could be given by making the translation from the global system (x_1^0, x_2^0) to the local system $(x_1^{(k)}, x_2^{(k)})$, *i.e.*, equation (2.1):

$$\begin{aligned} M_{\Gamma_0}(x_1^0, x_2^0) &= M_{\Gamma}(x_1^0, x_2^0) = \sum_{k=1}^{N} M_{\Gamma_k}(x_1^0, x_2^0) \\ &= \sum_{k=1}^{N} \left\{ \oint_{\Gamma_k} (W x_i^0 n_i - T_l u_{l,i} x_i^0) ds \right\} \\ &= \sum_{k=1}^{N} \left\{ \oint_{\Gamma_k} \left[W(x_i^{(k)} + \xi_i^{o(k)}) n_i - T_l u_{l,i}(x_i^{(k)} + \xi_i^{o(k)}) \right] ds \right\} \\ &= \sum_{k=1}^{N} \left\{ \oint_{\Gamma_k} \left[W x_i^{(k)} n_i - T_l u_{l,i} x_i^{(k)} \right] ds + \xi_i^{o(k)} \oint_{\Gamma_k} (W n_i - T_l u_{l,i}) ds \right\} \\ &= \sum_{k=1}^{N} \left\{ M_{\Gamma_k}(x_1^{(k)}, x_2^{(k)}) + \xi_1^{o(k)} J_1^{(k)} + \xi_2^{o(k)} J_2^{(k)} \right\}, \end{aligned} \tag{2.35}$$

where $M_{\Gamma_k}(x_1^{(k)}, x_2^{(k)})$ is evaluated along Γ_k in the local system $(x_1^{(k)}, x_2^{(k)})$, $J_1^{(k)}$ and $J_2^{(k)}$ are contributed by the kth microcrack, which could be evaluated by using the projected relations (see equation (2.5) in section 2.2 or see the work by Zhao and Chen (1997) and Chen (2001b), which are cited here for convenience

$$\begin{aligned} J_1^{(k)} &= J_{1*}^{(k)} \cos\phi_k - J_{2*}^{(k)} \sin\phi_k, \\ J_2^{(k)} &= J_{1*}^{(k)} \sin\phi_k + J_{2*}^{(k)} \cos\phi_k, \end{aligned} \tag{2.36}$$

and

$$
\begin{aligned}
J_{1*}^{(k)} &= \frac{\kappa_2 + 1}{8\mu_2} \left[(K_{IR}^{(k)})^2 + (K_{IIR}^{(k)})^2 - (K_{IL}^{(k)})^2 - (K_{IIL}^{(k)})^2 \right], \\
J_{2*}^{(k)} &= \frac{\kappa_2 + 1}{4\mu_2} \left[K_{IL}^{(k)} K_{IIL}^{(k)} - K_{IR}^{(k)} K_{IIR}^{(k)} \right] + F_{2*}^{(k)},
\end{aligned}
\tag{2.37}
$$

where the subscript star refers to the quantities in the local system $(x_{1*}^{(k)}, x_{2*}^{(k)})$ (see Figure 2.12), the subscripts R and L denote the right and left tips of the kth microcrack, the subscripts I and II denote the Mode I and Mode II stress intensity factors, respectively, κ_2 and μ_2 are elastic constants of the brittle phase, *i.e.*, the material 2, and $F_{2*}^{(k)}$ denotes the contribution induced from the traction-free faces of the kth microcrack to the second component of the J_k vector (see equation (1.30) in Chapter 1).

It is seen from (2.35) that the formulation of the M integral in the present case is similar to (2.7) in section 2.2 for homogeneous brittle solids. It could also be divided into two distinct parts. One of them, *i.e.*, the net part denotes by M_N, is contributed by the crack tip stress intensity factors of each sub-interface crack:

$$
\begin{aligned}
M_N &= \sum_{k=1}^{N} M_{\Gamma_k}(x_1^{(k)}, x_2^{(k)}) \\
&= \sum_{k=1}^{N} \frac{\kappa_2 + 1}{8\mu_2} \left\{ (K_{IR}^{(k)})^2 + (K_{IIR}^{(k)})^2 + (K_{IL}^{(k)})^2 + (K_{IIL}^{(k)})^2 \right\} a_k,
\end{aligned}
\tag{2.38}
$$

where a_k refers to the half length of the kth microcrack. The other part, *i.e.*, the additional part denoted by M_A, is induced from the coordinates of the kth microcrack center and the J_k vector evaluated in the local system $(x_{1*}^{(k)}, x_{2*}^{(k)})$ and along the closed contour Γ_k:

$$
M_A = \sum_{k=1}^{N} \left\{ \xi_1^{o(k)} J_1^{(k)} + \xi_2^{o(k)} J_2^{(k)} \right\},
\tag{2.39}
$$

or in the local system $(x_{1*}^{(k)}, x_{2*}^{(k)})$:

$$
M_A = \sum_{k=1}^{N} \left\{ \xi_1^{o(k)} (J_{1*}^{(k)} \cos\phi_k - J_{2*}^{(k)} \sin\phi_k) + \xi_2^{o(k)} (J_{1*}^{(k)} \sin\phi_k + J_{2*}^{(k)} \cos\phi_k) \right\}.
\tag{2.40}
$$

However, it should be emphasized that all the physical quantities in equations (2.36)–(2.40) have been influenced by the existence of the interface. In other words, the values of the J_k vector and the M integral are quite different from those in homogeneous cases owing to the interaction between the interface and the sub-interface cracks. A fundamental understanding of this influence will be discussed in the next section.

Since the origin of the system (x_1^o, x_2^o) lies on the interface, the crack center coordinate $\xi_2^{o(k)}(k = 1, 2, \cdots, N)$ could not vary when the origin moves a distance, say η_0, from one point to other point on the interface (See Figure 2.8). What is more, after the translation with the distance η_0 the formulation (2.39) becomes

$$M_A^* = \sum_{k=1}^{N} \left\{\xi_1^{o(k)} J_1^{(k)} + \xi_2^{o(k)} J_2^{(k)}\right\} + \eta_0 \sum_{k=1}^{N} J_1^{(k)}, \tag{2.41}$$

where the superscript star denotes the quantities in a new system (x_{1*}^o, x_{2*}^o) whose origin also lies on the interface, but whose vertical axis translates the distance η_0 (see Figure 2.12).

As discussed above, the last term in the right hand side of (2.41) vanishes since the interface is parallel to the x_1-axis. Therefore the independence of the M integral from the global system selection has been proved since the interface segment $\overline{AB}$ has no contribution to the J_1 integral and in turn has no contribution to the additional part of the M integral:

$$\begin{aligned} M_A &= M_A^*, \\ M_N &= M_N^*. \end{aligned} \tag{2.42}$$

(B) The origin of the coordinate system not located on the interface

Reconsider the microcrack damage problem shown in Figure 2.12 but take another coordinate system (x_1, x_2), whose origin is not located on the interface. After performing some manipulations, the following formulation is given:

$$\begin{aligned} M_{\Gamma_o}(x_1, x_2) &= M_{\overline{AB}}(x_1, x_2) + M_{\Gamma}(x_1, x_2) \\ &= \sum_{k=1}^{N} \left\{ M_{\Gamma_k}(x_1^{(k)}, x_2^{(k)}) + \xi_1^{(k)} J_1^{(k)} + \xi_2^{(k)} J_2^{(k)} \right\} \\ &+ M_{\overline{AB}}(x_1, x_2), \end{aligned} \tag{2.43}$$

where the last term $M_{\overline{AB}}(x_1, x_2)$ is induced from the segment $\overline{AB}$ which should be specially considered with caution in the (x_1, x_2) system:

$$\begin{aligned} M_{\overline{AB}}(x_1, x_2) &= \int_A^B (W x_i n_i - T_l u_{l,i} x_i) ds \\ &= \eta_2 \int_{x_{1A}}^{x_{1B}} \Big[(W^+ - W^-) - (\sigma_{l2}^+ u_{l,2}^+ - \sigma_{l2}^- u_{l,2}^-)\Big] dx, \end{aligned} \tag{2.44}$$

in which η_2 is the vertical distance from the origin O' of (x_1, x_2) to the interface, $x_{1A}$and x_{1B} are the horizontal coordinates of points A and B on the interface (see Figure 2.12), respectively, and the superscripts $+$ and $-$ refer to the upper and lower boundary values on the interface, respectively.

Obviously, the path dependency of the M integral could be seen in (2.44) since different segment $\overline{AB}$ will lead to different value of $M_{\overline{AB}}(x_1, x_2)$. However, when setting the closed contour Γ_0 to be infinite large so that the whole interface is enclosed, equations (2.43) and (2.44) will then become:

$$M_{\Gamma_\infty}(x_1, x_2) = \sum_{k=1}^{N} \left\{ M_{\Gamma_k}(x_1^{(k)}, x_2^{(k)}) + \xi_1^{(k)} J_1^{(k)} + \xi_2^{(k)} J_2^{(k)} \right\} + M_\infty(\eta_2) \tag{2.45}$$

and

$$M_\infty(\eta_2) = \eta_2 J_{2\infty} = \eta_2 \int_{-\infty}^{\infty} \Big[(W^+ - W^-) - (\sigma_{l2}^+ u_{l,2}^+ - \sigma_{l2}^- u_{l,2}^-)\Big] ds, \tag{2.46}$$

where (2.46) could be evaluated after the interaction problem shown in Figure 2.12 is solved (Zhao and Chen, 1997; Chen, 2001b). Obviously, the value of $J_{2\infty}$ does not depend on the selection of the coordinates although the value of $M_\infty(\eta_2)$ may depend on η_2.

It should be emphasized that equation (2.28) has shown the vanishing nature of the total contribution arising from all the microcracks and the whole interface to the J_2 integral. When making a coordinate transformation from the (x_1, x_2) system to the (x_1^0, x_2^0) system whose origin lies on the interface, equation (2.45) yields that

$$
\begin{aligned}
M_{\Gamma_\infty} &= \sum_{k=1}^{N}\left\{M_{\Gamma_k}(x_1^{(k)},x_2^{(k)}) + (\xi_1^{0(k)}+\eta_1)J_1^{(k)} + (\xi_2^{0(k)}+\eta_2)J_2^{(k)}\right\} \\
&+ \eta_2 J_{2\infty} = \sum_{k=1}^{N}\left\{M_{\Gamma_k}(x_1^{(k)},x_2^{(k)}) + \xi_1^{0(k)}J_1^{(k)} + \xi_2^{0(k)}J_2^{(k)}\right\} \\
&+ \eta_1 \sum_{k=1}^{N} J_1^{(k)} + \eta_2 \sum_{k=1}^{N} J_2^{(k)} + J_{2\infty}.
\end{aligned}
\tag{2.47}
$$

Obviously the second term in the right hand side of (2.47) vanishes owing to the validity of (2.5) in section 2.2 (since the interface is parallel to the x_1-axis). What is more, the third term vanishes too owing to the new conservation law of the J_k vector (2.28) established above. Therefore equation (2.47) directly leads to the following conclusion:

$$
M_{\Gamma_\infty}(x_1,x_2) = M_{\Gamma_0}(x_1^0,x_2^0), \tag{2.48}
$$

which confirms the independence of the M integral from the selection of the origin of the global coordinate system, whatever the origin lies on the interface or not.

It should be emphasized that equation (2.48) is conditional, that is, the closed contour should be infinitely large so that the whole interface is enclosed in it, otherwise the finite segment $\overline{AB}$ on the interface leads to the path dependency of the $M_{\Gamma_\infty}(x_1,x_2)$ integral. In the forthcoming manipulations, it is always assumed that the origin of the global coordinate system lies on the interface and the closed contour be infinite large.

As regards the rotation of the coordinates, it is easy to prove directly from the definition of the M integral that its value remains unaltered when the rotation of the coordinate system occurs (see Park and Earmme, 1986). Consequently the value of the M integral in bimaterials does not depend on the selection of the coordinates, either on the translation or on the rotation. The assumption is that the closed contour has been chosen as infinitely large and not only all the microcracks, but also the whole interface are enclosed in it. Therefore the confusion mentioned above which has been bothering many people for a long time is clarified.

4.3 *M* integral analysis for microcrack damage in the brittle phase

Numerical results of the M integral analysis are given in this section for multiple microcracks beneath the interface in the brittle phase of metal/ceramic bimaterial solids. Here the global coordinate system is always originated from a certain point on the interface so that the contribution of the microcracks to the M integral could be evaluated by (2.38) and (2.40) with no regards to those induced from the interface. The numerical results for crack tip *SIF's* as well as those contributed by the traction-free faces of the microcracks to the J_2 integral are computed by using the technique proposed by Zhao and Chen (1997). The detailed derivation procedure is no longer repeated here. The bimaterial solids considered below are loaded by the remote tensile stress σ^∞ parallel to the vertical axis x_2^0. Moreover, two tensile stresses σ_{1xx}^∞ and σ_{2xx}^∞ are aided such that the following relation is met (Isida and Noguch, 1994):

$$\sigma_{2xx}^\infty + \sigma^\infty \sin^2\psi = \frac{1}{1+\kappa_2}\left\{\frac{\mu_1}{\mu_2}(1+\kappa_2)(\sigma_{1xx}^\infty + \sigma^\infty \sin^2\psi)\right\}$$
$$+\frac{1}{1+\kappa_2}\left[3-\kappa_2-\frac{\mu_2}{\mu_1}(3-\kappa_1)\right]\sigma_{yy}^\infty, \tag{2.49}$$

where

$$\sigma_{yy}^\infty = \sigma_{1yy}^\infty = \sigma_{2yy}^\infty = \sigma^\infty \cos^2\psi, \tag{2.50}$$

and μ_1 and κ_1 are material constants of the upper material (see Figure 2.12).

In the numerical results presented below, the stress σ_{2xx}^∞ is always assumed to be zero, while the stress σ_{1xx}^∞ does not appear in the manipulations and the calculations although equation (2.49) is actually met (Zhao and Chen, 1997).

(A) Two microcracks beneath interface

Four kinds of two microcrack configurations are considered as shown in Figures 2.14 –2.17, respectively. Here d denotes the distance between the two centers of the microcracks, h denotes the distance from the center of one microcrack to the interface, and α_0 and ϕ_0 denote the location angle and the oriented angle, respectively. The lengths of the two microcracks are assumed to be equal with each other and all numerical results are confirmed by the consistency check (2.28) in section 2.4.2 of this chapter. Here Cu/Al_2O_3 and Ni/MgO bimaterial solids as well as a homogeneous elastic solid are considered for making comparisons. The ductile phase, *i.e.,* Cu or Ni, always occupies the upper half plane, while the brittle phase, *i.e.,* Al_2O_3 or MgO, always occupies the lower half plane. the Dundur parameters α, β, and the oscillatory index ε for the two bimaterial solids are shown in Table 2.2 (Hutchinson *et. al,* 1987). It is noted that α is always negative, while β is positive for Ni/MgO and negative for Cu/Al_2O_3, and ε is negative for Ni/MgO and positive for Cu/Al_2O_3. Obviously, α, β, and ε are equal to zero in homogeneous cases.

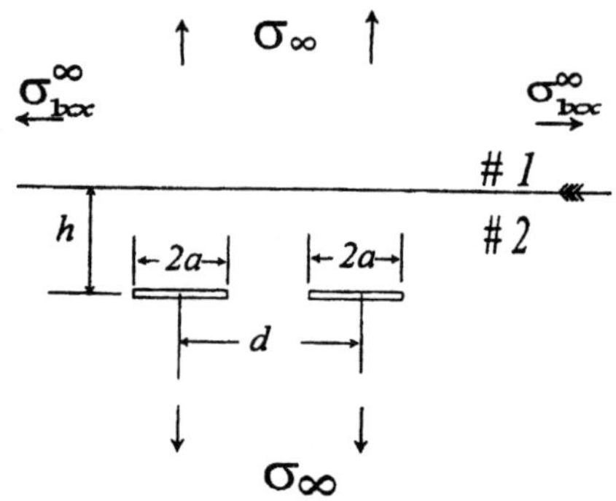

Figure 2.14. First example of two microcrack configuration beneath interface.

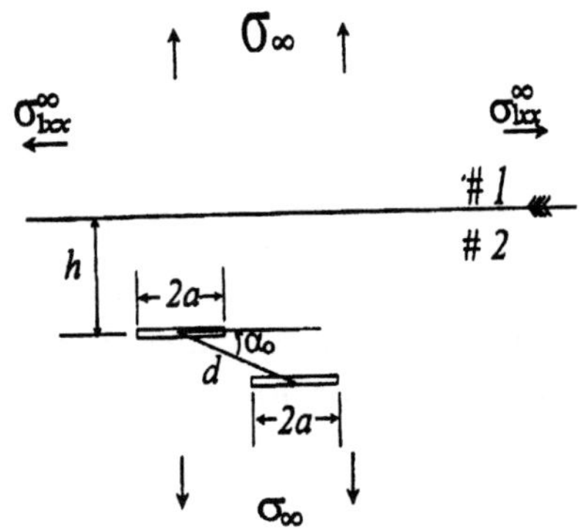

Figure 2.15. Second example of two microcrack configuration beneath interface.

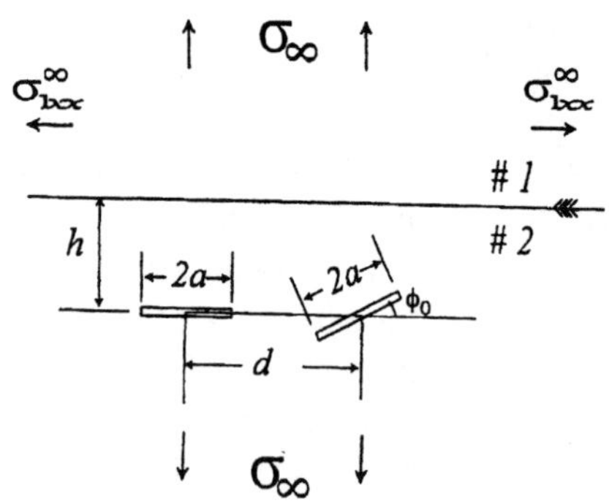

Figure 2.16. Third example of two microcrack configuration beneath interface.

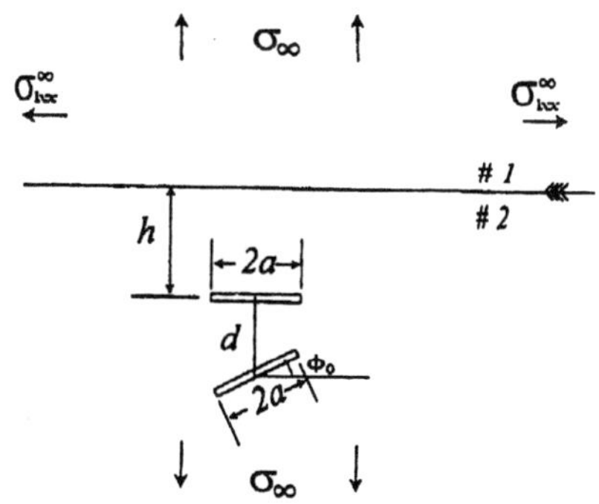

Figure 2.17. Fourth example of two microcrack configuration beneath interface.

Table 2.2. Dundurs' parameters α and β and the oscillation index ε.

Parameters	α	β	ε
Air/Al_2O_3	-1.000	-1.000	infinite
Cu/Al_2O_3	-0.5118	-0.089	0.028
Ni/MgO	-0.138	0.015	-0.00049
homogeneous	0.000	0.000	0.000

All the computed values of the M integral for the four examples of two-crack configurations are normalized by

$$M_o = \frac{\sigma_\infty^2 \pi (1-\nu_2) a^2}{2\mu_2,} \tag{2.51}$$

where a refers to the half length of each microcrack and σ_∞ is the remote tensile stress which is always loaded along the direction perpendicular to the interface.

Nevertheless, the sub-interface cracks are assumed to be stationary or no crack is allowed to grow. This could be ensured by controlling the remote tensile loading σ^∞ together with the fracture criterion (2.14) in section 2.3 of this chapter.

Computed values of M_N,M_A and M against the normalized distance d/a with *h/a=1.25* for the first example (Figure 2.14) are plotted in Figures 2.18 (a, b, c), respectively. All the values are normalized by (2.13) or (2.51) for making comparisons. The real curves refer to the results in homogeneous solids, the imaginary curves refer to the results in Cu/Al_2O_3 bimaterial solids, while the dotted curves refer to the results in Ni/MgO bimaterial solids. It is seen that, when d/a decreases for the collinear microcracks beneath the interface, the net part of the M integral, M_N, always increases and the additional part of the M integral, M_A, always decreases, whilst their summation, $M = M_N + M_A$, always increases, no matter what kind of metal/ceramic bimaterial solids is considered. This means that in the present microcrack arrangement the M integral is mainly governed by the net part. Particularly, when $d/a > 4$, M_A is within ten percent of M and it could be neglected indeed. However, when $d/a < 4$, specially when $d/a < 3$, the contribution of M_A to M becomes remarkable which could no longer be neglected. It could be concluded from Figure 2.18(c) that the M integral represents the level of the interaction effect and the coalescence tendency between the two collinear microcracks although it has been proved to be the summation of the two parts, *i.e.,* M_N and M_A. The larger interaction effect between the two microcracks is, the larger value of M in magnitude is. Of course, when the interaction effect becomes smaller corresponding to the case that the two collinear microcracks are located far apart, the net-part of the M integral, *i.e.,* M_N, provides a good approximation to the M integral. Moreover, it is seen in Figures 2.18(a,b,c) that the imaginary curves Cu/Al_2O_3 and the dotted curves Ni/MgO have signif icant divergences from the real curves corresponding to homogeneous cases. Obviously, this is owing to the mismatch nature of the bimaterial solids, Ni/MgO and Cu/Al_2O_3.

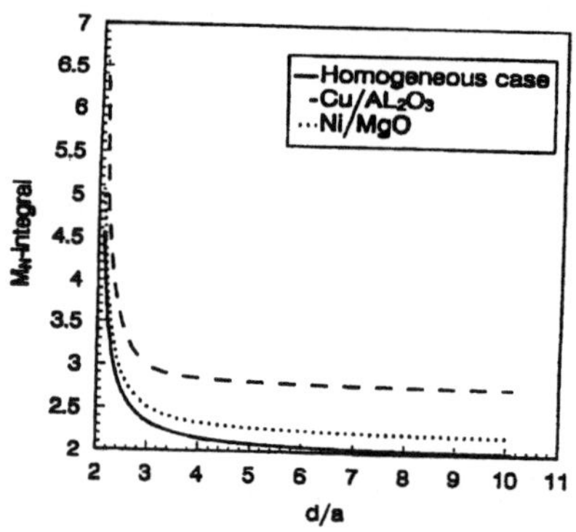

(a) The net part M_N.

What is more, it could be seen that the existence of the ductile phase, Cu or Ni, always leads to increase M_N and decrease M_A, and in turn to increase M. It is concluded that this kind of collinear microcrack arrangement in the brittle phase of the metal/ceramic bimaterial solids shows less stable nature and larger coalescence tendency than the corresponding one in homogeneous solids owing to the existence of ductile phases and that the M integral actually plays

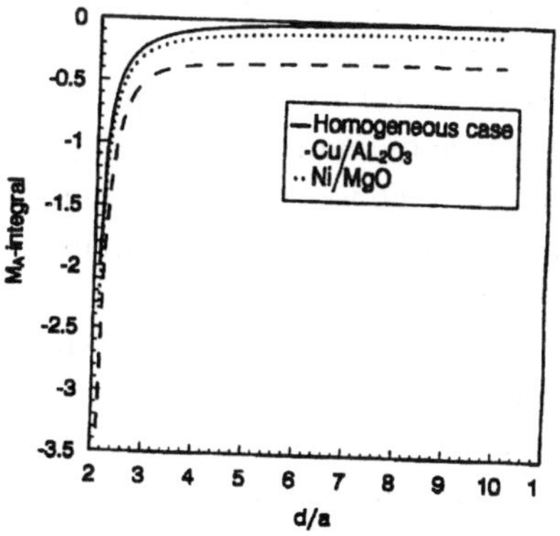

(b) The additional part M_A.

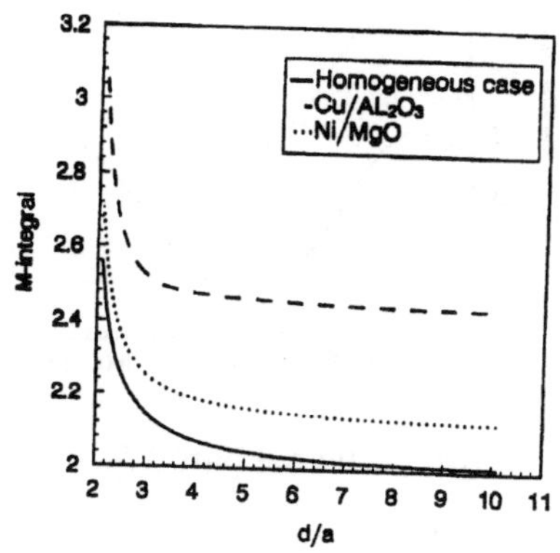

(c) The total $M = M_N + M_A$.

Figure 2.18. M against d/a for the first example of two crack configuration

an important role to describe microcrack damage in bimaterial solids. Of great interest is the influence of the ductile phase on the abovementioned divergence of M_N,M_A, and M which is obviously governed by the first Dundurs parameter α with no regards to the second parameter β. As shown in Table 2.2, $\alpha = -0.138$ for Ni/MgO and $\alpha = -0.5118$ for Cu/Al_2O_3 (see Hutchinson *et. al.*, 1987; Fleck *et al.*, 1991), whilst $\alpha = 0$ for homogeneous solids. This means that the value of α for Cu/Al_2O_3 shows larger deviation from zero (corresponding to homogeneous solids) than the value of α for Ni/MgO. Indeed, it is this reason that the computed values of M_N, M_A, and M for Cu/Al_2O_3 bimaterial show larger divergence from the values for homogeneous solids than those for Ni/MgO bimaterial. Nevertheless to say, the second parameter β has no such a nature (see Table 2.2). In other words, it is the first parameter α rather than the second parameter β that dominates the divergence of the imaginary curves and the dotted curves from the real curves in Figures 2.18(a, b, c). Moreover, as the computed values of M for metal/ceramic bimaterial solids are always much larger than those for homogeneous solids, it is concluded that a certain microcrack array in the brittle phase of metal/ceramic bimaterial solids shows lower stability and larger coalescence tendency than the same microcrack array in homogeneous brittle solids.

In order to avoid coincidence and make the above conclusions to be much clearer, we consider another microcrack arrangement shown in Figure 2.15. Computed values of M_N, M_A, and M against the location angle α_0 with $d/a = 4.0$ and $h/a = 1.25$ are plotted in Figures 2.19(a, b, c), respectively.

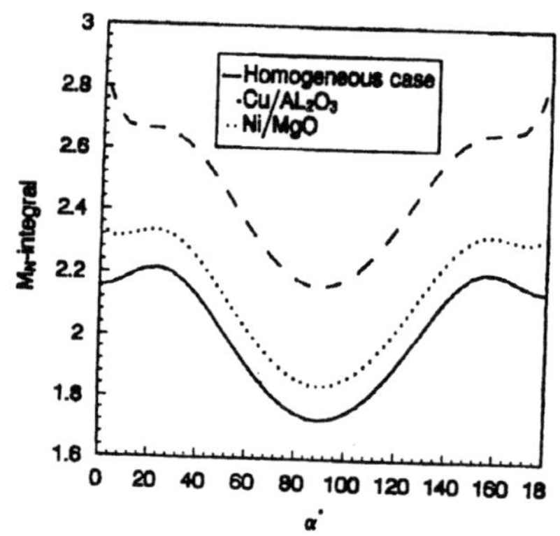

(a) The net part M_N.

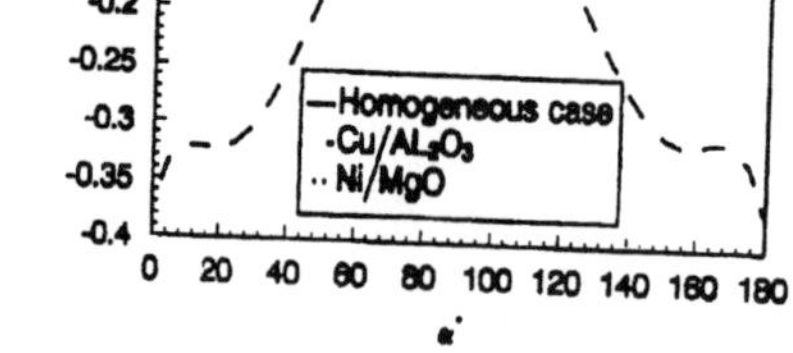

(b) The additional part M_A

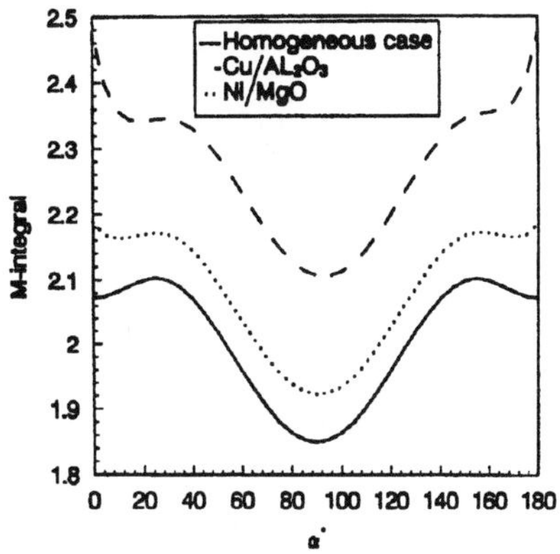

(c) The total $M = M_N + M_A$.

Figure 2.19. M against α_0 for the second example of two crack configuration.

It is seen that the similar conclusions could be given as those mentioned above for Figure 2.14. Indeed, the values of M_N, M_A, and M for Cu/Al_2O_3 bimaterial (imaginary curves) and those for Ni/MgO bimaterial (dotted curves) show large divergence from those for homogeneous solids (real curves). This divergence is also governed by the first Dundurs parameter α rather than β. It is also seen that, to a certain degree, for example, when the location angle α_0 in Figure 2.15 is between 40 and 140 degrees, the net part of the M integral, *i.e.,* M_N overwhelming the additional part M_A becomes a good approximation of the M integral with errors less than ten percent. However, when taking $\alpha_0 < 40$ degrees or $\alpha_0 > 140$ degrees, especially when taking $\alpha_0 < 30$ degrees or $\alpha_0 > 150$ degrees, the contribution of M_A to the M integral could no longer be neglected. Of great interest is that the values of the M integral in Figure 2.19(c) for metal/ceramic bimaterial solids are always larger than the corresponding values for homogeneous brittle solids. This confirms again that a certain microcrack array in the brittle phase of metal/ceramics bimaterial solids shows lower stability and larger coalescence tendency than the same microcrack array in homogeneous brittle solids.

Let us consider the third microcrack arrangement as shown in Figure 2.16. Computed values of M_N, M_A, and M against the oriented angle ϕ_0 with $h/a = 1.25$ and $d/a = 4.33$ are plotted in Figures 2.20(a, b, c), respectively. Although the microcrack arrangement is quite different from the two examples discussed above, the similar conclusions could still be given. The values of M_N in metal/ceramic bimaterial solids are always larger than those derived in homogeneous solids. Quite contrary, the values of M_A in metal/ceramic bimaterial solids are always smaller than those derived in homogeneous solids.

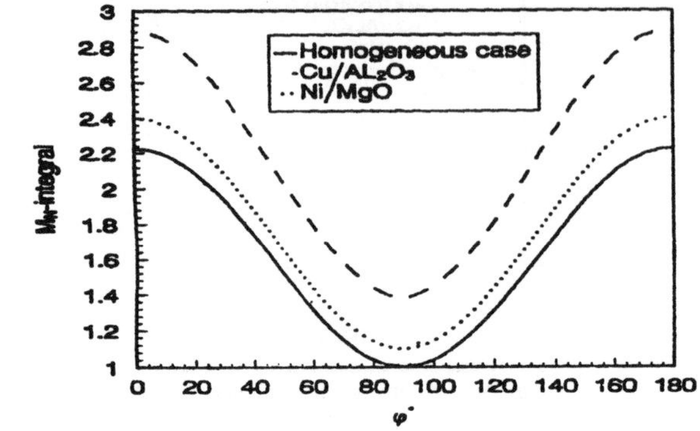

(a) The net part M_N.

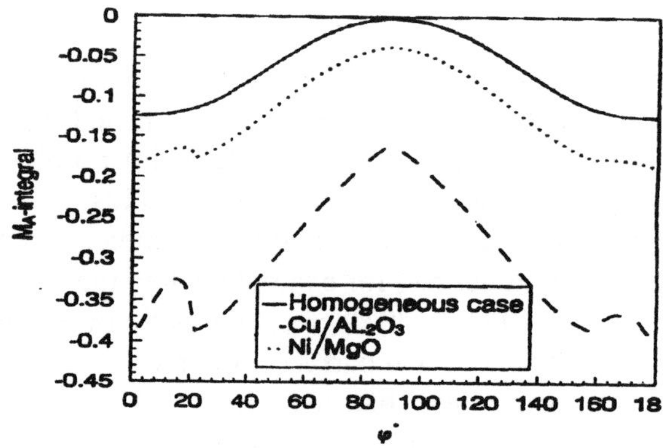

(b) The additional part M_A

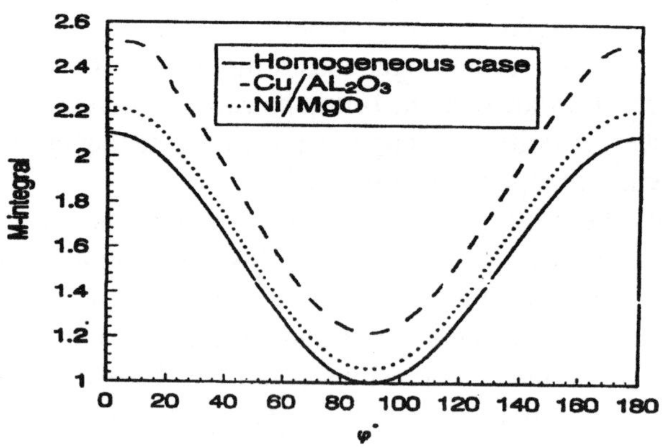

(c) The total $M = M_N + M_A$.

Figure 2.20. M against ϕ_0 for the third example of conf iguration.

There exist apparently significant divergences of the values of M in bimaterial solids from those in homogeneous brittle solids. The divergences are actually governed by the first Dundurs parameter α with no regards to the second parameter β. What is more, the larger divergence of α from zero, the larger divergences of M from those for homogeneous brittle solids.

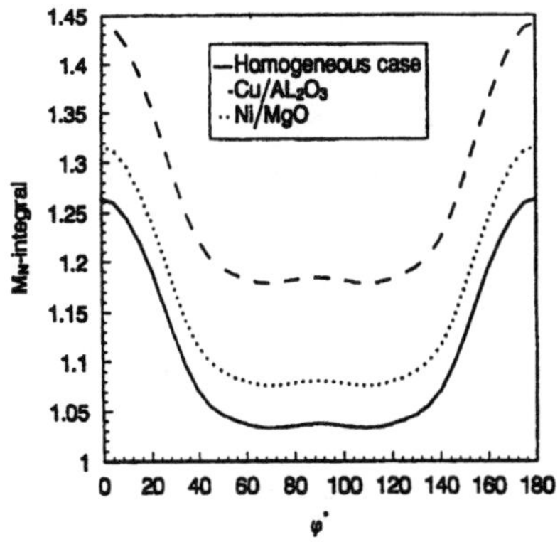

(a) The net part M_N.

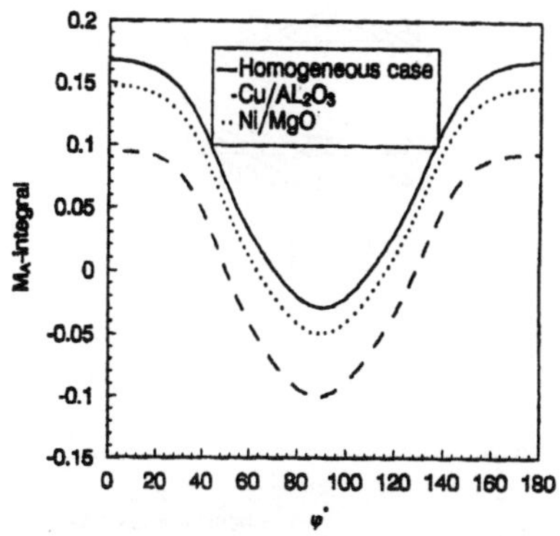

(b) The additional part M_A

Finally, consider the fourth microcrack arrangement shown in Figure 2.17. Computed values of M_N, M_A, and M against the oriented angle ϕ_0 with $h/a = 2.0$ and $d/a = 1.2$ are plotted in Figures 2.21(a, b, c), respectively. Again, readers can see that although the microcrack arrangement is quite different from the three examples discussed above, the similar conclusions could be given, which are no longer repeated here for shortening the length of this section.

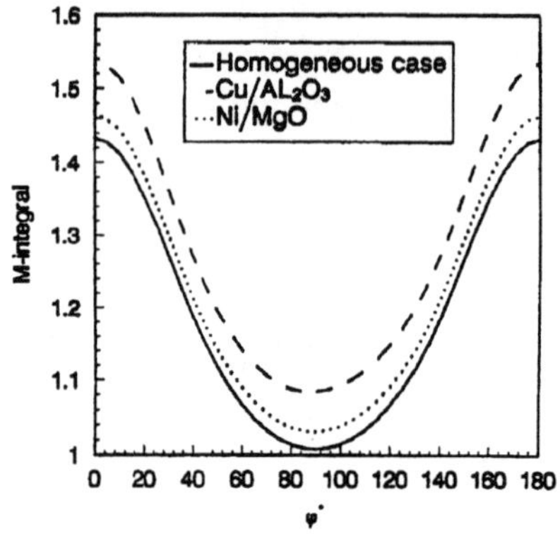

(c) The total $M = M_N + M_A$.

Figure 2.21. M against ϕ_0 for the fourth example of configuration.

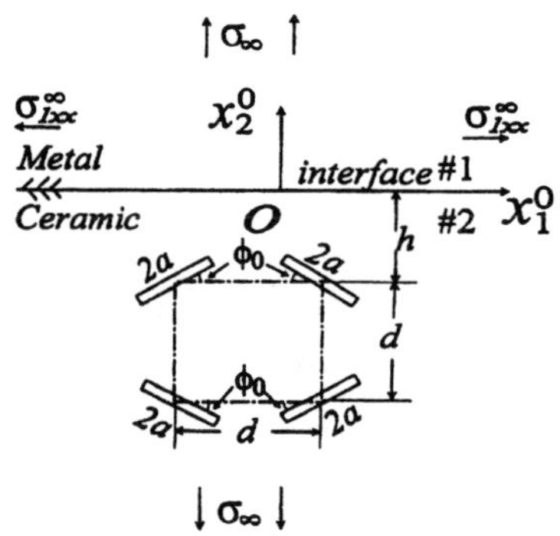

Figure 2.22. Four microcracks beneath the interface.

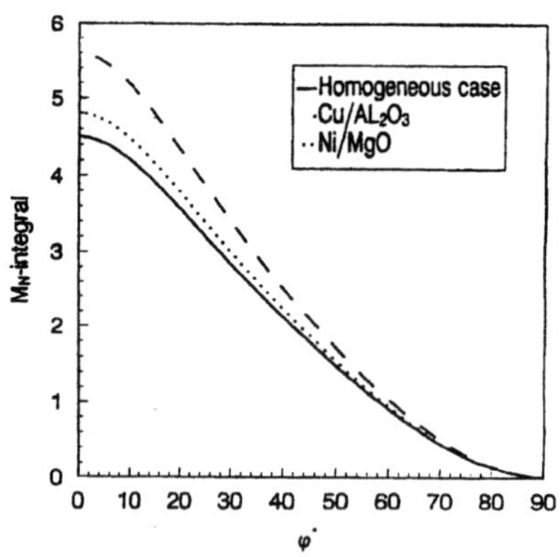

(a) The net part M_N.

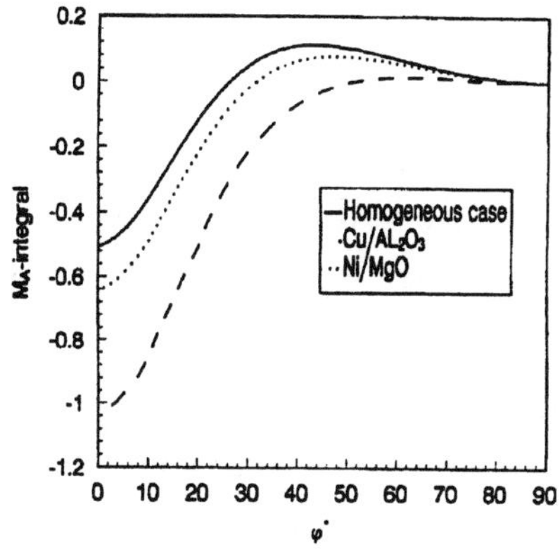

(b) The additional part M_A

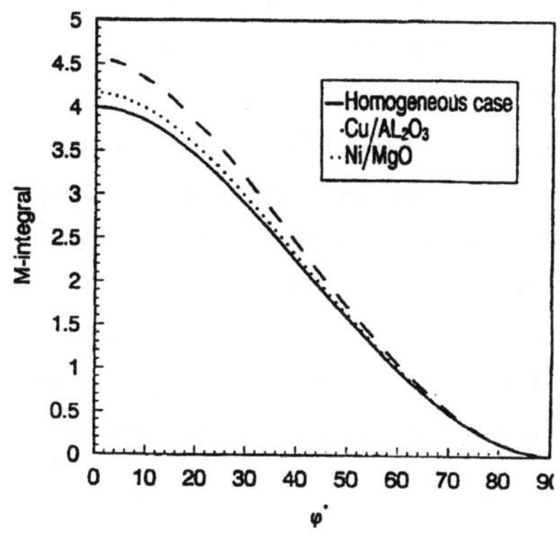

(c) The total $M = M_N + M_A$.

Figure 2.23. M against ϕ_0 for the four cracks.

(B) Four microcracks beneath interface

Four interacting microcracks beneath the interface in the brittle phase of bimaterial solids are shown in Figure 2.22. For convenience, we take the lengths of the microcracks to be the same and take the normalized center to center distances to be equal with each other d/a=2.25. Whilst the oriented angle ϕ_0 is taken to be a variable from 0 to 90 degrees and the normalized distance between the interface and the neighbouring microcracks to be h/a=1.2 (see Figure 2.22). Computed values of M_N, M_A and M against the oriented angle ϕ_0 are plotted in Figure 2.23(a, b, c), respectively. It is seen that the conclusions derived in the two microcracks arrangements could still be given in the present four microcracks configuration. It is concluded that the conclusions given in this chapter are inherent in the M integral analysis no matter what kind of microcrack configurations in the brittle phase is considered, providing that the microcracks are stationary, fully open, and not intersected.

5. A half-plane brittle solid containing multiple cracks

In order to show the potential applications of the present investigation, an extension of the present investigation to treat the multiple crack problem in a half-plane brittle solid is discussed briefly (see Figure 2.24). Here the outside boundary of the solid is taken along the x_1-axis, which is considered as a special kind of interface between air and the solid with the first Dundurs parameter $\alpha = -1$. The global coordinate system is originated at a fixed point on the boundary. The remote loading is preferred along the x_1-axis only. A two-crack array is considered. The left crack is taken as always perpendicular to the boundary, while the right crack is oriented by an angle ϕ_0, which is taken as a variable from 5 degrees to 175 degrees. The normalized distance h/a=1.25 and d/a=1.25$\sqrt{3}$. It is found that the M integral takes the same variable tendencies for the three different kinds of material combinations (Air/Al_2O_3 with $\alpha = -1$, Cu/Al_2O_3 with $\alpha = -0.5118$, and Ni/MgO with $\alpha = -0.138$). However, there exist apparent divergences among the three curves, respectively, corresponding to the real curve, the dotted curve, and the imaginary curve. As discussed above, the divergences are dominated by the first Dundurs parameter α. The larger divergence of the parameter is from zero, the larger value of the M integral is deduced. It is concluded that the outside boundary of the half-plane solid could actually be treated as the special kind of interface between air and the brittle solid.

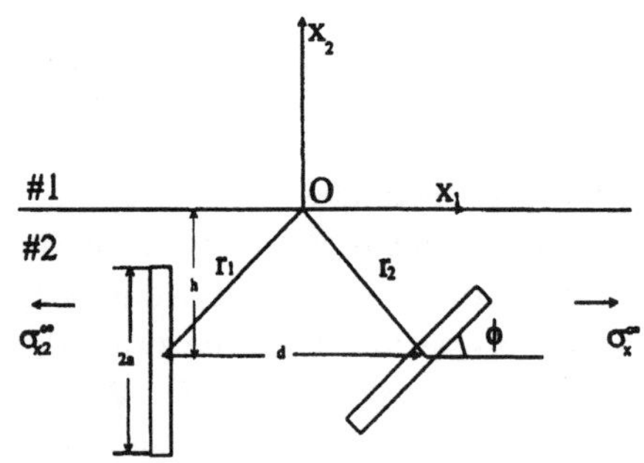

Figure 2.24. A half-plane containing multiple cracks.

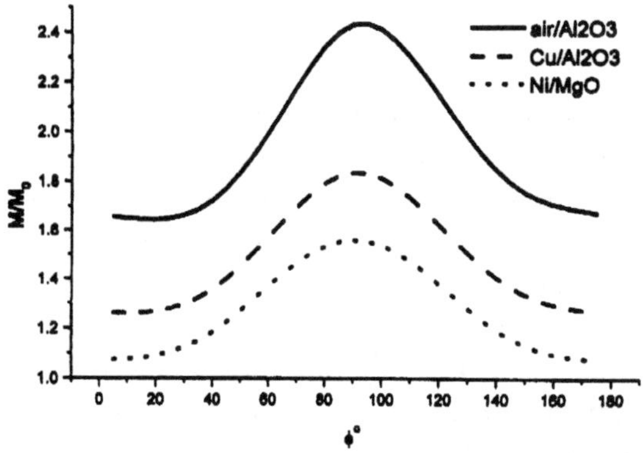

Figure 2.25. Variable tendencies of the M integral for the three kinds of material combinations.

Therefore the role that the outside boundary plays in the M integral analysis has been clarified. The present discussion could then easily be extended to treat a finite plane solid containing multiple cracks without any difficulty. The major idea is that the outside boundaries of the solid should be treated as a special kind of interface between air and the solid (or between a rigid body and the solid).

6. Brief summary

From the discussions of the M integral analysis in this chapter, the following conclusions can be summarized:

(A) Although the existence of interface gives considerable trouble in calculating the M integral in the brittle phase of a metal/ceramic bimaterial solid with strongly interacting microcracks, the M integral could still be divided into two distinct parts. As in homogeneous cases, the first part is induced from the stress intensity factors at all microcrack tips although the existence of the interface disturbs the values of the *SIF's* significantly. Of great interest is that the second part has been proved to arise only from the global coordinates of each microcrack center and the J_k vector of each crack irrespective of the contribution induced from the existence of the interface. This is because the M integral is independent from the selection of the origin of the global coordinate system when the closed integral contour is chosen to be infinitely large, *i.e.,* not only all the microcracks, but also the whole interface are enclosed in it. Therefore, the origin of the system could always be chosen on a certain point of the interface. Moreover, it is found that, generally speaking, the first part of the M integral (*i.e.,* the net part M_N) does not make overwhelming superiority over the second part (*i.e.,* the additional part M_A) for a certain microcrack array in the brittle phase.

(B) The M integral plays an important role for microcrack damage in the brittle phase of metal/ceramic bimaterial solids. As in homogeneous brittle solids, it could be considered as a measure of the damage level, the microcrack stability, and the microcrack coalescence tendency. The present discussion does confirm the conclusions given in the above sections of this chapter and does extend the M integral analysis to treat multiple crack problems in much more complicated cases. It actually provides a new way to describe the microcrack damage phenomena, at least, in brittle solids, in the brittle phase of metal/ceramic bimaterial solids, and in half plane brittle solids. It is found that the restriction arising from the discussion in section 2.3, that is, the mathematical treatments are only valid in an infinite case, could be removed without doubt.

(C) There always exist significant divergences of the M integral values in metal/ceramic bimaterial solids from those for the same microcrack array in homogeneous brittle solids. The divergences are governed by the first Dundurs parameter α with no regards to the second parameter β. The larger divergence of α from zero, the larger divergences of the values of the M integral in metal/ceramic solids from those in homogeneous brittle solids, providing that the same microcrack configuration is considered in both kinds of solids.

(D) For the same kind of microcrack array the values of the M integral in the brittle phase of a metal/ceramic bimaterial are always larger than those in a homogeneous brittle solid. This is owing to the existence of the ductile phase or owing to the interaction of the microcrack array with the interface. Therefore it is concluded that a certain microcrack array in the former case shows lower stability and larger coalescence tendency than the same array in the latter case. Of course, the largest value of the M integral occurs for the same microcrack array in air/brittle bimaterial since in this bimaterial the first Dundur parameter takes -1 corresponding to the largest divergence from zero (homogeneous cases).

(E) The present discussion for a half-plane brittle solid containing multiple interacting cracks shows that the M integral analysis could be extended to treat microcrack damage problems in a finite brittle solid. The only correct way is to consider the outside boundaries of the finite solid as a special kind of interface between air and the solid (or between a rigid body and the solid) so that the contribution arising from the outside boundaries to the M integral can be taken into account.

(F) Finally, it should be emphasized that the discussions and conclusions presented in this chapter have not been covered in other books yet. Although the M integral was established many years ago (Eshelby, 1970, 1975; Budiansky and Rice, 1973), Chen (2001a,b) was the first to recognize its potential use in microcrack damage mechanics. However, how to use the conservation laws to replace the conventional effective elastic moduli in general cases remains to be further investigated. Using the Boundary Element Methods (Crouch and Starfield, 1983), Wang and Chen (2001a, b) performed an attempt to treat a finite microcracking brittle solid.

References for Chapter 2

Bazant, Z.P., (1986) Mechanics of distributed cracking, *Applied Mechanics Reviews.*, **39**, 675 – 705.

Bazant, Z.P., and Cedolin, L., (1991) Stability of structures: Elastic, inelastic, fracture and damage theories, *Oxford University Press*, New York, Chapter 12.

Benveniste, T., (1986) On the Mori-Tanaka's Method in Cracked Bodies, *Mech. Res. Comm.*, **13**, 193 – 201.

Budiansky, B., and Rice, J. R., (1973) Conservation laws and energy-release rates, *ASME Journal of Applied Mechanics*, **40**, 201 – 203.

Budiansky, B., and O'Connell, R.J., (1976) Elastic Moduli of a Cracked Solid, *International Journal of Solids and Structures*, **12**, 81 – 97.

Chaboche, J.L., (1988a) Continuum Damage Mechanics: Part I-General Concepts, *ASME Journal of Applied Mechanics,* **55**, 59 – 64.

Chaboche, J.L., (1988b) Continuum Damage Mechanics: Part II-Damage Growth, Crack Initiation, and Crack Growth, *ASME Journal of Applied Mechanics,* **55**, 65 – 72.

Chen, Y. Z., (1984) General Case of Multiple Crack Problems in an Infinite Body, *Engineering Fracture Mechanics,* **20**, 591 – 597.

Chen, Y. H., and Hasebe, N., (1998) A consistency check for strongly interacting multiple crack problem in isotropic, anisotropic, and bimaterial solids, *International Journal of Fracture*, **89**, 333 – 353.

Chen, Y.H. (2001a) M-integral analysis for two-dimensional solids with strongly interacting microcracks, Part I. In an infinite brittle solid, *International Journal of Solids and Structures,* **38**, 3193 – 3212.

Chen, Y.H. (2001b) M-integral analysis for two-dimensional solids with strongly interacting microcracks, Part II. In the brittle phase of an infinite metal/ceramic bimaterial, *International Journal of Solids and Structures,* **38**, 3213 – 3232.

Christensen, R.M. (1990) A Critical Evaluation for a Class of Micromechanics Models, *J. Mech. Phys. Solids,* **38**, 379 – 404.

Crouch, S.L., and Starfield, A.M., (1983) *Boundary Element Methods in Solids Mechanics, George Allen & Unwin,* London.

Eshelby, J.D., (1970) *The Energy Momentum Tensor in Continuum Mechanics, Inelastic Behavior of Solids*, edited by Kanninen, M.F., *et al.,* McGraw-Hill, New York.

Eshelby, J. D., (1975) The elastic energy momentum tensor, *J. Elasticity* **5**, 321 – 335.

Fleck, N. A., Hutchinson, J. W., and Suo, Z. (1991) Crack path selection in a brittle adhensive layer, *International Journal of Solids and Structures*, **27**, 1683 – 1703.

Freund, L.B. (1978) Stress Intensity Factor Calculation Based on a Conservation Integral, *International Journal of Solids and Structures,* **14**, 241 – 250.

Gross, D., (1982) Spannungsintensitaetsfaktoren von Risssystemen (Stress Intensity Factors of System of Cracks), *Ing.-Arch*, **51**, 301 – 310 (in German).

Herrmann, G.A., and Hermann, G. (1981) On Energy Release Rates for a Plane Cracks, *ASME Journal of Applied Mechanics,* **48**, 525 – 530.

Horii, H., and Nemat Nasser, S. (1983) Overall Moduli of Solids with Microcracks: Load-Induced Anisotropy, *Journal of Mechanics and Physics of Solids*, **33**,155 – 171.

Horii, H., and Nemat Nasser, S. (1985) Elastic Fields of Interacting Inhomogeneities, *International Journal of Solids and Structures*, **21**, 731 – 745.

Hutchinson, J.W., Mear, M.E., and Rice, J.R. (1987) Crack paralleling an interface between dissimilar materials, *ASME Journal of Applied Mechanics*, **54**, 828 – 832.

Isida, M., and Noguchi, H. (1994) Distributed cracks and kinked cracks in bonded dissimilar planes with an interface, *International Journal of Fracture*, **66**, 313 – 337.

Jun, J.W. (1991) On Two-Dimensional Self-Consistent Micromechanical Damage Models for Brittle Solids, *International Journal of Solids and Structures,* **27**, 227 – 258.

Jun, J.W., and Lee, X. (1991) On Three-Dimensional Self-Consistent Micromechanical Damage Models for Brittle Solids, Part I. Tensile Loadings,'*ASCE J. Eng. Mech.,* **117**, 1495 – 1515.

Jun, J.W., and Chen, T.M. (1994a) Effective Elastic Moduli of Two-Dimensional Brittle Solids with Interacting Microcracks, Part I: Stationary, *ASME Journal Applied Mechanics*, **61**, 349 – 357.

Jun, J.W., and Chen, T.M. (1994b) Effective Elastic Moduli of Two-Dimensional Brittle Solids with Interacting Microcracks, Part II: Evolutionary Damage Models, *ASME Journal of Applied Mechanics*, **61**, 358 – 366.

Kachanov, L.M. (1958) Time of the Rupture Process Under Creep Conditions, *Izv. Akad. Nauk. SSR. Otd Tekh. Nauk* . No.8, 26 – 31.

Kachanov, M. (1992) Effective Elastic Properties of Cracked Solids: Critical Review of Dome Basic Concepts, *Applied Mechanics Reviews,* **45**, 304 – 335.

Kachanov, M. (1993) Elastic Solids of Many Cracks and Related Problems, *Advances of Applied Mechanics*, **30**, 259 – 428.

Kachanov, M., and Laures, J.P. (1989) Three-Dimensional Problems of Strongly Interacting Arbitrarily Located Penny-Shaped Cracks, *International Journal of Solids and Structures*, **23**, 11 – 21.

Kanninen, M.F., and Popelar, C.F. (1985) *Advanced Fracture Mechanics,* Oxford University Press, New York.

King, R.B., and Herrmann, G. (1981) Nondestructive Evaluation of the J-and M-integrals, *ASME Journal Applied Mechanics*, **48,** 83 – 87.

Knowles, J.K., and Sternberg, E. (1972) On a Class of Conservation Laws in Linearized and Finite Elastostatics, *Arch. Rat. Mech. Anal.,* **44**, 187 – 211.

Krajcinovic, D., (1989) *Damage Mechanics, Mechanics of Materials,* **8**, 117 – 197.

Mori, T., and Tanaka, K. (1973) Average Stress in Matrix and Average Elastic Energy of Materials with Misf itting Inclusions, *Acta Metallurgica*, **21**, 571 – 574.

Nemat-Nasser, S., and Hori, M. (1990) Elastic Solids with Microdefects, *Micromechanics and Inhomogeneity*, Edited by G. J. Weng, M. Taya, and H. Abe., *Springer-Verlag*, New York, 137 – 159.

Park, J.H., and Earmme., Y.Y. (1986) Application of Conservation Integrals to Interfacial Crack Problems, *Mechanics of Materials,* **5**, 261 – 276.

Smelser R. E., and Gurtin E. (1977) On the J-integral for bi-material bodies., *International Journal of Fracture*, **13**, 382 – 384.

Wang, D.F., and Chen, Y.H. (2001a), Evolutionary microcrack damage by the M integral description in a f inite brittle solid, Part I, submitted.

Wang, D.F., and Chen, Y.H. (2001b), Evolutionary microcrack damage by the M integral description in a f inite brittle solid, Part II, submitted.

Zdenek, P., and Bazant, P. (1992) Nonlocal damage theory based on micromechanics of crack interactions, *ASCE J. Eng. Mech.*, **120**, 593 – 617.

Zhao, L.G., and Chen, Y.H. (1997) Further Investigation of Sub-interface Cracks, *Archive of Applied Mechanics*, **67**, 393 – 406.

Appendix of Chapter 2

As discussed by Herrmann and Hermann (1981), the traction-free surfaces of a crack have some contribution to the second component of the J_k vector. For the kth microcrack shown in Figure 2.1, this contribution could be calculated in the local system $(x_{1*}^{(k)}, x_{2*}^{(k)})$

$$F_{2ak} = -\int_{-a_k}^{a_k} (w^+ - w^-) dx_{1*}^{(k)},$$

(A2.1)

where W^+ and W^- indicate the upper and lower boundary values along the crack surfaces of the strain energy which could be given as

$$W^+ = \frac{1-\nu^2}{2E}(\sigma_{11*}^+(t))^2,$$

$$W^- = \frac{1-\nu^2}{2E}(\sigma_{11*}^-(t))^2,$$

(A2.2)

where t denotes $x_{1*}^{(k)}$, $\sigma_{1*}^+(t)$ and $\sigma_{1*}^-(t)$ are the normal stresses parallel to the upper and lower crack surfaces, respectively.

A special numerical technique was introduced by Zhao and Chen (1997), from which a deduced singular integral, *i.e.*,

$$I = \frac{1}{2\pi i\sqrt{a_k^2 - t^2}} \int_{-a_k}^{a_k} \frac{P(s) - iQ(s)}{t - s} \sqrt{a_k^2 - s^2} ds,$$

(A2.3)

could be calculated. Here $P(s)$ and $Q(s)$ are the so called pseudo-tractions acting on both surfaces of the kth microcrack.

Equation (A2.3) is normalized as

$$I = \frac{1}{2\pi i\sqrt{1 - (t/a_k)^2}} \int_{-1}^{1} \frac{P(u) - iQ(u)}{(t/a_k - u)} \sqrt{1 - u^2} du,$$

(A2.4)

where

$$u = s/a_k,$$

(A2.5)

Expanding *P(u)-iQ(u)* into the second kind of the Chebyshev polynomial U_j(u) as follows

$$P(u) - iQ(u) = \sum_{j=1}^{M-1} B_j U_j,$$

(A2.6)

with

$$B_j = \frac{2}{\pi} \int_{-1}^{1} \sqrt{1-u^2}(P(u) - iQ(u))U_j(u)du, \tag{A2.7}$$

and substituting equation (A2.6) into (A2.4), the following numerical technique is obtained

$$\begin{aligned} I &= \frac{1}{2\pi i\sqrt{1-(t/a_k)^2}} \int_{-1}^{1} \frac{P(u) - iQ(u)}{(t/a_k - u)} \sqrt{1-u^2} du \\ &= \frac{1}{2i\sqrt{1-(t/a_k)^2}} \sum_{j=0}^{M-1} B_j T_{j+1}(t/a_k), \end{aligned} \tag{A2.8}$$

where $T_{j+1}(t/a_k)$ is the first kind of the Chebyshev polynomial.

Chapter 3

THE PROJECTED CONSERVATION LAW OF J_K VECTOR IN MICROCRACK SHIELDING PROBLEMS OF BRITTLE SOLIDS

This chapter deals with the conservation laws in microcrack shielding problems of brittle solids. Firstly, the engineering background of the so called microcrack shielding problems is discussed briefly. Two analytical techniques: the continuum mechanics approach (Ortiz, 1987, 1989) and the discrete model approach (Gong and Horii, 1989) are discussed also. Secondly, attention is focused on the discrete model, in which a semi-infinite crack interacting with single or many microcracks in the near-tip process zone is atressed in detail. The pseudo-traction method proposed by Horii and Nematt Nasser (1985, 1987) and Chen (1984a,b) is used here, from which a system of the Fredholm integral equations with infinite integral intervals are deduced. A special technique is proposed for substituting the infinite integral intervals into finite integral intervals, and in this way the Fredholm integral equations can be solved numerically by using the Chebyshev numerical integration technique (Fleck *et al.,* 1992; Chen and Hasebe, 1994; Chen, 1996; Zhao and Chen 1998). In particular, the projected conservation law of the J_k vector in the discrete model is established. Manipulations and numerical results reveal that there is a redistribution of the remote J integral when it transmits across the microcrack zone from infinity to the macrocrack tip. In other words, the remote J integral is divided into two parts as a result of microcracking. One of them is contributed from the macrocrack tip and the other is contributed from the formation of the near-tip microcracks. The latter is then evaluated by using the projected relations between the values of the J_k vector in local coordinate systems and those in a global coordinate system originated at the macrocrack tip. It is shown that in this circumstance the J_2 integral plays a similar role as the J_1 integral. The calculation of the J_2 integral is discussed in detail, which is not only concerned with the contributions arising from crack tips but also concerned with those from the traction-free surfaces of cracks. Although the previous investigations have shown the redistribution of stress when microcracks are formed in the near-tip process zone of a macrocrack and Hutchinson (1987) has pointed out the two sources for microcrack shielding effect, the discussion performed in this chapter verifies that it is this redistribution of stress which yields the redistribution of the remote J integral. Finally, the effect of the T stress on the redistribution is discussed in detail too (Zhao and Chen, 1998).

1. Microcrack shielding problems

From the engineering point of view, there now exists experimental evidence that stress-induced microcracking near a macrocrack tip enhances the fracture toughness of brittle materials such as ceramics and rocks (Hoagland, 1973; Clausen, 1976; Wu, *et al.,* 1978; Hoagland and Embury, 1980; Evans and Faber, 1981; 1984; Budiansky, *et al.,* (1983); Hutchinson, 1983, 1987; Ortiz, 1987, 1988, 1989; Faber and Fu, 1985; Fu and Evans, 1985; Rose, 1986a,b; Montagut and Kachcnov, 1988; Bowling, *et al.,* 1987; Ruhle, *et al.,* 1987; Clarke, 1984; Kachanov and Montagut, 1986; Chudnovsky, *et al.,* 1987a,b; Gong and Horii, 1989; Rubinstein, 1985, 1986, 1988; Laws and Brockenbrough, 1988; Charalambides and McMeeking, 1987a,b; Cai, *et al.,* 1990; Faber, *et al.,* 1991; Gong and Meguid, 1991; Chudnovsky and Wu, 1990; Wu and Chudnovsky, 1991; Cai and Faber, 1992; Chen and Hasebe, 1995a,b,c; Wang, *et al.,* 1996; Chen, 1996; Zhao and Chen, 1998). Experimental studies carried out by Ruhle *et al.,*(1987) have provided conclusive evidence of stress-induced microcracking toughening in a zirconia toughened alumina. After their attempt, Faber *et. al.* (1991) have shown a relationship between microcrack formation and an increase in toughness in $SiC - TiB_2$ composites with phases of different thermal expansion coefficients. Indeed, there is a commonly recognized opinion that the stress-induced microcracks near a macrocrack tip shield the macrocrack from the applied stress, thereby increasing the fracture toughness (Hutchinson, 1987; Ortiz, 1987).

However, it is critical for microcracking to provide effective shielding that microcracks remain stable following nucleation. In ceramics and rocks, this is accomplished by microcracks forming at grain boundary facets and remaining confined to them thereafter. Under increasing loads the number of available nucleation sites is eventually exhausted and the material asymptotically attains some reduced elastic moduli (Ortiz, 1987, 1989). The presence of softer material around the crack tip has the effect of screening the remote loads, thereby reducing the level of stress. Furthermore, newly formed microcracks partially relieve residual stresses that develop in the macroscopic scale as a transformation strain and constitutes another source of toughening. Just as pointed out by Hutchinson (1987), there are two sources of the redistribution of stresses in the near-tip stress field of a macroscopic crack. One is owed to the reduction in the effective elastic moduli resulting from microcracking. The other is the strain arising from the release of the residual stresses when microcracks are formed. The residual stresses in question develop in the fabrication of polycrystalline or multi-phase materials owed to thermal mismatches between phases or thermal anisotropies of the single crystals. The spatial variation of these stresses is set by the grain size or by the scale of second phase particles. These residual stresses play an important role in determining the onset and extent of microcracking. Moreover, the microcracks partly relieve the residual stresses producing strains, which are manifest on the macroscopic scale as inelastic or transformation strains. Ultimately, the macrocrack advances by coalescence with microcracks ahead of the tip, a mechanism which is promoted by a lot of microcracking (Ortiz, 1987, 1988, 1989). In addition to the experimental studies mentioned above, microcrack toughening has also been the subject of numerous modeling studies. The two basic approaches are continuum modeling and discrete modeling. Based on the work by Ortiz (1987), we will first discuss the continuum modeling, in which the conservation law of the J integral is adopted. And then, based on the work by Chen (1996) and Chen and Ma (1997), we will focus our attention on the discrete modeling and on what role the J_k vector plays.

2. A continuum theory of microcrack shielding

In this section a continuum mechanics model proposed by Ortiz (1987) is discussed. In this model the microcracking region is described by a different constitutive equation with a reduced effective modulus. See Figure 3.1. The near tip process zone contains many miscrocracks. Three different subregions are introduced: The saturated zone where the existing microcracks do not grow and new microcracks are not allowed to create, the transition zone where the existing microcracks may grow and new microcracks may be created, and the elastic zone where no microcracks are created. The so-called small scale approach (Horii and Nemat Nasser, 1987) is assumed, that is, either the process zone or the microcrack lengths are much less than the macrocrack length. Therefore the macrocrack can be considered as a semi-infinite crack.

This reduction in modulus in the saturation zone has two counteracting effects. On the one hand, it contributes to the reduction in the near tip stresses thus leading to a shielding effect. On the other hand, it contributes to the reduction of the resistance of the brittle material to fracture. The resulting enhancement or degradation of the toughness of the material would ultimately depend upon the net outcome of the two effects mentioned above (Ortiz, 1987). Our attention is not focused on the detailed account induced from these two counteracting effects, rather, it is focused on the conservation theory of the J integral as a basic assumption adopted in the continuum mechanics model (Ortiz, 1987, 1988,

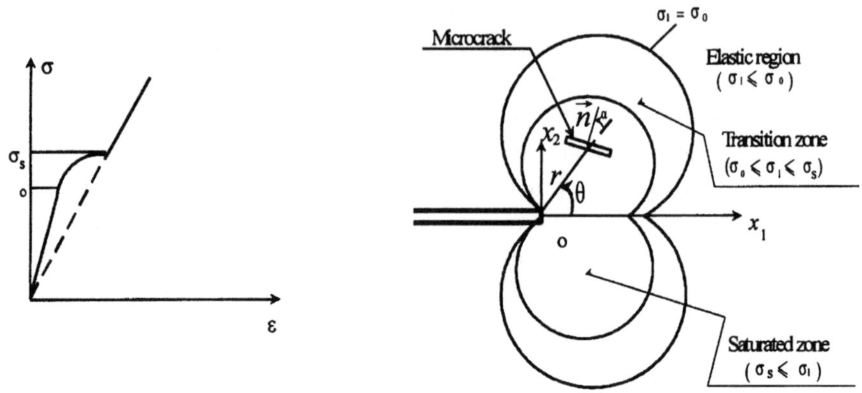

Figure 3.1. Small scale microcracking problem for a stationary crack.

1989; Hutchinson, 1987). In almost all investigations of continuum modelling it is assumed that

$$J_{\infty} = J_t, \tag{3.1}$$

where J_{∞} is evaluated along a remote contour lying entirely with the undamaged region, while J_t is evaluated along a contour contained within the saturated region surrounding a macrocrack.

The formulation (3.1) implies that the microcracks formed in the saturated region have no contribution to the J integral. We will show a contrary conclusion in the next section.

The continuum theory of microcrack shielding starts from a phenomenological constitutive model in which stresses and strains are assumed to be linearly related according to

$$\varepsilon_{ij} = C_{ijkl}\sigma_{kl}, \tag{3.2}$$

where the elastic compliances C_{ijkl} are regarded as internal variables which evolve as a result of damage processes taking place at a microstructural level. A threshold is postulated below which no further damage can occur. For the class of material under consideration, the onset of damage is assumed to occur when the maximum tensile stress σ_1 reaches a critical value $\sigma_c \geq 0$, *i.e.*,

$$\varphi(\sigma, \sigma_c) \equiv \sigma_1 - \sigma_c = 0. \tag{3.3}$$

The direction of incremental damage is given by a damage rule

$$\dot{C}_{ijkl} = \dot{\lambda} n_i n_j n_k n_l, \tag{3.4}$$

where $n_i, \sqrt{n_i n_i} = 1$, is the direction of maximum tensile stress and the multiplier λ can be regarded as an effective added flexibility owing to damage. Implicit in equation (3.4) is the assumption that the newly nucleated microcracks tend to be preferentially oriented normal to the direction of maximum tensile stress and thus most loss of stiffness occurs in that direction.

The evaluation of the critical stress σ_c is assumed to be governed by a damage rule of the type

$$\dot{\sigma} = h(\sigma_c) \cdot \dot{\lambda} \tag{3.5}$$

for some modulus $h(\sigma_c)$. In this simple model the dependence of h on σ_c can be determined from the uniaxial tension stress strain curve alone.

For the purpose of the asymptotic analysis that follows it proves more convenient to employ a deformation type constitutive theory rather than the incremental model introduced above. A deformation theory of damage can be readily obtained by integrating the incremental constitutive equations along proportional stress paths. This results in the following stress strain relation

$$\varepsilon_{ij} = (C^0_{ijkl} + \lambda(\sigma_1) n_i n_j n_k n_l)\sigma_{kl}, \tag{3.6}$$

where C^0_{ijkl} are the isotropic elastic moduli of the undamaged material, and the function $\lambda(\sigma_1)$ can be determined directly from the uniaxial tension test. A typical uniaxial tension stress strain curve has already been shown in Figure 3.1. It can be seen that the material initially exhibits an elastic domain after which damage starts to accumulate. Eventually a saturation stage is reached in which no further damage takes place. In this saturation stage equation (3.6) simplifies to

$$\begin{aligned} \varepsilon_{ij} &= (C^0_{ijkl} + \lambda_s n_i n_j n_k n_l)\sigma_{kl} \\ &= C^0_{ijkl} \cdot \sigma_{kl} + \lambda_s \sigma_1 n_i n_j, \end{aligned} \tag{3.7}$$

where λ_s is a constant of value

$$\lambda_s = \frac{1}{E_s} - \frac{1}{E_0}. \tag{3.8}$$

Here E_0 is the initial Young's modulus of the material and E_s is the slope of the uniaxial tension stress strain curve in the saturation stage (See Figure 3.1).

The stress strain relation (3.6) has a hyperelastic structure

$$\varepsilon_{ij} = \frac{\partial \chi(\sigma)}{\partial \sigma_{ij}}, \tag{3.9}$$

where the complementary energy potential $\chi(\sigma)$ takes the form

$$\chi(\sigma) = \frac{1}{2} C^0_{ijkl} \sigma_{ij} \sigma_{kl} + f(\sigma_1) \tag{3.10}$$

and the function $f(\sigma_1)$ and $\lambda(\sigma_1)$are related by $f(\sigma_1) = \lambda(\sigma_1)$.

In the saturation stage equation (3.6) reduces to (3.7) and $f(\sigma_1)$ reduces to $\lambda_s \sigma_1^2/2$, which renders the complementary energy potential (3.10) a homogeneous function of degree two of the stress tensor. Thus the strain energy potential $W(\varepsilon)$ is also a homogeneous function of degree two and satisfies the identity

$$W(\varepsilon) = \chi(\sigma) = \frac{1}{2} \sigma_{ij} \varepsilon_{ij} \tag{3.11}$$

when equation (3.9) is valid.

This means that the strain energy potential and the complementary energy potential take the same numerical value when evaluated at stresses and strains which satisfy the stress strain relation (3.9).

To present the asymptotic field for a stationary macrocrack, it is assumed that the process zone around a macrocrack tip to which microcracking is confined is small compared with the length of the crack but orders of magnitude larger than the characteristic size of the microcracks and their mean separation. The first condition leads to formulate an asymptotic problem for a semi-infinite crack as shown in Figure 3.1, while the second condition is needed for the constitutive model presented above to apply. Here, attention is confined to plane strain conditions and Mode I loading, *i.e.*, applied loads are symmetric with respect to the plane of the macrocrack. In Figure 3.1 three well differentiated regions surrounding the crack tip can be identified. The innermost region the strains are large enough so that the material can be assumed to be in the saturation stage of microcracking. The middle region is the transition zone in which the maximum tensile stress σ_1 is larger than σ_0 and smaller than σ_s, while the surrounding region is purely linear isotropic elastic without damage in which the state of stress is given by the well known expression

$$\sigma_{ij}(r, \theta) = \frac{K_\infty}{\sqrt{2\pi r}} \bar{\sigma}^0_{ij}(\theta), \tag{3.12}$$

where K_∞ is the remote stress intensity factor which can be considered as the measure of the remote Mode I loading level, and the universal angular distributions $\overline{\sigma}^0_{ij}(\theta)$ for a linear isotropic elastic material are given in the polar-coordinate system by

$$\left\{ \begin{array}{c} \bar{\sigma}^0_{rr}(\theta) \\ \bar{\sigma}^0_{\theta\theta}(\theta) \\ \bar{\sigma}^0_{r\theta}(\theta) \end{array} \right\} = \left\{ \begin{array}{c} (5/4)\cos(\theta/2) - (1/4)\cos(3\theta/2) \\ (3/4)\cos(\theta/2) + (1/4)\cos(3\theta/2) \\ (1/4)\sin(\theta/2) + (1/4)\sin(3\theta/2). \end{array} \right\} \tag{3.13}$$

The behavior of the material surrounding the crack tip, *i.e.*, the saturation zone, is assumed to be described by the stress strain law (3.6). Since these relation are derived from a complementary energy potential which is homogeneous of degree two, a classical argument shows that the leading term in the asymptotic expansion of the stress field has to be the from

$$\sigma_{ij}(r,\theta) = \frac{K_t}{\sqrt{2\pi r}}\tilde{\sigma}_{ij}(\theta), \tag{3.14}$$

where K_t is the local stress intensity factor of the near-tip stress fields and the angular distributions $\tilde{\sigma}_{ij}(\theta)$ are unknown to be determined. Generally speaking, K_t is different from K_∞ because the crack tip is shielded from the remote loads by the intervening microcracks. If assuming that crack growth is controlled by the value of K_t, a principal objective of the present analysis is to establish a relation between K_t and K_∞, and the extent to which K_t is reduced below K_∞ provides a quantitative measure of shielding.

Because the constitutive behavior of the damaged material surrounding the crack tip as shown in Figure 3.1 is nonlinear and strongly anisotropic one would expect singular stress fields which substantially depart from those in linear isotropic elasticity. It is shown next that this is not the case. In fact, the linear isotropic elastic stress field provides the exact asymptotic solution of the present problem, *i.e.*,

$$\tilde{\sigma}_{ij}(\theta) = \tilde{\sigma}^0_{ij}(\theta), \tag{3.15}$$

as anticipated by Hutchinson (1987), under certain circumstances, a similar situation is encountered in materials exhibiting linear creep and grain boundary cavitation. On the basis of this observation he aided the result stated above, namely, that the constitutive relation (3.7) implies the same angular distribution of stresses as those in the linear elastic solution.

To prove equation (3.15) Ortiz (1987) started by noting that the stress field (3.14), (3.15) satisfies equilibrium and traction-free boundary conditions on the crack faces. Thus it only remains to be shown that the corresponding strains satisfy compatibility. With reference to Figure 3.1 the direction of maximum tensile stress is computed to be

$$n_i = (\cos\alpha, \sin\alpha),$$
$$\alpha = \begin{cases} \pi/4 - \theta/4 & 0 < \theta \le \pi, \\ -\pi/4 - \theta/4 & -\pi \le \theta < 0. \end{cases} \tag{3.16}$$

Therefore it is seen that the angle α made by n_i and the radial direction varies linearly with the polar angle θ from a value of $\alpha = 45°$ at $\theta = 0^+$ to $\alpha = 0$ on the crack face $\theta = \pi$. The vector n_i can be regarded as giving an indication of the preferred orientation of the microcracks. In this light it is interesting to note that for $\theta = 0$ one has $\tilde{\sigma}_{rr} = \tilde{\sigma}_{\theta\theta}$ and $\tilde{\sigma}_{r\theta} = 0$. This would appear to render α indeterminate on the plane of the crack. However, this indeterminacy can be resolved by computing the limiting values of α as the plane $\theta = 0$ is approached from above and below. This operation yields two values of $a = 45°$, respectively. Thus, the model predicts two families of perpendicular microcracks at $\theta = 0$ symmetrically distributed with respect to the plane of the crack.

Substituting equation (3.16) into equation (3.7) the asymptotic strain field is computed to be

$$\begin{Bmatrix} \varepsilon_{rr}(r,\theta) \\ \varepsilon_{\theta\theta}(r,\theta) \\ \varepsilon_{r\theta}(r,\theta) \end{Bmatrix} = \frac{K_t}{\sqrt{2\pi r}} \cdot \frac{1}{2G_0} \begin{Bmatrix} (5/4 - 2v_0) \cdot \cos(\theta/2) - (1/4)\cos(3\theta/2) \\ (3/4 - 2v_0) \cdot \cos(\theta/2) + (1/4)\cos(3\theta/2) \\ (1/2)\sin(\theta/2) + (1/2)\sin(3\theta/2) \end{Bmatrix}$$

$$+ \frac{K_t}{\sqrt{2\pi r}} \cdot \lambda_s \cdot (\cos(\theta/2) + (1/2)\sin\theta) \begin{Bmatrix} (1 + \sin(\theta/2))/2 \\ (1 - \sin(\theta/2)/2 \\ \cos(\theta/2) \end{Bmatrix}, \quad 0 \le \theta \le \pi, \tag{3.17}$$

where E_0, v_0, and $G_0 = E_0/2(1 + v_0)$ are the initial Young's modulus, Poisson's ratio, and shear modulus of the undamaged material. It is noted that the first term is the isotropic linear elastic solution corresponding to a stress intensity factor K_t. The second term represents the effect of damage and vanishes identically for $\lambda_s = 0, i, e., E_s = E_0$. The strains in the lower half plane $-\pi \le \theta < 0$ are obtained from the symmetrical conditions

$$\begin{aligned} \varepsilon_{rr}(r,-\theta) &= \varepsilon_{rr}(r,\theta), \\ \varepsilon_{\theta\theta}(r,-\theta) &= \varepsilon_{\theta\theta}(r,\theta), \\ \varepsilon_{r\theta}(r,-\theta) &= -\varepsilon_{r\theta}(r,\theta). \end{aligned} \tag{3.18}$$

Lengthy but straightforward algebra shows that the strain field (3.17) identically satisfies the compatibility equation. Hence, equations (3.14) and (3.15) do indeed provide a closed form asymptotic solution of the present problem. The displacement field can be computed from the strain displacement relations

$$\left\{ \begin{array}{c} u_r(r,\theta) \\ u_\theta(r,\theta) \end{array} \right\} = \frac{K_t}{4G_0} \cdot \sqrt{\frac{r}{2\pi}} \cdot \left\{ \begin{array}{c} (2\kappa_0 - 1)\cos(\theta/2) - \cos(3\theta/2) \\ -(2\kappa_0 + 1)\sin(\theta/2) + \sin(3\theta/2 \end{array} \right\}$$

$$+K_t\lambda_s\sqrt{\frac{r}{2\pi}} \cdot \left\{ \begin{array}{l} (\cos(\theta/2) + (1/2)\sin\theta)(1 + \sin(\theta/2)) \\ (\cos\theta - \sin(\theta/2) - \sin(3\theta/2)) - 1 \end{array} \right\} \quad , 0 \le \theta \le \pi, \tag{3.19}$$

where $\kappa_0 = 3 - 4v_0$. The displacement in the lower half plane follows from the symmetrical conditions

$$\begin{array}{l} u_r(r_1, -\theta) = u_r(r,\theta), \\ u_\theta(r_1, -\theta) = -u_\theta(r,\theta), \end{array} \tag{3.20}$$

of particular interest is the crack opening profile

$$\delta(r) = -u_\theta(r_1, -\pi) + u_\theta(r_1, +\pi) = 8K_t \left[\frac{1 - v_0^2}{E_0} + \lambda_s \right] \sqrt{\frac{r}{2\pi}}. \tag{3.21}$$

As can be expected, the opening profile is parabolic, as in the linear elastic solution.

To have a complete asymptotic solution, it remains to determine the value of the local stress intensity factor K_t as a function of the applied stress intensity factor K_∞. This relation follows simply from an application of the J integral of Rice (1968a.b). Under monotonic loading the stress path at all points is nearly proportional and the response of the material is indistinguishable from that of a small strain, nonlinear elastic solid with the complementary energy potential (3.10). Under these conditions, the J integral is defined as (3.1) and

$$J_\infty = \frac{1 - v_0^2}{E_0} K_\infty^2. \tag{3.22}$$

For a contour shrunk down to the crack tip (3.1) can be evaluated from equations (3.14), (3.15), and (3.18). For the particular material model under consideration the complementary potential $\chi(\sigma)$ coincides with $W(\varepsilon)$ as computed from equations (3.7), (3.14) and (3.15). A lengthy but straightforward computation yields

$$J_t = \frac{1 - v_0^2}{E_0} K_t^2 + \beta \lambda_s K_t^2, \tag{3.23}$$

where the numerical constant β takes the value $\beta = 1.0942$. Finally, the conservation law of the J integral (3.1) yields

$$\frac{K_t}{K_\infty} = \frac{1}{\sqrt{1 + \frac{\beta}{1-v_0^2}(E_0/E_s - 1)}}. \tag{3.24}$$

Obviously the value of K_t/K_∞ is always less than unity since the value of $E_0/E_s$is generally larger than unity. Consequently, a substantial reduction in K can be expected even for moderate deviations from elastic behavior. This reduction represents the shielding effect owing to microcracking (Ortiz, 1987, 1988, 1989). However, it should be emphasized that the analysis resented above has been based on the path independence of the J integral, *i.e.,* equation (3.1), which means that the microcracks formed in the near-tip process zone have no contributions to the J integral. Moreover, the second source, *i.e.,* the strain arising form the release of the residual stresses induced from microcracking has not been taken into account. It is not clear, at least, not fully clear at present, whether or not the release of the residual stresses plays an important role in the problem under consideration. We will discuss this topic in the forthcoming sections. It should be emphasized also that the net toughness enhancement cannot be expected to be as substantial as equation (3.24) would tend to suggest, because the microcracks created ahead of the main crack degrade the crack extension resistance of the material (Ortiz, 1989; Gong and Meguid, 1991).

3. A discrete modelling of shielding problems

As another theoretical estimation of the microcrack shielding effect, the discrete modeling has also been widely used in the literature (see Kachanov and Montagut, 1986; Rose 1986a; Chudnovsky, *et al.,* 1987a,b; Rubinstein, 1985, 1986, 1988; Cai and Faber, 1992; Gong and Meguid, 1991; Chen, 1996; Zhao and Chen, 1998). The goal of this section is not supply the detailed results in this topic, rather, it is aimed at providing the projected conservation theory of J_k vector for the discrete model in brittle solids. The basic idea starts from the author's recent work (Chen, 1996), in which the contribution of microcracks in a near-tip stress field to J integral is studied in detail. The J integral defined by Rice (1968a,b) and the related J_k vector discussed by Budiansky and Rice (1973), Bergez (1974), Cherepanov (1979) and Herrmann and Herrmann (1981) are evaluated and re-examined, respectively, along three different closed contours. The first of them surrounds the tip of a semi-infinite crack only, the second encloses the microcracks completely, whilst the third encloses both the tip and all the microcracks completely. Unlike equations (1.33) and (1.34) or (1.39) and (1.40) discussed in Chapters 1 and 2, another simple, but universal relation is found among three values of the J integral corresponding to the contributions induced from the semi-infinite crack tip, the microcracks, and the remote stress field, respectively. In Section 3.3.1 fundamental solutions for a semi-infinite crack and a finite crack are given. In Section 3.3.2 the pseudo-traction methods proposed by Horii and Nemat Nasser (1987) and Fredholm integral equations proposed by Chen (1984a,b) are solved with the aid of the variable change (Fleck *et al.,* 1991) and the Chebyshev numerical integral method. In Section 3.3.3 the projected conservation theory of the J_k vector is presented and discussed in detail from which the following questions could be answered:

(i) What happens to the J integral when taking into account microcracks in a near-tip stress field?

(ii) What is the contribution arising from the existence of microcracks to the J integral?

(iii) Whether a relation exists among the three values of the J integral contributed respectively from the semi-infinite crack tip, the microcracks, and the remote stress field?

(iv) What role the components of the J_k-vector play in a discrete model although the second component of J_k was previously proved to be significant in Chapters 1 and 2?

(v) Whether or how the T stress influences the three values of the J integral?

Detailed examinations of numerical results are shown in Tables and Figures, which reveal that shielding or antishielding effects in the discrete model can be considered as a redistribution of the J integral arising from the existence of the microcracks. Moreover, the present examinations imply that the total effect of a cloud of microcracks in the near-tip stress field is dominated by a single parameter. Indeed, there exists an inherent and universal relation among the three values of the J integral and the second component in the discrete model for microcrack shielding problems.

3.1 Fundamental solutions

(A) Consider a semi-infinite crack along $(-\infty, 0)$ of the negative x-axis in the (x, y) plane (see Figure 3.2a,b). The crack faces are subjected to a pair of concentrated normal or a pair of tangential traction, respectively. The well known Muskhelishivili (1953) representation of two-dimensional elastic displacement and stress fields can be put in the following form:

$$\begin{aligned} \sigma_x + \sigma_y &= 2\left[\Phi(z) + \overline{\Phi(z)}\right], \\ \sigma_y - \sigma_x + 2i\sigma_{xy} &= 2\left[(\overline{z} - z)\Phi'(z) - \Phi(z) + \overline{\Omega}(z)\right], \quad (3.25) \\ 2\mu(u_x + iu_y) &= \kappa\varphi(z) - z\overline{\varphi'(z)} - \overline{\omega(z)}, \quad (3.26) \end{aligned}$$

where $i = \sqrt{-1}$, $\Phi(z)$ and $\Omega(z)$ are complex potentials, $\Phi(z) = \varphi'(z)$ and $\Omega(z) = \omega'(z)$, μ is the shear modulus of the brittle material under consideration, $\kappa = 3 - 4\nu$ for plane strain and $\kappa = (3 - \nu)/(1 + \nu)$ for generalized plane stress.

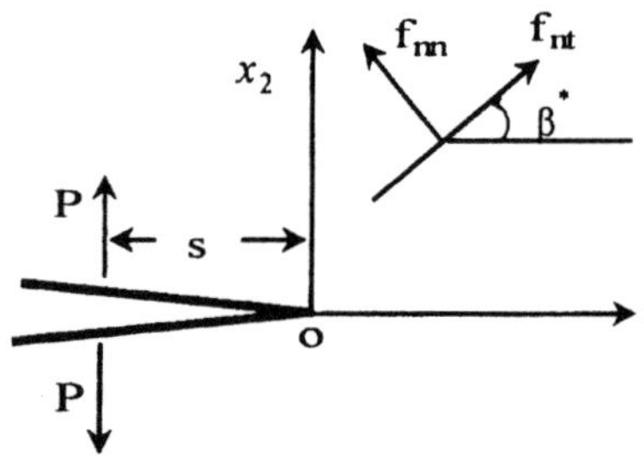

(a) under the normal traction P.

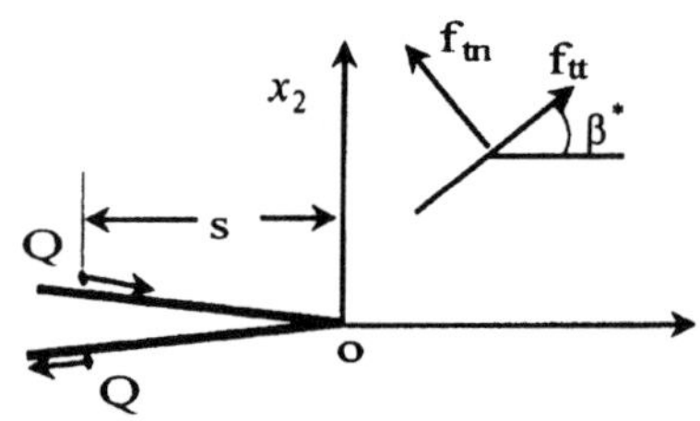

(b) under the tangential traction Q.

Figure 3.2. A semi-infinite crack under the normal traction P and under the tangential traction Q.

The problems shown in Figure 3.2(a) and 3.2(b) can be reduced to the Riemann–Hilbert problems. Without going into the detail, the corresponding complex potentials are given in a compact form:

$$\Phi(z) = \Omega(z) = \Phi_P(z) + \Phi_Q(z) = \frac{P - iQ}{2\pi(z + s)}\sqrt{\frac{s}{z}}, \quad (3.27)$$

where the subscripts P and Q denote the normal and tangential tractions, respectively, and s is the distance from the loaded point to the origin.

The tractions induced from the concentrated loads P and Q at any point in the (x, y)-plane along any orientation β can be formulated as follows:

$$\begin{aligned} f_{nn} - if_{nt} &= \Phi_P(z) + \overline{\Phi_P(z)} + e^{-2i\beta}\left(\Omega_P(\overline{z}) - \overline{\Phi_P(z)} + (z - \overline{z})\overline{\Phi_P'(z)}\right), \\ f_{tn} - if_{tt} &= \Phi_Q(z) + \overline{\Phi_Q(z)} + e^{-2i\beta}\left(\Omega_Q(\overline{z}) - \overline{\Phi_Q(z)} + (z - \overline{z})\overline{\Phi_Q'(z)}\right), \end{aligned} \quad (3.28)$$

where f_{nn}, f_{nt}, f_{tn} and f_{tt} are the so called influence coefficients in the boundary element methods (*BEM*) or the kernel functions in Fredholm integral equations (*FIE*); $\Phi_P(z) = \Omega_P(z)$ and $\Phi_Q(z) = \Omega_Q(z)$ are the complex potentials induced from P and Q, respectively; the subscripts n and t have definite meanings which indicate the corresponding quantities in the normal direction and tangential direction, respectively; for example, f_{nt} denotes the contribution induced from the normal traction P acting on the crack faces to the tangential direction at any point in the (x, y) plane (see Figure 3.2a).

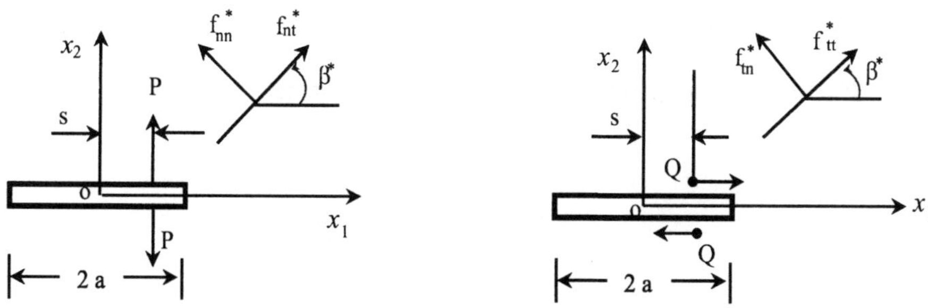

(a) under the normal traction $P*$. (b) under the tangential traction $Q*$.

Figure 3.3. a finite crack under concentrated tractions $P*$ and $Q*$.

(B) Consider a finite crack of length 2a along the x^*- axis in a local coordinate system $z^* = x^* + iy^*$(see Figure 3.3a, b). Assume that a pair of normal traction $P*$ and a pair of tangential traction $Q*$ are acting on any point t of both faces of the crack, respectively.

The problem shown in Figures 3.3(a) and 3.3(b) can be reduced to the Riemann–Hilbert problems (Chen, 1984a,b). Without going into the detail, the corresponding complex potentials are given in a compact form in the local coordinate system (x^*,y^*).

$$\Phi(z^*) = \Omega(z^*) = \Phi_P(z^*) + \Phi_Q(z^*) = -\frac{P*-iQ*}{2i\pi}\sqrt{\frac{t^2-a^2}{z^{*2}-a^2}}\frac{1}{z^*-t}, \tag{3.29}$$

where t denotes the distance from the loaded point to the origin.

The tractions at any point in the (x^*, y^*) plane along any orientation $\beta*$ induced form $P*$ and $Q*$ can then be formulated as follows

$$f_{nn}* - if_{nt}* \;=\; \Phi_P(z^*) + \overline{\Phi_P(z^*)}$$

$$
\begin{aligned}
&+ \quad e^{-2i\beta*}\left(\Omega_P\left(\overline{z^*}\right) - \overline{\Phi_P\left(z^*\right)} + \left(z^* - \overline{z^*}\right)\overline{\Phi_P\left(z^*\right)}\right),\\
f_{tn} * -if_{tt}* \quad &= \quad \Phi_Q\left(z^*\right) + \overline{\Phi_Q\left(z^*\right)}\\
&+ \quad e^{-2i\beta*}\left(\Omega_Q\left(\overline{z^*}\right) - \overline{\Phi_Q\left(z^*\right)} + \left(z^* - \overline{z^*}\right)\overline{\Phi_Q\left(z^*\right)}\right),
\end{aligned}
\tag{3.30}
$$

where f^*_{nn}, f^*_{nt} f^*_{tn} and f^*_{tt} are the so called influence coefficients in *BEM* or the kernel functions in *FIE*. It is noted that for large $|z^*|$, the four kernel functions in equation (3.30) have an asymptotical nature of $|z^*|^{-2}$ which will be adopted in sequel.

3.2 Pseudo-traction methods and integral equations

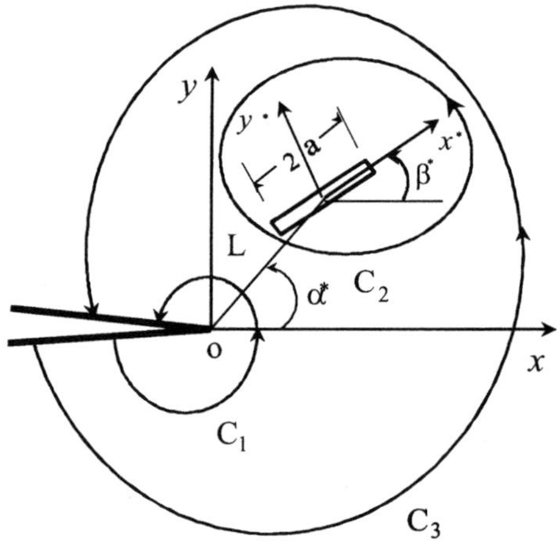

Figure 3.4. A semi-infinite crack interacted with a finite crack in a discrete model.

We study a discrete model shown in Figure 3.4, in which a semi-infinite crack is interacted with a finite crack (microcrack). Three closed contours are introduced which enclose the semi-infinite crack tip, the finite crack, and both, namely C_1, C_2 and C_3, respectively.

Although the interaction problem shown in Figure 3.4 has been well studied in the literature (see, *e.g.*, Rose, 1986a; Horii and Nemat Nasser, 1987; Gong and Horii, 1989), an alternative method is presented below for giving readers the necessary understandings of the mathematical treatments and numerical examinations to the projected conservation theory of the J_k vector in Figure 3.4. The method is based on the well known superimposing technique proposed firstly by Gross (1982), and then by Chen (1984a,b), Horii and Nemat Nasser (1987),

Gong and Horii (1989), Gong and Meguid (1991), Rubinstein (1985, 1986, 1988) and many others. Using the 'Pseudo-traction' method (Horii and Nemat Nasser, 1987), the present problem could be decomposed into two subproblems, each of which contains one single crack (see Figure 3.5). The unknown normal traction $P_0(s)$ and tangential traction $Q_0(s)$ are assumed to be applied on the faces of the semi-infinite crack. Similarly, the unknown normal traction $P_1(t)$ and tangential traction $Q_1(t)$ are assumed to be applied on the faces of the finite crack. Using the fundamental solutions presented in Section 3.1 and the superimposing technique, the following integral equations with infinite integral interval are reduced:

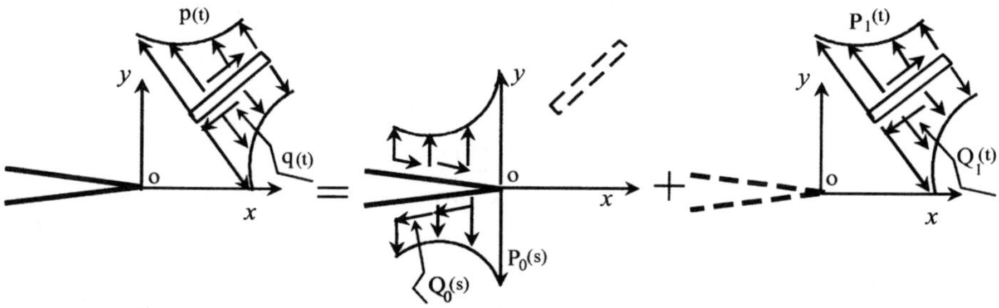

Figure 3.5. Decomposition of the discrete model into two subproblems.

$$P_0(s) + \int_{-a}^{a} P_1(t)\, f_{nn,10}*(t,s)\,dt + \int_{-a}^{a} Q_1(t)\, f_{tn,10}*(t,s)\,dt = 0, \quad (-\infty \prec s \prec 0),$$

$$Q_0(s) + \int_{-a}^{a} P_1(t)\, f_{nt,10}*(t,s)\,dt + \int_{-a}^{a} Q_1(t)\, f_{tt,10}*(t,s)\,dt = 0, \quad (-\infty \prec s \prec 0),$$

$$P_1(t) + \int_{-\infty}^{0} P_0(s)\, f_{nn,01}(s,t)\,dt + \int_{-\infty}^{0} Q_0(s)\, f_{tn,10}(t,s)\,dt = p(t), \quad (-a \prec t \prec a),$$

$$Q_1(t) + \int_{-\infty}^{0} P_0(s)\, f_{nt,01}(s,t)\,dt + \int_{-\infty}^{0} Q_0(s)\, f_{tt,01}(s,t)\,ds = q(t), \quad (-a \prec t \prec a), \tag{3.31}$$

where the eight kernel functions $f_{nn,10}*$,$f_{nt,10}*$, $f_{tn,10}*$, $f_{tt,10}*$, $f_{nn,01}$,$f_{nt,01}$, $f_{tn,01}$ and $f_{tt,01}$ are given by equations(3.30) and (3.28); the subscripts n, t, 0, 1 indicate the normal quantities, the tangential quantities, the semi-infinite crack and the finite crack, respectively; $p(t)$ and $q(t)$ are the known residual stresses to be released at the microcrack location which could be expressed by using the near-tip stress field at the semi-infinite crack tip

$$\begin{aligned} p(t) &= \tfrac{\sigma_x^t+\sigma_y^t}{2} - \tfrac{\sigma_x^t-\sigma_y^t}{2}\cos 2\beta* - \sigma_{xy}^t \sin 2\beta*, \\ q(t) &= -\tfrac{\sigma_x^t-\sigma_y^t}{2}\sin 2\beta* + \sigma_{xy}^t \cos 2\beta*, \end{aligned} \tag{3.32}$$

where the formulations of σ_x^t, σ_y^t and σ_{xy}^t are well known as follows

$$\begin{aligned} \sigma_x^t &= \frac{K_I^\infty}{\sqrt{2\pi r}}\cos\frac{\theta}{2}\left[1-\sin\frac{\theta}{2}\sin\frac{3\theta}{2}\right] - \frac{K_{II}^\infty}{\sqrt{2\pi r}}\sin\frac{\theta}{2}\left[2+\cos\frac{\theta}{2}\cos\frac{3\theta}{2}\right], \\ \sigma_y^t &= \frac{K_I^\infty}{\sqrt{2\pi r}}\cos\frac{\theta}{2}\left[1+\sin\frac{\theta}{2}\sin\frac{3\theta}{2}\right] + \frac{K_{II}^\infty}{\sqrt{2\pi r}}\sin\frac{\theta}{2}\cos\frac{\theta}{2}\cos\frac{3\theta}{2}, \\ \sigma_{xy}^t &= \frac{K_I^\infty}{\sqrt{2\pi r}}\sin\frac{\theta}{2}\cos\frac{\theta}{2}\cos\frac{3\theta}{2} + \frac{K_{II}^\infty}{\sqrt{2\pi r}}\cos\frac{\theta}{2}\left[1-\sin\frac{\theta}{2}\sin\frac{3\theta}{2}\right], \end{aligned} \tag{3.33}$$

where r and θ are the polar coordinates, and $K_I^\infty$and K_{II}^∞ are the remote applied stress intensity factors of Mode I and Model II, respectively.

Obviously, as pointed out by Gong and Meguid (1991), the residual stresses will be released when the microcrack is formed and shielding effects may also be induced from the release of these residual stresses. The aggregate effect of these localized residual stresses leads to a macroscopic residual strain, which according to Hutchinson (1987) and Evans and Faber (1984) could represent the prime source of microcrack toughening. However, the integral equation (3.31) can not be solved by using the traditional numerical integral method since the infinite integral intervals are involved. The following changes of variables proposed by Fleck *et al.*, (1991),

$$\begin{aligned} s &= (u-1)/(u+1), \\ ds &= 2du/(u+1)^2, \end{aligned} \tag{3.34}$$

allow equation (3.31) to be then reduced to the following equations without infinite intervals:

$$P_0(u) + \textstyle\int_{-a}^a P_1(t)\, f_{nn,10}*(t,u)\,dt + \int_{-a}^a Q_1(t)\, f_{tn,10}*(t,u)\,dt = 0, \quad (-1<u<1),$$

$$Q_0(u) + \textstyle\int_{-a}^a P_1(t)\, f_{nt,10}*(t,u)\,dt + \int_{-a}^a Q_1(t)\, f_{tt,10}*(t,u)\,dt = 0, \quad (-1<u<1),$$

$$
\begin{aligned}
P_1(t) &+ \int_{-1}^{1} P_0(u)\, f_{nn,01}(u,t) \frac{2du}{(u+1)^2} \\
&+ \int_{-1}^{1} Q_0(u)\, f_{tn,10}(u,t) \frac{2du}{(u+1)^2} = p(t)\,, \quad (-a < t < a)\,, \\
Q_1(t) &+ \int_{-1}^{1} P_0(u)\, f_{nt,01}(u,t) \frac{2du}{(u+1)^2} \\
&+ \int_{-1}^{1} Q_0(u)\, f_{tt,10}(u,t) \frac{2du}{(u+1)^2} = q(t)\,, \quad (-a < t < a)\,.
\end{aligned}
\tag{3.35}
$$

Here, for the sake of convenience, the following notations are used:

$$
\begin{aligned}
f^*_{nn,10}(t,s) &= f^*_{nn,10}\left(t, \frac{u-1}{u+1}\right) \equiv f^*_{nn,10}(t,u)\,, \\
f^*_{nt,10}(t,s) &\equiv f^*_{nt,10}(t,u)\,, \\
f^*_{tn,10}(t,s) &\equiv f^*_{tn,10}(t,u)\,, \\
f^*_{tt,10}(t,s) &\equiv f^*_{tt,10}(t,u)\,,
\end{aligned}
\tag{3.36}
$$

$$
\begin{aligned}
f_{nn,01}(s,t) &= f_{nn,01}\left(\frac{u-1}{u+1}, t\right) \equiv f_{nn,01}(u,t)\,, \\
f_{nt,01}(s,t) &\equiv f_{nt,01}(u,t)\,, \\
f_{tn,01}(s,t) &\equiv f_{tn,01}(u,t)\,, \\
f_{tt,10}(s,t) &\equiv f_{tt,10}(u,t)\,,
\end{aligned}
\tag{3.37}
$$

and

$$
P_0(s) \equiv P_0(u)\,, Q_0(s) \equiv Q_0(u) \tag{3.38}
$$

However, the factor of $(u+1)^2$ in the second and third terms of equation (3.35) presents a new problem in evaluation of the integrals near $u = -1$. Calling the asymptotic behaviors of $P_0(u)$ and $Q_0(u)$ (see Section 3.1), it follows that

$$
P_0(u) \sim O\left((u+1)^2\right), Q_0(u) \sim O\left((u+1)^2\right), \tag{3.39}
$$

as $u \to -1$, and equation (3.35) can then reduced to the normal Fredholm integral equations without further problems:

$$
\begin{aligned}
F_0(u)\frac{(u+1)^2}{2} &+ \int_{-a}^{a} P_1(t)\, f_{nn,10}*(t,u)\,dt \\
&+ \int_{-a}^{a} Q_1(t)\, f_{tn,10}*(t,u)\,dt = 0, \quad (-1 \prec u \prec 1), \\
G_0(u)\frac{(u+1)^2}{2} &+ \int_{-a}^{a} P_1(t)\, f_{nt,10}*(t,u)\,dt \\
&+ \int_{-a}^{a} Q_1(t)\, f_{tt,10}*(t,u)\,dt = 0, \quad (-1 \prec u \prec 1), \\
P_1(t) &+ \int_{-1}^{1} F_0(u)\, f_{nn,01}(u,t)\,du \\
&+ \int_{-1}^{1} G_0(u)\, f_{tn,01}(u,t)\,du = p(t), \quad (-a \prec t \prec a), \\
Q_1(t) &+ \int_{-1}^{1} F_0(u)\, f_{nt,01}(u,t)\,du \\
&+ \int_{-1}^{1} G_0(u)\, f_{tt,01}(u,t)\,du = q(t), \quad (-a \prec t \prec a),
\end{aligned}
\tag{3.40}
$$

where

$$
\begin{aligned}
F_0(u) &= 2P_0(u)\,/\,(u+1)^2, \\
G_0(u) &= 2Q_0(u)\,/\,(u+1)^2.
\end{aligned}
\tag{3.41}
$$

After doing so, a linear system with $4M$ unknown $F_0(u_i), G_0(u_i)$ $P_1(t_i)$ and $Q_1(t_j)$ $(i, j$=1,2,3,...,$M)$ is given by using the Chebyshev numerical integral (Chen, 1996)

$$
\begin{aligned}
\frac{(u_i+1)^2}{2}F_0(u_i) &+ \sum_{j=1}^{M} P_1(t_j)\, f_{nn,10}*(t_j,u_i)\delta_j \\
&+ \sum_{j=1}^{M} Q_1(t_j)\, f_{tn,10}*(t_j,u_i)\delta_j = 0, \\
\frac{(u_i+1)^2}{2}G_0(u_i) &+ \sum_{j=1}^{M} P_1(t_j)\, f_{nt,10}*(t_j,u_i)\delta_j \\
&+ \sum_{j=1}^{M} Q_1(t_j)\, f_{tt,10}*(t_j,u_i)\delta_j = 0,
\end{aligned}
$$

$$
\begin{aligned}
P_1(t_j) &+ \sum_{i=1}^{M} F_0(u_i)\, f_{nn,01}(u_i, t_j)\, \delta_i \\
&+ \sum_{i=1}^{M} G_0(u_i)\, f_{tn,01}(u_i, t_j)\, \delta_i = p(t_j)\,, \\
Q_1(t_j) &+ \sum_{i=1}^{M} F_0(u_i)\, f_{nt,01}(u_i, t_j)\, \delta_i \\
&+ \sum_{i=1}^{M} G_0(u_i)\, f_{tt,01}(u_i, t_j)\, \delta_i = q(t_j)\,,
\end{aligned}
\tag{3.42}
$$

where

$$
\begin{aligned}
u_i &= \cos\tfrac{(2i-1)\pi}{2M}, \\
t_j &= a \cdot \cos\tfrac{(2j-1)\pi}{2M}, \\
\delta_i &= \tfrac{\pi}{M}\sin\tfrac{(2i-1)\pi}{2M}, \\
\delta_j &= \tfrac{\pi}{M}\sin\tfrac{(2j-1)\pi}{2M}.
\end{aligned}
\tag{3.43}
$$

The changes of *SIF's* at the semi-infinite crack tip owing to microcracking and those at both tips of the microcrack can then be evaluated by using the quadrature rule (Erdogan, 1978):

$$
\begin{aligned}
\Delta K_I &= \tfrac{\sqrt{2\pi}}{M} \sum_{m=1}^{M} F_0\left(\cos\tfrac{(2m-1)\pi}{2M}\right)\left(1+\cos\tfrac{(2m-1)\pi}{2M}\right), \\
\Delta K_{II} &= \tfrac{\sqrt{2\pi}}{M} \sum_{m=1}^{M} G_0\left(\cos\tfrac{(2m-1)\pi}{2M}\right)\left(1+\cos\tfrac{(2m-1)\pi}{2M}\right),
\end{aligned}
\tag{3.44}
$$

and

$$
\begin{aligned}
K_I^R &= \tfrac{\sqrt{\pi a}}{M} \sum_{m=1}^{M} P_1\left(a\cdot\cos\tfrac{(2m-1)\pi}{2M}\right)\left(1+\cos\tfrac{(2m-1)\pi}{2M}\right), \\
K_{II}^R &= \tfrac{\sqrt{\pi a}}{M} \sum_{m=1}^{M} Q_1\left(a\cdot\cos\tfrac{(2m-1)\pi}{2M}\right)\left(1+\cos\tfrac{(2m-1)\pi}{2M}\right), \\
K_I^L &= \tfrac{\sqrt{\pi a}}{M} \sum_{m=1}^{M} P_1\left(a\cdot\cos\tfrac{(2m-1)\pi}{2M}\right)\left(1-\cos\tfrac{(2m-1)\pi}{2M}\right), \\
K_{II}^L &= \tfrac{\sqrt{\pi a}}{M} \sum_{m=1}^{M} Q_1\left(a\cdot\cos\tfrac{(2m-1)\pi}{2M}\right)\left(1-\cos\tfrac{(2m-1)\pi}{2M}\right).
\end{aligned}
\tag{3.45}
$$

Finally, the local stress intensity factors at the semi-infinite crack tip are given

$$K_I^t = K_I^\infty + \Delta K_I,$$
$$K_{II}^t = K_{II}^\infty + \Delta K_{II}. \tag{3.46}$$

As discussed above, the principal objective of analytical treatments to shielding problems is to establish a relation between the local stress intensity factor K_I^t and the remote 'applied' stress intensity factors K_I^∞. The extent to which K_I^t is reduced below K_I^∞ provides a quantitative measure of shielding (Hutchinson, 1987; Ortiz, 1987).

3.3 Numerical examinations

Although the foregoing formulations apply to the combined Mode I and Mode II loading conditions at infinity, numerical results presented below and in the forthcoming sections are limited to purely Mode I loading conditions only, such that all the stress intensity factors expressed by equations (3.44),(3.45) and (3.46) could be normalized by the remote factor K_{I}^∞.

Table 3.1. The normalized stress intensity factor K_I^t/K_I^∞ for collinear cases with $L/2a$ being a variable.

$L/2a$	Gong and Horii (1987)	Present Solution (3.46)
0.55	1.652	1.651
0.60	1.387	1.386
0.65	1.274	1.273
0.70	1.209	1.209
0.75	1.167	1.167
1.00	1.067	1.067

Table 3.1 and 3.2 show that the numerical results of normalized stress intensity factors derived by using the above mentioned technique for collinear cases with $\alpha* = 0$ and $\beta* = 0$ in Figure 3.4 coincide well with those presented by Gong and Horii (1987). Here, $L/2a$ denotes the normalized distance between the semi-infinite crack tip and the microcrack center.

Table 3.2. Normalized stress intensity factors K_I^R/K_I^∞ and K_I^L/K_I^∞ for collinear cases with $L/2a$ being a variable.

$L/2a$	Gong and Horii (1987)		Present Solutions (3.45)	
	K_I^R/K_I^∞	K_I^L/K_I^∞	K_I^R/K_I^∞	K_I^L/K_I^∞
0.55	0.655	1.469	0.655	1.474
0.60	0.610	1.138	0.610	1.138
0.65	0.579	0.979	0.579	0.978
0.70	0.555	0.877	0.554	0.877
0.75	0.534	0.805	0.534	0.805
1.00	0.464	0.611	0.463	0.611

4. The J integral analysis: the projected conservation law of the Jk vector

Reconsider the discrete model shown in Figure 3.4 in which three closed contours are specially introduced. The first, C_1, encloses the semi-infinite crack tip only. The second, C_2, encloses the microcrack completely. The third, C_3, encloses both the microcrack and semi-infinite crack tip.

According to the discussion for the J integral presented in Chapter 1 we know that

$$\begin{aligned} J_t &= \oint_{C_1} [W dx_2 - T_i u_{i,1} ds] \\ &= \tfrac{1-\nu^2}{E}\left[(K_I^t)^2 + (K_{II}^t)^2\right] \end{aligned} \tag{3.47}$$

and

$$\begin{aligned} J_\infty &= \oint_{C_3} [W dx_2 - T_i u_{i,1} ds] \\ &= \tfrac{1-\nu^2}{E}\left(K_I^\infty\right)^2, \end{aligned} \tag{3.48}$$

where K_I^t and K_{II}^t are the local *SIF's* at the macrocrack tip, whilst K_I^∞ is the Mode I remote loading level; while plane strain conditions are considered and (x_1,x_2) denotes (x, y) coordinate system (Figure 3.4).

Since the local coordinate system (x^*,y^*) does not coincide with the global coordinate system (x, y), the value of the J integral could not be evaluated directly along C_2 in the global coordinate system. For doing so, the J_k^* integral vector in the local coordinate system (x^*,y^*) or (x_1^*, x_2^*) should be introduced (see equation (1.5) in Chapter 1):

$$J_k* = \oint_{C_2} (W * n_k * -T_i * u_{i,k}*)\, ds, \ (k = 1, 2), \tag{3.49}$$

where the over star $*$ denotes the corresponding quantities in the (x_1^*, x_2^*) system.

As discussed in Chapter 1, the first component of J_k^* is path independent, which has the same expression as the J^* integral in the (x_1^*, x_2^*) system. Since the contour C_2 encloses the microcrack completely, the relation between the J_1^* and microcrack tip parameters should be as

$$\begin{aligned} J_1^* = J^* &= \oint_{C_2} \left(W^* n_1^* - T_i^* u_{i,1}^*\right) ds \\ &= \frac{1-\nu^2}{E}\left[\left(K_I^R\right)^2 + \left(K_{II}^R\right)^2 - \left(K_I^L\right)^2 - \left(K_{II}^L\right)^2\right], \end{aligned} \tag{3.50}$$

whereK_I^R, K_{II}^R,K_I^L, and K_{II}^L are stress intensity factors at both tips of the microcrack shown in Figure 3.4.

Also, as discussed in Chapter 1, the second component of J_k^* is path independent too at the present case since the contour C_2 encloses the microcrack completely. Thus the following relation should be valid:

$$\begin{aligned} J_2^* &= \oint_{C_2} \left(W^* n_2^* - T_i^* u_{i,2}^*\right) ds \\ &= \frac{2\left(1-\nu^2\right)}{E}\left[\left(K_I^L K_{II}^L\right) - \left(K_I^R K_{II}^R\right)\right] + F_{2a}^*, \end{aligned} \tag{3.51}$$

where F_{2a}^* is the additional contribution arising from the traction-free faces of the microcrack to the J_2^* integral (Herrmann and Herrmann, 1981). We have discussed well this topic in Chapter 1.

Readers should remember equations (1.16) and (1.17) in Chapter 1. We recall them and then the last term in equation (3.51), *i.e.*, expression of F_{2a}^* is given as

$$\begin{aligned} F_{2a}^* &= -\int_{-a}^{a} \left(W_*^+ - W_*^-\right) dx_1^* \\ &= -\frac{1-\nu^2}{2E}\int_{-a}^{a} \left[\left(\sigma_x^{*+}\right)^2 - \left(\sigma_x^{*-}\right)^2\right] dx_1^*, \end{aligned} \tag{3.52}$$

where $W = W^*$ is the strain energy density; σ_x^* is the in-face normal stress component along the traction-free faces of the microcrack; superscripts $+$ and $-$

refer to the upper face and lower face of the microcrack, respectively (Herrmann and Herrmann, 1981).

The stress component σ_x^* along the microcrack faces is contributed by two parts, one of which is induced from the macrocrack tip stress field and the pseudo-tractions $P_0(s)$ and $Q_0(s)$ (see Figure 3.5) and the other is the influence arising from the macrocrack itself with the pseudo-traction $P_1(t)$ and $Q_1(t)$ (see Figure 3.5). Once the linear system (3.42) is solved and the pseudo-traction $P_0(s)$, $Q_0(s)$, $P_1(t)$ and $Q_1(t)$ are derived numerically, the value of F_{2a}^* can then be evaluated without any difficulty. We will discuss the detailed process in Chapter 4 (or see, Zhao and Chen, 1997).

As discussed in Chapter 1, the J_k^* vector is dependent on the coordinate rotation. It has already been pointed out in equation (3.49) that the J_k^* vector, in the present circumstance, is defined in the local coordinate system (x_1^*, x_2^*) and that, when performing the J analysis in the global system (x_1, x_2), a coordinate system transformation should be made, form which we derive

$$\begin{aligned} \Delta J &= \oint_{C_2} (W n_1 - T_i u_{i,1})\, ds \\ &= J_1^* \cos\beta^* - J_2^* \sin\beta^*, \end{aligned} \tag{3.53}$$

where J_1^* and J_2^* are formulated by equations(3.50) and (3.51); ΔJ is the contribution of the microcrack to the J integral in the global coordinate system(x_1, x_2).

Introducing a supplementary contour AB and adopting the path independent nature of the J integral in the global coordinate system (x_1, x_2), we know that (see Figure 3.6)

$$\begin{aligned} &J_t + \Delta J = J_\infty, \\ &(J_t + \Delta J)\,/J_\infty = 1. \end{aligned} \tag{3.54}$$

Unlike the new conservation laws (1.39) and (1.40) established in Chapter 1, equation (3.54) together with equation (3.53) falls in the conventional concepts of invariant integrals established by Rice (1968a,b) and others. Figure 3.6 shows a closed contour specially introduced, which contains a segment *AB*. Readers can see that according to the conventional path independent concept, the J integral value calculated along the closed contour $C_3 + AB - C_2 + BA + CD - C_1 + FE$ vanishes since there is no discontinuity within the region enclosed by the closed contour $C_3 + AB - C_2 + BA + CD - C_1 + FE$. This understanding directly yields equation (3.54).

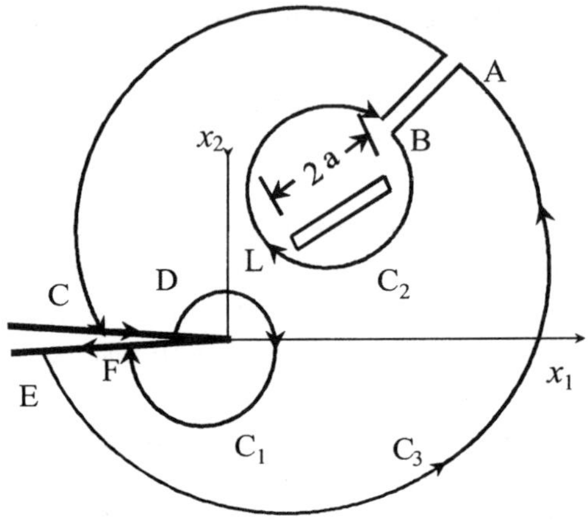

Figure 3.6. The path independent nature of the J-integral with a supplementary contour $\overline{AB}$.

However, equation (3.54) still constructs a projected conservation law of the J_k integral vector for the discrete model of microcrack shielding problems in brittle solids. This is because, physically, this law describes the redistribution aspect of the remote applied J integral when it transmits across the microcrack zone from infinity to the macrocrack tip. In other words, the remote J integral is divided into two distinct parts owing to microcracking. One of them corresponding to the first term in (3.54) is contributed from the macrocrack tip and the other corresponding to the second term in (3.54) is contributed from the formation of the near-tip microcracks. What is more, the latter should be evaluated by using the projected relations between the values of the J_k vector in local coordinate systems and those in a global coordinate system originated at the macrocrack tip. It is shown that in this circumstance the J_2 integral plays a role similar to the J_1 integral. Although the previous investigations have shown the redistribution of stress when microcracks are formed in the near-tip process zone of a macrocrack and Hutchinson (1987) has pointed out the two sources for microcrack shielding effect, the discussion performed in this chapter verifies that it is this redistribution of stress that does yields the redistribution of the remote J integral. Obviously equation (3.54) provides a quite different aspect in a discrete model from equation (3.1) in the continuum modeling by Ortiz (1987).

5. Numerical results and discussions

Numerical examinations are shown in Figure 3.7. As could be expected, equation (3.54) is always valid, no matter how the location angle and the normalized distance are.

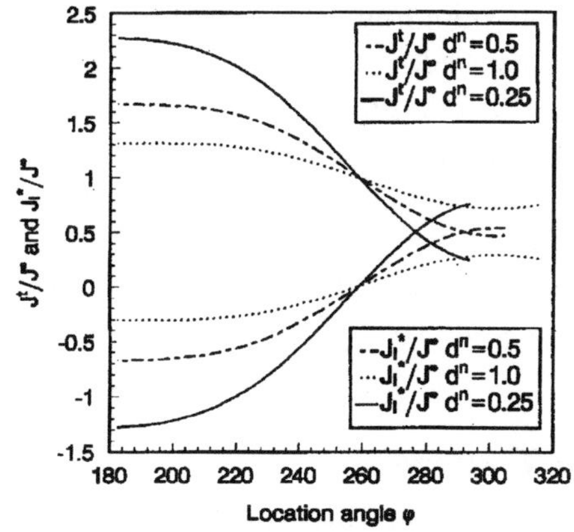

Figure 3.7. Numerical examinations for the projected conservation theory of the J_k vector (single microcrack).

Table 3.3. Numerical examinations for the projected conservation theory of the J_k vector (two microcracks).

$\alpha_1 = 30°$ $\alpha_2 = 100°$	$\beta_1 = 20°$ $\beta_2 = 20°$	$\beta_1 = 100°$ $\beta_2 = 90°$	$\beta_1 = 160°$ $\beta_2 = 120°$
K_I^t/K_I^∞	1.0659	1.0649	1.0201
K_{II}^t/K_I^∞	−0.0397	−0.1300	−0.1055
Microcrack 1			
K_I^R/K_I^∞	0.5724	0.2929	0.8414
K_{II}^R/K_I^∞	0.1973	−0.1145	−0.1891
K_I^L/K_I^∞	0.6522	0.5188	0.6609
K_{II}^L/K_I^∞	0.3484	−0.0657	0.0383
F_{2n}^*/J^∞	0.0773	0.0796	0.0222
Microcrack 2			
K_I^R/K_I^∞	0.5154	0.2358	0.1624
K_{II}^R/K_I^∞	0.1649	0.1926	−0.0429
K_I^L/K_I^∞	0.5811	0.2578	0.2799
K_{II}^L/K_I^∞	−0.0854	0.3446	−0.3408
F_{2n}^*/J^∞	−0.1085	0.0170	−0.1915
J_t/J_∞	1.1377	1.1509	1.0531
$\Delta J/J_\infty$	−0.1377	−0.1509	−0.0531
$(J_t + \Delta J)/J_\infty$	1.0000	1.0000	1.0000

The numerical examinations confirm the conclusions presented by equation (3.54). Indeed, the remote J integral (J_∞) is redistributed and divided into two parts owing to the existence of microcracks in the near-tip stress field of a macrocrack. We note that the signs of the ΔJ integral, *i.e.*, equation (3.53), govern the so called anti-shielding effect and the shielding effect. Obviously the former effect is corresponding to the negative value of ΔJ, whilst the latter is to the positive values of ΔJ. Obviously the so called neutral angle at which the transition from an anti-shielding effect to a shielding effect occurs can be recognized when a microcrack in the near-tip process zone has no contribution to the J integral. Two microcracks configurations are also calculated. Numerical results are listed in Table 3.3. It is no doubt that the redistribution is still seen although both macrocrack microcrack and microcrack microcrack interactions are taken into account.

6. Effect of the T stress

Early in the 1970s Larsson and Carlsson (1973) and Rice (1974) showed that the sign and magnitude of the T stress substantially alter the size and shape of the plane-strain crack tip plastic zone at finite load level. Bilby *et al.*, (1986) succeeded in indicating that the T stress extends the range of validity of the small scale yielding conditions at finite strain. Betegon and Hancock (1991) characterized the elastic-plastic crack tip fields by using a two-parameter law, *i.e.*, J integral and T stress. Another significance of the T stress in fracture analysis is that it governs the stability of a straight crack path (Cotterell and Rice, 1980; Fleck *et al.*, 1991) under mode I loading conditions. Tvergaad and Hutchinson (1994) studied the effect of the T stress on mode I crack growth resistance in a ductile solid.

6.1 What is the T stress?

The definition and notation of the T stress was done by Rice (1974) using the standard Williams expansion (1957) and taking the first and the second terms into account:

$$\begin{vmatrix} \sigma_{11} & \sigma_{12} \\ \sigma_{21} & \sigma_{22} \end{vmatrix} = \frac{K}{\sqrt{2\pi r}} \begin{vmatrix} f_{11}(\theta) & f_{12}(\theta) \\ f_{21}(\theta) & f_{22}(\theta) \end{vmatrix} + \begin{vmatrix} T & 0 \\ 0 & 0 \end{vmatrix} + O(\sqrt{r}). \quad (3.55)$$

The first term in the above expansion is singular at the crack tip, whereas the second term is the finite and bounded T stress in the notation of Rice (1974), which can be regarded as a stress acting parallel to the crack plane.

Zhao and Chen (1998) studied the effect of the T stress in microcrack shielding problems by adopting the discrete model requiring the consideration of the interaction of a macrocrack with microcracks. The numerical scheme has been well addressed in the above sections or thoroughly by Chen (1996) and Zhao and Chen (1997).

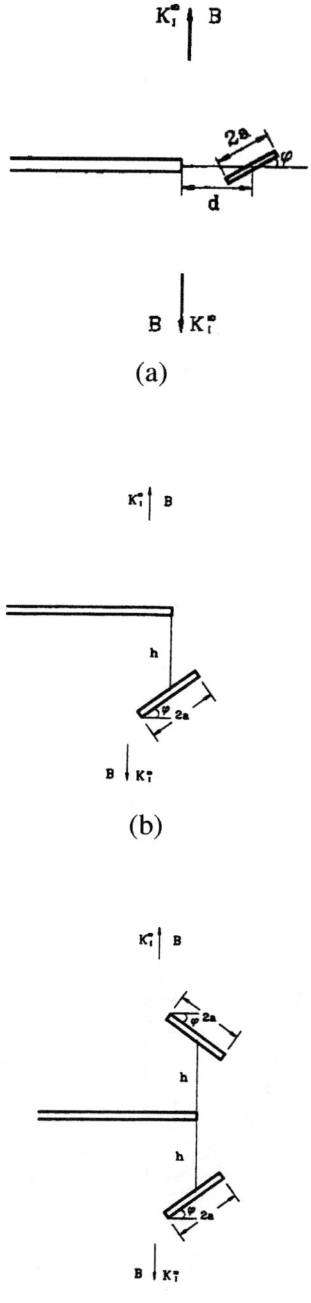

Figure 3.8. Discrete model of microcrack shielding problems with one or two microcracks.

Readers can easily extend the scheme to the current problem shown in Figure 3.8(a,b,c) taking the T stress into account in the remote stress field. Here the magnitude of the remote T stress, T^∞, is defined through a biaxial parameter B, introduced by Leevers and Radon (1983) and Betegon and Hancock (1991)

$$B = \frac{T^\infty \sqrt{\pi a}}{K_I^\infty}, \tag{3.56}$$

where a is half of the mean length of the N-mocrocracks in the near-tip process zone.

For a finite crack of length $2a$ in an infinite plate, the T stress is dependent on the applied remote loading. For example, when the plate is loaded by a stress σ normal to the crack and $R\sigma$ parallel to it, we have (Cotterell and Rice, 1980),

$$T = (R-1)\sigma, \qquad K_I = \sigma\sqrt{\pi a}, \qquad B = \frac{T\sqrt{\pi a}}{K_I} = R - 1. \tag{3.57}$$

We can see that $-1 \le B \le 1$ for $0 \le R \le 2$, so it is appropriate to take some values of B in the interval $(-1, 1)$ in the following calculation.

6.2 What role does the T stress play in microcrack shielding problems?

Numerical results for interacting effects as influenced by the T stress are plotted in Figures 3.9(a,b,c,d) for the configuration shown in Figure 3.8(a) with the variation of the microcrack orientation angle φ.

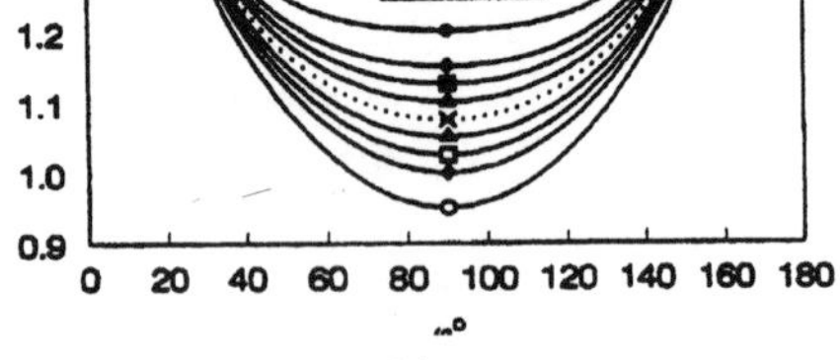

(a) $d/a = 1.1$

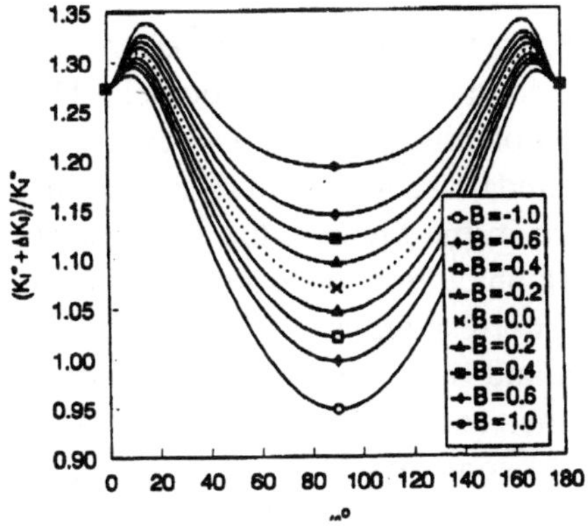

(b) $d/a = 1.3$

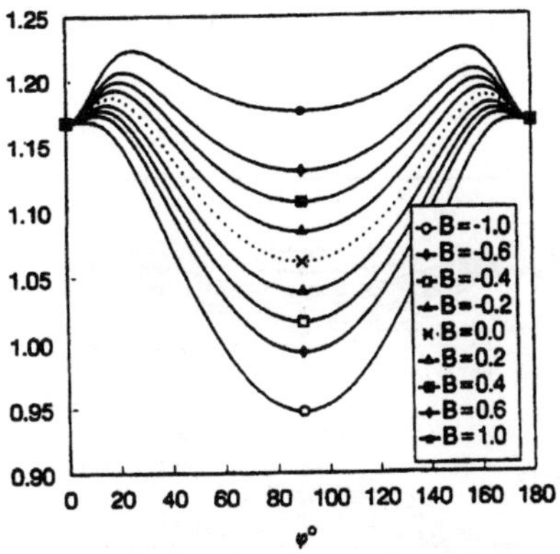

(c) $d/a = 1.5$

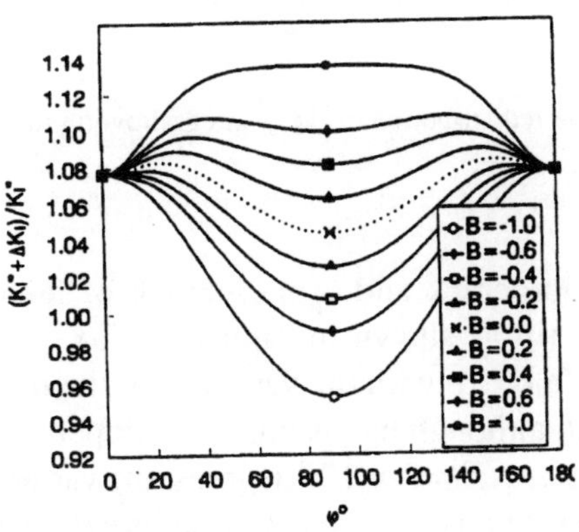

(d) $d/a = 2.0$

Figure 3.9. Normalized Local *SIF* of the macrocrack vs. the microcrack orientation angle φ.

It can be seen that when $\varphi = 0$ or 180 degrees, *i.e.,* the collinear microcrack, the T stress has no effect on the results. However, the sign and magnitude of the T stress alter the results for the oriented microcrack. There exist obvious differences between the results obtained with the T stress (see the continuous curves in the figures for $B \neq 0$) and those without it (see the dotted curves in the figures for $B = 0$). As shown in the figures, the T stress substantially influences the oriented microcrack shielding corresponding to $(K_I^\infty + \Delta K)/K_I^\infty < 1$ or amplification effect corresponding to $(K_I^\infty + \Delta K)/K_I^\infty > 1$, especially for large values of d/a and $|B|$. The positive T stress ($B > 0$) increases the amplification effect, whereas the negative T stress ($B < 0$) decreases the amplification effect and, to some extent, it will cause a shielding effect. the T stress will produce the most significant effect when $\varphi = 90$ degrees. As regards the conservation law (3.54) together with (3.53), Figure 3.10 shows the results of the redistribution of the J integral for $d/a = 1.5$. We can see that the law is still valid when considering the T stress , although the T stress influences the values of J_t and ΔJ.

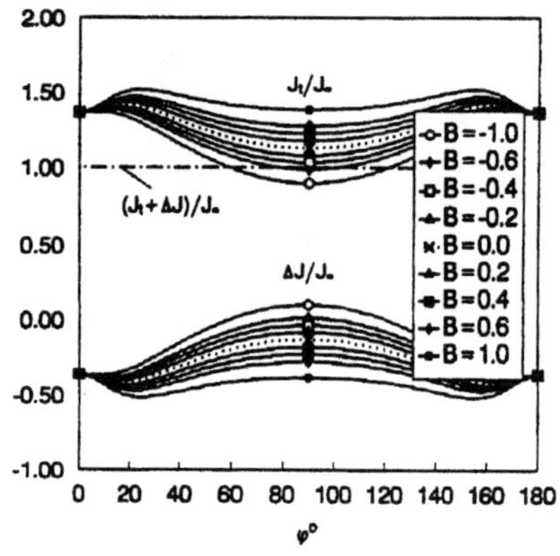

Figure 3.10. The redistribution of the J integral owing to microcracking.

Then consider a macrocrack and a microcrack of length $2a$ just centered below the macrocrack tip as shown in Figure 3.8(b). Numerical results are plotted in Figures 3.11(a,b,c,d) with the variation of the microcrack orientation angle φ^0. It is seen that, although the results are independent of the T stress for the parallel microcrack, *i.e.,* φ^0=0 or 180 degrees, the variable tendencies for the oriented microcrack as influenced by the T stress are quite different from those shown in Figures 3.9(a,b,c,d). The curves in each of Figures 3.11(a,b,c,d) are noted to cross at a point, which hints that the results are weakly dependent on the T stress for the specified microcrack orientation. Such specified microcrack orientation is denoted by φ_T in this chapter and φ_T =87.6 degrees for h/a =1.1, 84.3 degrees for h/a =1.3, 81.9 degrees for h/a =1.5 and 78.2 degrees for h/a =2.0. When $\varphi < \varphi_T$, the positive T stress ($B > 0$) increases the shielding effect or decreases the amplif ication effect and the negative T stress ($B < 0$) just leads to an opposite effect from the positive T stress.

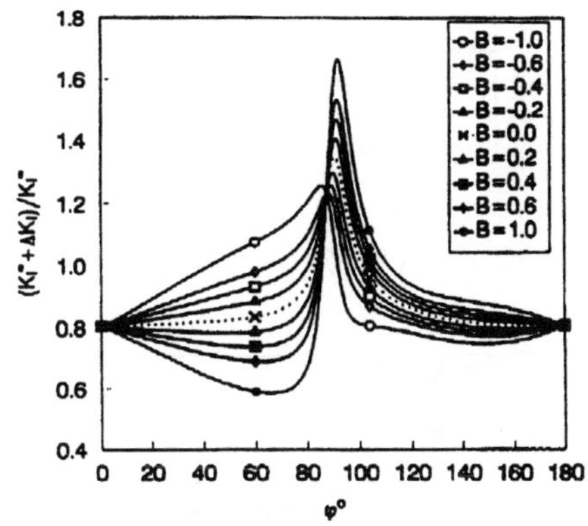

(a) $h/a = 1.1$

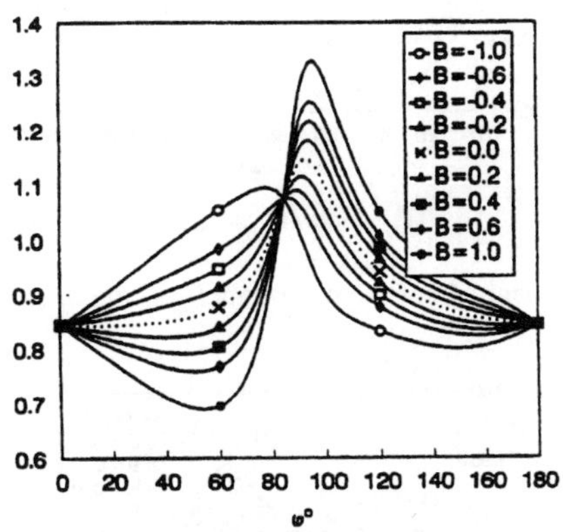

(b) $h/a = 1.3$

Of great interest is that the conservation law (3.54) is still valid in this case shown in Figure 3.12 no matter how the T stress is specif ied. In determining the shielding or amplif ication effect of the oriented microcrack, it is necessary to consider the effect of the T stress. This effect is substantially dependent on the sign and magnitude of the T stress as well as the geometry of the microcrack arrangement.

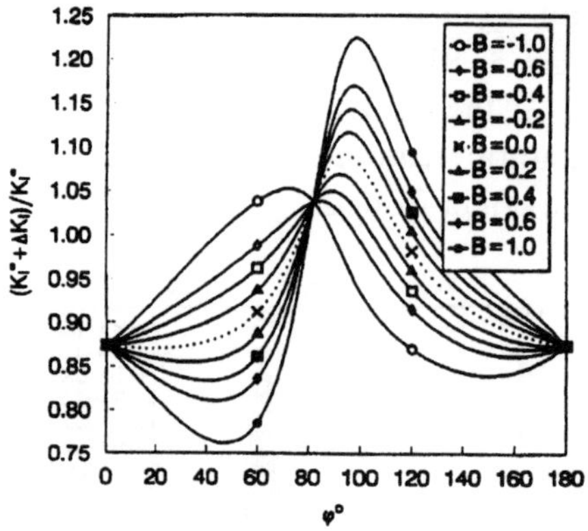

(c) $h/a = 1.5$

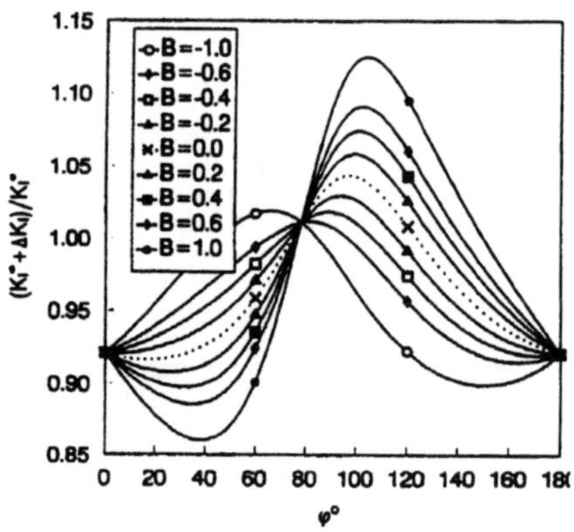

(d) $h/a = 2.0$

Figure 3.11. Normalized Local *SIF* of the macrocrack vs. the microcrack orientation angle φ.

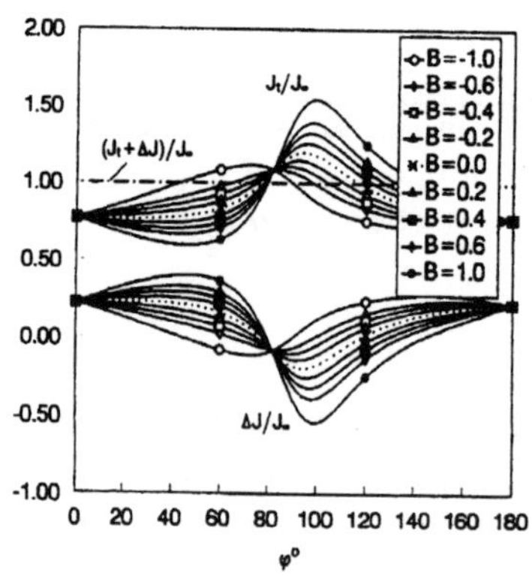

Figure 3.12. The redistribution of the J integral as influenced by the T stress.

Let us consider the more complicated example as shown in Figure 3.8(c), in which two symmetric microcracks are interacted with the macrocrack tip. Figure 3.13 shows the numerical results for the interacting effect. Similarly, the T stress influences the shielding effect and anti-shielding effect substantially. Whereas the conservation law (3.54) is re-examined in Figure 3.14, which shows that the validity of the conservation law is verified again.

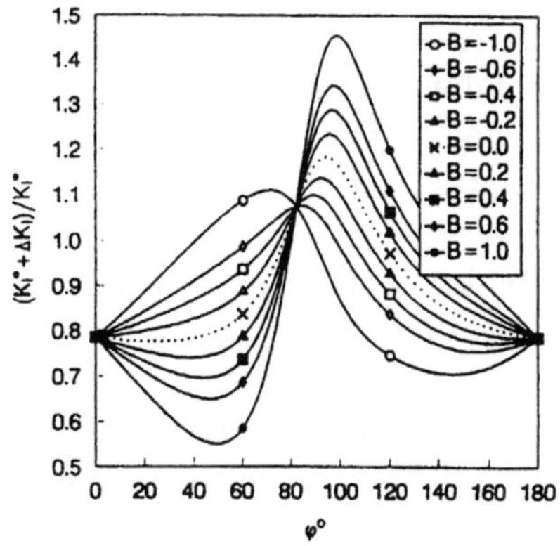

Figure 3.13. Normalized Local *SIF* of the macrocrack vs. the microcrack orientation angle φ with h/a=1.5.

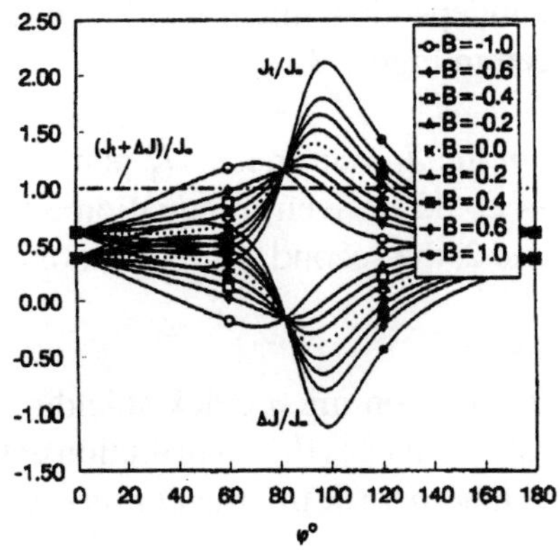

Figure 3.14. The redistribution of the J integral as influenced by the T stress.

It is no doubt that the law (3.54) does describe the key aspect of a discrete model in shielding problems. It is concluded too that the positive and negative values of the T stress always play opposite effects in any kind of macro-micro crack configuration in shielding problems. That is, if the positive T stress increases the shielding effect, the negative T stress does always decrease it, vice versa.

7. Brief summary

In this chapter we have shown that the projected conservation law of the J_k vector (3.53) and (3.54) is valid in a discrete model of microcrack shielding problems in brittle solids. Although the T stress does significantly influence the shielding and anti-shielding effects in the discrete model, the conservation law always remains unchanged. Thus, the questions cited in the beginning of section 3.2 can be answered at present.

(A) There is a redistribution of the J integral when taking into account microcracks in a near-tip stress field.

(B) The existence of microcracks does contribute to the J integral owing to the redistribution of stress in the near-tip field as a consequence of microcracking (Hutchinson, 1987).

(C) There exists a conservation law, *i.e.,* (3.54), amongst the three values of the J integral contributed respectively from the semi-infinite crack tip, the microcracks, and the remote stress field. Unlike the new laws (1.39) and (1.40) presented in Chapters 1 and 2, the present law is directly deduced from the conventional concept of path independent integral in Fracture Mechanics (Eshelby, 1970, 1975) as can be seen in Figure 3.6.

(D) Both components of the J_k vector, *i.e.,* J_1^* and J_2^*, play the similar roles in the present problem as could be seen in equation 3.53. In other words, as discussed in Chapters 1 and 2, the second component of the J_k vector could not be disregarded.

(E) The effect of the T stress on microcrack shielding problems is remarkable. Although the T stress influences the contribution to the J integral induced from microcracks and the macrocrack tip, the conservation law (3.54) remains unchanged.

As pointed out by Hutchinson (1987), the so called shielding effect by microcracks to a macrocrack tip is induced from the redistribution of stress in the near-tip stress field and it is concerned with two sources. The first one of them is the reduction in the effective moduli in the microcrack zone and the second is the release of residual stresses at the microcrack locations. Both sources are the energy dissipative process. Obviously it is this redistribution of stress or the energy dissipative process that does lead to the redistribution of the remote J integral mentioned above. Moreover, the discussion presented in this chapter reveals that the total effect of a cloud of microcracks in the near-tip stress field of a macrocrack is dominated by a single phenomenological parameter, *e.g.*, ΔJ the contribution arising from the existence of the cloud of microcracks to the J integral. The present discussion supports the so called Local *ERR* and Global *ERR* concepts.

We will show in the next chapter that the projected relation (3.53) together with the conservation law (3.54) is an inherent and universal one, which will be proved to be valid in some similar problems in much more complicated materials, *e.g.*, metal/ceramic bimaterials, dissimilar composite materials, and piezoelectric materials.

References for Chapter 3

Berges, D. (1974) Determination of stress intensity factors by use of path independent integrals, *Mechanics Research Communications*, **1**, 179 – 180.

Betegon, C., and Hancock, J.W. (1991) Two-parameter characterization of elastic plastic crack-tip fields, *ASME Journal of Applied Mechanics,* **58**, 104 – 110.

Bilby, B.A., Cardew, G.E., Goldthorpe, M.R., and Howard, I.C., (1986), A finite element investigation of the effect of specimen geometry on the fields of stress and strain at the tips of stationary cracks, *Size Effects in Fracture, Mechanical Engineering Publications, Ltd., London*, 37 – 46.

Bowling, G.D., Faber, K.T., and Hoagland, R.D., (1987) Computer simulations of R-curve behavior in microcracking materials , *J. Am. Ceram. Soc.*, **70**, 849 – 854.

Budiansky, B., and Rice, J.R.,(1973) Conservation laws and energy release rates, *ASME Journal of Applied Mechanics,* **40**, 201 – 203.

Budiansky, B., Hutchinson, J.W., and Lambropoulos, J.C., (1983) Continuum theory of dilatant transformation toughening in ceramics , *International Journal of Solids and Structures*, **19**, 337 – 355.

Cai, H., Gu, W.H., and Faber, K.T., (1990) Microcrack toughening in a Sic-TiB_2 Composites , Proceedings of the American Society for Composites, *Fifth Technical Conference on Composite materials in transition, Technomic Publishing Co. Lancaster. PA,* 892 – 901.

Cai, H., and Faber, T.K., (1992) On the use of approximation methods for microcrack shielding problems , *ASME Journal of Applied Mechanics*, **59**, 497 – 501.

Charalambides, P.G., and McMeeking, R. M., (1987a) Finite element method simulation of crack propagation in a brittle microcracking solid , *Mechanics of Materials*, **6**, 71 – 87.

Charalambides, P.G., and R. M.McMeeking, R. M., (1987b) Near-tip mechanics of stress-induced microcracking in brittle materials , *J. Am. Ceram. Soc.*, **71**, 465 – 472.

Chen, Y.H., and Hasebe, N., (1994) Interaction between an main-crack and a parallel microcrack in an orthotropic plane elastic solid , *International Journal of Solids and Structures*, **31**, 1877 – 1890.

Chen, Y.H., and Hasebe, N., (1995a) Investigation of main-micro crack interaction problems in a unidirectional f ibre reinforced composite , *Engineering Fracture Mechanics*, **50**, 283 – 293.

Chen, Y.H., and Hasebe, HN., (1995b) On the inf luence of orthotropic material constants on the neutral-shielding angle in main-micro crack interaction problems , *Engineering Fracture Mechanics*, **51**, 349 – 360.

Chen, Y. H., and Hasebe, N., (1995c) Interaction of two-off-axis cracks exhibiting orthotropic and general anisotropic behavior under arbitrary extension , *Theoretical and Applied Fracture Mechanics.*, **22**, 249 – 260.

Chen, Y.H., (1996) On the contribution of discontinuities in a near-tip stress f ield to the Jintegral , *International Journal of Engineering Science.*, **34**, 819 – 829.

Chen, Y.H., and Ma, H., (1997) Explicit formulations of the J_2-integral in anisotropic bodies and it's application , *Science in China, E*, **27**, 310 – 317, (in Chinese).

Chen, Y.Z., (1984a) General case of multiple crack problems in an inf inite plate , *Engineering Fracture Mechanics*, **20**, 591 – 597.

Chen, Y.Z., (1984b) Multiple crack problems of antiplane elasticity in an inf inite body , *Engineering Fracture Mechanics*, **20**, 767 – 776.

Cherepanov, G.P., *Mechanics of Brittle Fracture* , McGraw-Hill, New York (1979).

Chudnovsky, A., Dolgopolsky, A., and Kachanov, M., (1987a) Elastic interaction of a crack with a microcrack array-I. Formulation of problem and general form of the solutions , *International Journal of Solids and Structures*, **23**, 1 – 10.

Chudnovsky, A., Dolgopolsky, A., and Kachanov, M., (1987b) Elastic interaction of a crack with a microcrack array-II. Elastic solution for two crack conf iguration (piecewise constant and linear approximations) , *International Journal of Solids and Structures*, **23**, 11 – 21.

Chudnovsky, A., and Wu, S., (1990) Effect of crack-microcrack interaction on energy release rates , *International Journal of Fracture*, **44**, 43 – 56.

Clarke, D. R., (1984) A simple calculation of process-zone toughening by microcracking , *J. Am. Ceram. Soc.*, **67**, $C15 - C16$.

Clausen, N., (1976) Fracture toughness of Al_2O_3 with a unstabilized ZrO_2 dispersed phase , *J. Am. Ceram. Soc.*, **59**, 49 – 51.

Cotterell, B., and Rice, J.R., (1980) Slightly curved or kinked crack , *International Journal of Fracture*, **16**, 155 – 169.

Erdogen, F., (1978) *Mixed Boundary Value Problems* , Mechanics Today, **4**, Pergamen Press, Oxford.

Eshelby, J.D., (1970) *The Energy Momentum Tensor in Continuum Mechanics, Inelastic Behavior of Solids*, edited by Kanninen, M.F., *et al.*, McGraw-Hill, New York.

Eshelby, J. D., (1975) The elastic energy-momentum tensor, *Journal of Elasticity* **5**, 321 – 335.

Evans, A.G., and Faber, K.T., (1981) Toughening of ceramics by circumferential microcracking , *J. Am. Ceram. Soc.*, **64**, 394 – 398.

Evans, A.G., and Faber, K.T., (1984) Crack-growth resistance of microcracking brittle materials , *J. Am Ceram. Soc.*, **67**, 255 – 260.

Faber, A.G., and Fu, Y., (1985) Some effects of microcracks on the mechanical properties of brittle solids-II. Microcrack toughening , *Acta Metall.*, **33**, 1525 – 1531.

Faber, K.T., Gu, G.H., Cai, H., Winholtz, R.A., and Magley, D.J., (1991) Fracture properties of SiC-Based Particulate Composites , Proceedings of the NATO Advanced Research Workshop on Toughening Mechanisms in Quasi-Brittle Materials, *Kluwer Academic Publishers, The Netherlands*, 3 – 17.

Fleck, N.A., Hutchinson, J.W., and Suo, Z., (1991) Crack path selection in a brittle adhensive layer , *International Journal of Solids and Structures*, **27**, 1683 – 1703.

Fu, Y., and Evans, A. G., (1985) Some effects of microcracks on the mechanical properties of brittle solids-I. Stress, Strain relations , *Acta Metall.*, **33**, 1515 – 1523.

Gong, S.X., and Horii, H., (1989) General solution to the problem of microcracks near the tip of a main crack , *Journal of Mechanics and Physics of Solids*, **37**, 27 – 46.

Gong, S.X., and Meguid, S.A., (1991) On the effect of the release of residual stress Owing to near-tip microcracking , *International Journal of Fracture*, **52**, 257 – 274.

Gross, D., (1982) Stress intensity factors of systems of cracks , *Ing. Arch.*, **51**, 301 – 310.

Herrmann, A.G., and Herrmann, G., (1981) On energy-release rates for a plane crack , *ASME Journal of Applied Mechanics*, **48**, 525 – 528.

Hoagland, R.G., Hahn, R.T., and Rosenf ield, A.R., (1973) Inf luence of microstructure on the fracture propagation in rocks , *Rock Mech.*, **5**, 77 – 106.

Hoagland, R.G., and Embury, J.D., (1980) A treatment of inelastic deformation around a crack tip Owing to microcracking , *J. Am. Ceram. Soc.*, **63**, 404 – 410.

Horii, H., and Nemat Nasser, S., (1985) Elastic fields of interacting in inhomo-geneities, *International Journal of Solids and Structures*, **21**, 731−745.

Horii, H., and Nemat Nasser, S., (1987) Interacting microcracks near the tip in the process zone of a macrocrack , *Journal of Mechanics and Physics of Solids,* **35**, 601 − 629.

Hutchinson, J.W., (1983) Constitutive behavior and crack tip fields for materials undergoing creep-constrained grain boundary cavitation , *Acta Metall.*, **31**, 1079 − 1088.

Hutchinson, J.W., (1987) Crack tip shielding by microcracking in brittle solids , *Acta Metall.*, **35**, 1605 − 1619.

Kachanov, M., and Montagut, E., (1986) Interaction of a crack with certain microcrack array , *Engineering Fracture Mechanics.*, **25**, 625 − 636.

Krenk, S., (1975) On the use of the interpolation polynomial for solutions of singular integral equations , *Quarterly of Applied Mathematics*, **32**, 479 − 484.

Larsson, S.G., and Carlsson, A.J., (1973) Influence of non-singular stress terms and specimen geometry on small-scale yielding at crack tips in elastic-plastic materials, *Journal of Mechanics and Physics of Solids*, **21**, 263 − 277.

Laws, N., and Brockenbrough, J. R., (1988) Microcracking in Polycrystalline solids , *J. Eng. Mater. Tech.*, **110**, 101 − 104.

Leevers, P.S., and Radon, J.C., (1983) Inherent stress biaxiality in various fracture specimen geometries, *International Journal of Fracture*, **19**, 311−325.

Montagut, E., and Kachcnov, M., (1988) On modelling a microcracked zone by weakened elastic material and on statistical aspects of crack-microcrack interaction , *International Journal of Fracture*, **37**, $R55 - R62$.

Muskhelishvili, N.I., (1953) *Some Basic Problems of Mathematical Theory of Elasticity* , Noordhoff Publishers, Groningen.

Ortiz, M., (1987) A continuum theory of crack shielding in ceramics , *ASME Journal of Applied Mechanics*, **54**, 54 − 58.

Ortiz, M., (1988) Microcrack coalescence and macroscopic crack growth initiation in brittle solids , *International Journal of Solids and Structures*, **24**, 231 − 250.

Ortiz, M., and Ginnakopoulos, A.E., (1989) Maximal crack tip shielding by microcracking , *ASME Journal of Applied Mechanics*, **56**, 279 − 283.

Rice, J.R., (1968a) A path independent integral and the approximate analysis of strain concentration by notches and cracks , *ASME Journal of Applied Mechanics* **35**, 379 − 386.

Rice, J.R., (1968b) *Mathematical analysis in the mechanics of fracture* , In *Fracture* (Edited by H. Liebowitz), **2**, 191 − 231, Academic Press, New York.

Rice, J.R., (1974) Limitations to the small scale yielding approximation for crack tip plasticity, *Journal of Mechanics and Physics of Solids,* **22**, 17 − 26.

Rose, L.R.F., (1986a) Microcrack interaction with a main-crack , *International Journal of Fracture*, **31**, 233 − 242.

Rose, L.R.F., (1986b) Effective fracture toughness of microcracked materials , *J. Am. Ceram. Soc.*, **69**, 212 – 214.

Rubinstein, A.A., (1985) Macrocrack interaction with semi-infinite microcrack array , *International Journal of Fracture*, **27**, 113 – 119.

Rubinstein, A.A., (1986) Macrocrack-microdefect interaction , *ASME Journal of Applied Mechanics*, **53**, 503 – 510.

Rubinstein, A.A., (1988) Macrocrack interaction with transverse array of microcracks , *International Journal of Fracture*, **36**, 15 – 26.

Ruhle, M., Evans, A.G., McMeeking, R.M., Charalambides, P.G., and Hutchinson, J.W., (1987) Microcrack toughening in Alumina / Zirconia , *Acta Metall.*, **55**, 2701 – 2710.

Tvergaad, V., and Hutchinson, J.W., (1994) Effect of T-stress on mode I crack growth resistance in a ductile solid , *International Journal of Solids and Structures*, **31**, 823 – 833.

Wang, X.M., Gao, S., and Chen, Y.H., (1996) Further investigation for the macro-microcrack interaction, I-In an infinite isotropic body , *International Journal of Solids and Structures*, **33**, 4051 – 4064.

Williams, M.L., (1957) On the stress distribution at the base of a stationary crack , *ASME Journal of Applied Mechanics*, **24**, 111 – 114.

Wu, S., and Chudnovsky, A., (1991) Elastic interaction of a crack with a random array of microcracks , *International Journal of Fracture*, **49**, 123 – 140.

Wu, C.C., Freiman, S.W., Rice, R.W., and Mecholsky, J.J., (1978) Microstructural aspects of crack propagation in ceramics , *J. Mater. Sci.*, 13, 2659 – 2670.

Zhao, L.G., and Chen, Y.H., (1998) Effect of the T stress in microcrack shielding problems., *ASME Journal of Applied Mechanics*, **65**, 71 – 75.

Chapter 4

APPLICATION OF THE CONSERVATION LAWS IN METAL/CERAMIC BIMATERIALS

In this chapter the following three topics: (i) interface crack; (ii) macrocrack microcrack interaction; and (iii) conservation laws of the J_k vector are discussed. These topics are of great significance in both theoretical and applied fracture mechanics.

As is widely accepted, the elastic interface crack problem has received considerable attention in the past 40 years since Williams (1959) determined the well known characteristic oscillating stress singularity in the near-tip stress field. Besides Williams' work there are so many investigations concerned with the topic in the literature. Cherepanov (1979), Erdogan (1965), England (1965), Atkinson (1977), Comninou (1977a,b, 1978, 1979), Rice and Sih (1965) and Hong and Stern (1978) presented the earlier investigation in the field. Works of Knowles and Sternberg (1983), Park and Earmme (1986), Yau and Wang (1984), He and Hutchinson (1989), Hutchinson *et al.,* (1987), Cai and Evans (1989), Rice (1988), Suo and Hutchinson (1989a,b), Suo (1989), Wang and Suo (1990), Gao (1991), Symington (1987), and Chen and Hasebe (1994a,b) provided examples of the contributions in the past 15 years. From the physical point of view the most significant advance was made by Rice (1988), who re-examined the elastic fracture mechanics concept for an interface crack between dissimilar elastic solids and proved that the complex stress intensity factor at the crack tip is still valid as a crack tip characterizing parameter under the conditions of the small scale non-linear material behavior and/or small contact zone near the crack-tip. That is, similar values of the factor for two cracked bodies then imply similar states near the crack tip, so that conditions for crack growth could be phrased in terms of the factor reaching a critical failure locus in a complex plane. Moreover, he derived the form of stress and displacement fields in the vicinity of the crack tip by using complex potential theory, which is equivalent to complete Williams expansions of both inner and outer (external to a nonlinear or contact zone) type. Although the elastic interfacial crack solutions predict interpenetration of the crack walls near the tip and the solutions must be wrong in detail at the scale of the contact zone, they do, nevertheless, provide a proper characterizing parameter, *i.e.*, the complex stress intensity factor, for the near tip state in typical circumstances when the zone size is much smaller than crack length. Chen and Hasebe (1994a,b) studied the Rice solutions (1988) in more detail. They found two important properties of the solutions, *i.e.*, the differential property and the so called pseudo-orthogonal property, by using the Betti's reciprocal theorem in plane elasticity. Several path independent integrals and weight functions are then obtained from which the coefficients in Rice's solutions, including the complex stress intensity factor and the well known T term for a given geometrical configuration and under a given loading condition, could be calculated without any difficulty.

As discussed in Chapter 3, the microcrack shielding problem (including the so called discrete for the macrocrack microcrack interaction problem) in brittle solids such as ceramics and rocks has become popular in the past twenty years (Hoagland *et al.*, 1973, 1980; Clausen, 1976; Wu *et al.*, 1978; Fu and Evans, 1985; Faber and Fu, 1985). A growing body of observational evidence attributes a role to microcracking due to fracture of brittle solids. On the one hand, microcracking is thought to contribute to the stability of macroscopic cracks by softening the material surrounding the tip, thereby mitigating the effect of the applied loads (Evans and Faber, 1981, 1984). On the other hand, the microcracks created ahead of the main crack can be expected to reduce the resistance to fracture of the material. The next extent of toughening of embrittlement is the result of the counteracting effects of shielding and degradation. Earlier studies of this problem focused mainly on the toughening effect of microcracking. A principal objective of these analyses is to establish a relation between the local stress intensity factor at the crack tip K_t and the amplitude K_∞ of the surrounding K field, or the remote stress intensity factor (sometimes called the 'applied' stress intensity factor). The extent to which K_t is reduced below K_∞ provides a quantitative measure of shielding. Hutchinson (1983, 1987) presented an analysis of the topic that takes transformation strains into consideration for both stationary and steady growth conditions. A noteworthy outcome of his work is the observation that anisotropic microcracking screens the applied loads more effectively than isotropic damage.

However, relatively little attention has been given to the problem of estimating the shielding effect of a microcracking metal/ceramic bimaterial solid in which microcracks are formed in the ceramic phase.

As discussed in Chapter 3, the simple but universal relation in brittle microcracked solids among three values of the J integral was found corresponding to the contributions induced from a macrocrack tip, the microcracks near the tip and the remote stress field, respectively (Chen, 1996; Zhao and Chen, 1998).

The second component J_2^* of the J_k^* vector in a local coordinate system chosen for each of the microcracks is shown to play an important rote in the relation. Since the relation concerns only with the projected values of both $J_1^*$and J_2^*components in a global coordinate system, which are path independent for a contour enclosing each of the microcracks completely, the relation could then be called as the projected conservation law of the J_k vector in the brittle microcracked solids. However, it is not clear whether or not the projected conservation theory is valid in bimaterial solids.

In this chapter, based on the work by Zhao and Chen (1997), we will perform some mathematical manipulations and numerical examinations to show that the oscillatory stress singularity, as well as the dissimilar material combinations, have no influence on the law itself, although they have some influence on the redistributed values of the J integral induced from an interface crack tip and sub-interface microcracks near the tip. Three numerical examples are given and the numerical results are shown in Tables and Figures, which are concerned with two kinds of bimaterial combinations Cu/Al_2O_3 and Ni/MgO presented by Hutchinson *et al.*, (1987). It is concluded also that the interaction effect between the interface crack and the sub-interface microcracks, whatever it is shielding or anti-shielding, can be considered as the redistribution of the remote J integral, which is divided into two parts, one being induced by the existence of the microcracks in brittle phases of the bimaterial solids and the other being contributed by the interface macrocrack tip.

1. Fundamental solutions for an interface crack and a sub-interface crack

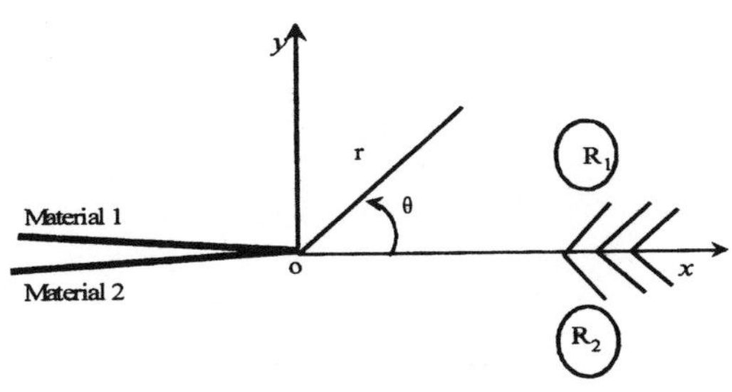

Figure 4.1. A semi-infinite crack along a bimaterial interface with coordinates (x, y) and (r,θ).

Consider an infinite bimaterial solid and let material 1 and material 2 occupy the upper and lower half-planes, respectively (see Figure 4.1). Assume that an interface crack has been created by thermal mismatch or other sources. In the coordinate system shown in Figure 4.1, the crack-tip field has a universal structure for each given material pair with the normal stress σ_{yy} and shear stress σ_{yx}along the interface ahead of the tip given by

$$(\sigma_{yy} + i\sigma_{yx})_{\theta=0} = \frac{k \cdot r^{i\varepsilon}}{\sqrt{2\pi r}}, \tag{4.1}$$

where $i = \sqrt{-1}$ and the 'oscillation index' ε is related to the material constants by

$$\varepsilon = \frac{1}{2\pi} \ln \frac{\kappa_1/\mu_1 + 1/\mu_2}{\kappa_2/\mu_2 + 1/\mu_1}. \tag{4.2}$$

Here subscripts 1 and 2 refer to the upper and lower materials, respectively, $k_j = 3 - 4\nu_j$ for plane strain and $k_j = (3 - \nu_j)/(1 + \nu_j)$ for plane stress, $j = 1, 2$, ν_j is Poisson's ratio, and μ_j is shear modulus.

The well known Dundurs' parameters (Rice, 1988) are also important to describe the mismatch nature of each given material pair:

$$\alpha = [\mu_1(\kappa_2 + 1) - \mu_2(\kappa_1 + 1)]/[\mu_1(\kappa_2 + 1) + \mu_2(\kappa_1 + 1)], \tag{4.3}$$

$$\beta = [\mu_1(\kappa_2 - 1) - \mu_2(\kappa_1 - 1)]/[\mu_1(\kappa_2 + 1) + \mu_2(\kappa_1 + 1)]. \tag{4.4}$$

Obviously it follows that

$$\varepsilon = \frac{1}{2\pi} \ln(\frac{1-\beta}{1+\beta}) \tag{4.5}$$

and

$$\frac{(1-\alpha)}{(1+\alpha)} = \frac{E_2'}{E_1'}, \tag{4.6}$$

where $E_j' = E_j$ for plane stress and $E_j' = E_j\big/(1 - \nu_j^2)$ for plane strain, j=1,2 (Fleck et al, 1991).

It is seen that there is a universal relation between ε and β and that α is an independent parameter which has no direct relation to ε and β.

The complete near-tip stress field in the *xy*-plane can be written in the from Rice's work (1988) or Gao (1991):

$$\sigma_{ij} = \frac{1}{2\pi r} \cdot [Re(kr^{i\varepsilon}) \cdot \Sigma^{I}_{ij}(\theta) + Im(kr^{i\varepsilon}) \cdot \Sigma^{II}_{ij}(\theta)], \qquad (i,j = 1,2; i.e., i,j = x,y). \tag{4.7}$$

In the sense of the crack-tip field structure expressed in (4.1) and (4.7), the angular functions $\sum^{I}_{ij}(\theta)$ and $\sum^{II}_{ij}(\theta)$ can be said to correspond to crack modes I and II that are inherently coupled together along the interface with a variable singularity strength $kr^{i\varepsilon}$.

The well known Muskhelishvili representation in the problem shown in Figure 4.1 can be put in the form (Muskhelishvili, 1953)

$$2\mu_j(u_j + iv_j) = \kappa_j\phi_j(Z) + (\overline{Z} - Z) \cdot \overline{\phi'_j(Z)} - \overline{\Omega_j(Z)}, \tag{4.8}$$

$$\sigma_{jxy} + \sigma_{jyy} = 2[\phi'_j(Z) + \overline{\phi'_j(Z)}], \tag{4.9}$$

$$\sigma_{jyy} - \sigma_{jxx} + 2i\sigma_{jxy} = 2[(\overline{Z} - Z) \cdot \phi_j(Z) - \phi'_j(Z) + \Omega'_j(Z)], \tag{4.10}$$

where $j = 1, 2$, $Z = x + iy$, $\phi_j(Z)$ and $\Omega_j(Z)$ are analytic in R_1 and R_2, respectively, $\phi'_j(Z) = d\phi_j/dZ$, and the over bar denotes complex conjugate.

The general solutions for $\phi'_1(Z), \Omega'_1(Z), \phi'_2(Z)$and $\Omega'_2(Z)$ are found by Rice (1988)

$$\phi'_1(Z) = e^{-\pi\varepsilon} \cdot Z^{-1/2-i\varepsilon} \cdot f(Z) + 2C_2 g(Z)/(C_1 + C_2), \tag{4.11}$$

$$\Omega'_1(Z) = e^{\pi\varepsilon} \cdot Z^{-1/2+i\varepsilon} \cdot \overline{f}(Z) - 2C_2\overline{g}(Z)/(C_1 + C_2), \tag{4.12}$$

$$\phi'_2(Z) = e^{\pi\varepsilon} \cdot Z^{-1/2-i\varepsilon} \cdot f(Z) + 2C_1 g(Z)/(C_1 + C_2), \tag{4.13}$$

$$\Omega'_2(Z) = e^{-\pi\varepsilon} \cdot Z^{-1/2+i\varepsilon} \cdot \overline{f}(Z) - 2C_1\overline{g}(Z)/(C_1 + C_2), \tag{4.14}$$

where

$$C_1 = (\kappa_1 + 1)/\mu_1, \qquad C_2 = (\kappa_2 + 1)/\mu_2, \tag{4.15}$$

and for a pure number ζ, $\zeta^{i\varepsilon}$ may be evaluated as $e^{i\varepsilon \ln \zeta}$, and

$$f(Z) = \sum_{n=0}^{\infty} a_n Z^n, \tag{4.16}$$

$$f(Z) = \sum_{n=0}^{\infty} b_n Z^n. \tag{4.17}$$

Then a_0 represents the strength of the crack tip singularity and may be written as

$$a_0 = \frac{\overline{k}}{(2\sqrt{2\pi}\cosh(\pi\varepsilon))}, \tag{4.18}$$

where k is the complex stress intensity factor which uniquely characterizes the oscillatory singular field (Rice, 1988).

The displacement jump along the crack faces is

$$(u_{1y} + iu_{1x})_{\theta=\pi} - (u_{2y} + u_{2x})_{\theta=\pi} = \frac{(C_1 + C_2)kr^{i\varepsilon}}{2\sqrt{2\pi}(1 + 2i\varepsilon)\cosh(\pi\varepsilon)}, \tag{4.19}$$

whereas the energy release per unit of new crack area is (Rice, 1988)

$$G = \frac{(C_1 + C_2)k\overline{k}}{16\cosh^2(\pi\varepsilon)}. \tag{4.20}$$

(A) Consider a semi-infinite interface crack under concentrated tractions acting on both face (See Figure 4.2(a) and 4.2(b)), the stresses at any point Z beneath the interface can be given as follows (Zhao and Chen, 1997):

$$\sigma_{\theta\theta} + i\sigma_{r\theta} = \phi_2'(Z) + \overline{\phi_2'(Z)} + e^{2i\theta}[\Omega_2'(Z) - \phi_2'(Z) - (Z - \overline{Z})\phi_2(Z)], \tag{4.21}$$

and

$$\phi_2'(Z) = h_2 F(Z), \qquad h_2 = \mu_1\mu_2, /(\mu_2 + \mu_1\kappa_2), \tag{4.22}$$

$$\Omega_2'(z) = h_1\overline{F}(z), \quad h_1 = \mu_1\mu_2/(\mu_1 + \mu_2\kappa_1), \tag{4.23}$$

$$F(z) = F_P(z) = \frac{P}{2\pi h_1 e^{\pi\varepsilon}} \cdot \frac{1}{z + b}\left(\frac{b}{z}\right)^{1/2+i\varepsilon} \tag{4.24}$$

for the normal concentrated traction P shown in Figure 4.2(a);

$$F(z) = F_q(z) = \frac{iQ}{2\pi h_1 e^{\pi\varepsilon}} \cdot \frac{1}{z + b}\left(\frac{b}{z}\right)^{1/2+i\varepsilon} \tag{4.25}$$

for the shear concentrated traction Q shown in Figure 4.2(b); b is the distance between the traction acting point and the original point, $\alpha°$is the oriented angle.

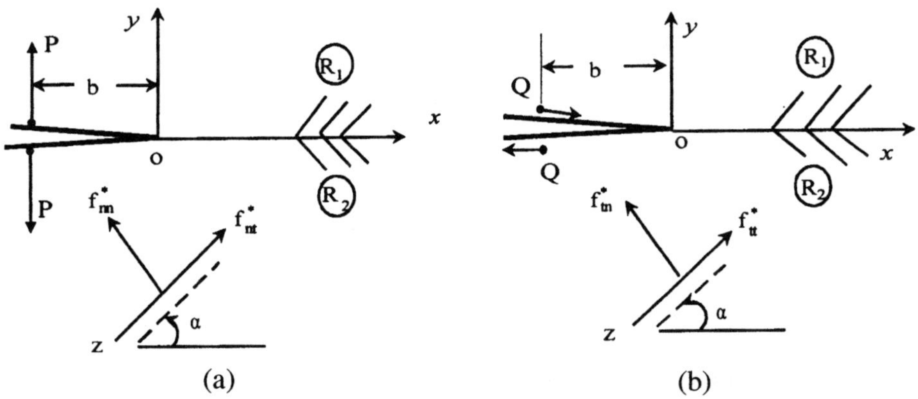

(a) (b)

Figure 4.2. An interface crack under normal concentrated tractions P or under shear concentrated tractions Q.

Thus the so called influence coefficients induced from the normal concentrated traction P at any point z beneath the interface are derived as follows

$$f^*_{nn}+if^*_{nt} = h_2F_P(z)+h_2\overline{F_P(z)}+e^{2i\alpha^*}[h_1\overline{F_P}(z)-h_2F_P(z)-(z-\overline{z})F'_P(z)], \tag{4.26}$$

which will be used in the following sections as the kernel functions in the deduced integral equations.

Similarly, the so called influence coefficients induced from the shear concentrated traction Q at any point z beneath the interface are derived as follows

$$f^*_{tn}+if^*_{tt} = h_2F_q(z)+h_2\overline{F_q(z)}+e^{2i\alpha^*}[h_1\overline{F_q}(z)-h_2F_q(z)-(z-\overline{z})F'_q(z)]. \tag{4.27}$$

The complex stress intensity factor for the interface crack induced from P and Q is

$$k = K_1 - iK_2 = \frac{2(P-iQ)}{\sqrt{2\pi b}}\cosh(\pi\varepsilon)b^{i\varepsilon}. \tag{4.28}$$

(B) Consider a finite crack of length $2a$ under concentrated tractions P and Q on the crack faces in an infinite homogeneous solid. The corresponding complex potentials denoted by $\Phi_{1P}(z)$ and $\Psi_{1P}(z)$ as well as $\Phi_{1q}(z)$ and $\Psi_{1q}(z)$ are (Chen, 1986):

$$\Phi_{1P}(z) = \frac{P}{2\pi}\cdot\frac{1}{z-b}\cdot\sqrt{\frac{a^2-b^2}{z^2-a^2}}, \tag{4.29}$$

$$\Psi_{1P}(z) = \overline{\Phi}_{1P}(z) - \Phi_{1P}(z) - z\cdot\Phi'_{1P}(z), \tag{4.30}$$

and

$$\Phi_{1q}(z) = -\frac{iQ}{2\pi}\cdot\frac{1}{z-b}\cdot\sqrt{\frac{a^2-b^2}{z^2-a^2}}, \tag{4.31}$$

$$\Psi_{1q}(z) = \overline{\Phi}_{1q}(z) - \Phi_{1q}(z) - z\cdot\Phi'_{1q}(z). \tag{4.32}$$

Putting the crack beneath the interface in a bimaterial solid as shown in Figures 4.3(a) and 4.3(b), Chen (1986) derived the corresponding complex potentials:

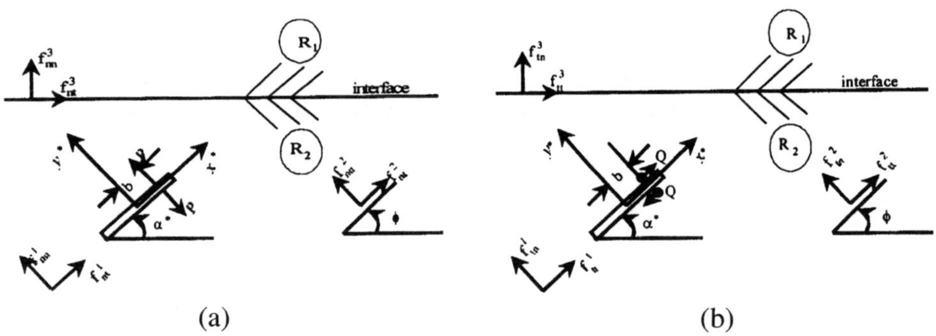

Figure 4.3. A sub-interface crack under normal concentrated tractions P (a) or under shear concentrated tractions Q (b).

$$\begin{aligned}
\Phi_1(z) &= (1+\Lambda)\Phi_{1P}(z),\\
\Psi_1(z) &= (\Pi-\Lambda)[p\cdot\Phi_{1P}(z) + (pz+q)\Phi'_{1P}(z)] + (1+\Pi)\Psi_{1P}(z),\\
&\qquad \text{in } R_1,
\end{aligned} \tag{4.33}$$

$$\begin{aligned}
\Phi_2(z) &= \Phi_{1P}(z) + \Phi_{2C}(z),\\
\Psi_2(z) &= \Psi_{1P}(z) + \Psi_{2C}(z), \qquad \text{in } R_2,
\end{aligned} \tag{4.34}$$

for the normal concentrated tractions P shown in Figure 4.3(a); and

$$\begin{aligned}
\Phi_1(z) &= (1+\Lambda)\Phi_{1q}(z), \\
\Psi_1(z) &= (\Pi-\Lambda)[p\cdot\Phi_{1q}(z)+(pz+q)\Phi'_{1q}(z)]+(1+\Pi)\Psi_{1q}(z), \\
&\qquad \text{in } R_1, \qquad (4.35) \\
\Phi_2(z) &= \Phi_{1q}(z)+\Phi_{2d}(z), \\
\Psi_2(z) &= \Psi_{1q}(z)+\Psi_{2d}(z), \qquad \text{in } R_2,
\end{aligned}$$

(4.36)

where

$$z = x^* + iy^*, \quad p = e^{2i\alpha^*}, \quad q = -2hie^{i\alpha^*}, \qquad (4.37)$$

$$\begin{aligned}
\Lambda &= \frac{\mu_1\kappa_2-\mu_2\kappa_1}{\mu_1+\mu_2\kappa}, \\
\Pi &= \frac{\mu_1-\mu_2}{\mu_2+\mu_1\kappa_2},
\end{aligned}$$

(4.38)

$$\Phi_{2C}(z) = \Pi\{\overline{\Phi}_{1P}(pz+q)+pz\cdot\Phi'_{1p}(pz+q)+p\overline{\Psi}_{1p}(pz+q)\}, \qquad (4.39)$$

$$\begin{aligned}
\Psi_{2C}(z) &= \Lambda p\overline{\Phi}_{1P}(pz+q)-\Pi p\{\overline{\Phi}_{1P}(pz+q)+(3pz+2q)\overline{\Phi'}_{1p}(pz+q) \\
&+pz(pz+q)\overline{\Phi}_{1p}(pz+q)+p\overline{\Psi}_{1p}(pz+q)+p(pz+q)\overline{\Psi'}_{1p}(pz+q)\},
\end{aligned} \qquad (4.40)$$

$$\Phi_{2d}(z) = \Pi\{\overline{\Phi}_{1q}(pz+q)+pz\cdot\Phi'_{1q}(pz+q)+p\overline{\Psi}_{1q}(pz+q)\}, \qquad (4.41)$$

$$\begin{aligned}
\Psi_{2d}(z) &= \Lambda p\overline{\Phi}_{1q}(pz+q)-\Pi p\{\overline{\Phi}_{1q}(pz+q)+(3pz+2q)\overline{\Phi'}_{1q}(pz+q) \\
&+pz(pz+q)\overline{\Phi}_{1q}(pz+q)+p\overline{\Psi}_{1q}(pz+q)+p(pz+q)\overline{\Psi'}_{1q}(pz+q)\}.
\end{aligned} \qquad (4.42)$$

The influence coefficients along the crack line, the dashed line, and the interface Owing to the normal concentrated traction P in Figure 4.3(a) are denoted by the superscripts 1, 2 and 3, respectively, which are given as follows (Chen, 1986):

$$\begin{aligned} f_{nn}^1 + i f_{nt}^1 &= -P\delta(t-b) + \Phi_{2C}(t) + \overline{\Phi_{2C}(t)} + t \cdot \Phi_{2C}'(t) + \Psi_{2C}(t), \\ f_{nn}^2 + i f_{nt}^2 &= \Phi_2(z) + \overline{\Phi_2(z)} + e^{2i(\phi-\alpha^*)}[\overline{z} \cdot \Phi_2'(z) + \Psi_2(z)], \\ f_{nn}^3 + i f_{nt}^3 &= \Phi_1(z) + \overline{\Phi_1(z)} + e^{-2i\alpha^*}[\overline{z} \cdot \Phi_1'(z) + \Psi_1(z)], \end{aligned} \tag{4.43}$$

where equations (4.33) and (4.34) are used.

Similarly, the influence coefficients along the crack line, the dashed line and the interface Owing to the shear concentrated traction Q in Figure 4.3(b) are given as follows

$$\begin{aligned} f_{tn}^1 + i f_{tt}^1 &= -iQ\delta(t-b) + \Phi_{2d}(t) + \overline{\Phi_{2d}(t)} + t \cdot \Phi_{2d}'(t) + \Psi_{2d}(t), \\ f_{tn}^2 + i f_{tt}^2 &= \Phi_2(z) + \overline{\Phi_2(z)} + e^{2i(\varphi-\alpha^*)}[\overline{z} \cdot \Phi_2'(z) + \Psi_2(z)], \\ f_{tn}^3 + i f_{tt}^3 &= \Phi_1(z) + \overline{\Phi_1(z)} + e^{-2i\alpha^*}[\overline{z} \cdot \Phi_1'(z) + \Psi_1(z)], \end{aligned} \tag{4.44}$$

where equations (4.35) and (4.36) are used.

The complex stress intensity factor is then given as follows

$$K^R = K_1^R - iK_2^R = \frac{P - iQ}{\sqrt{\pi a}}\sqrt{\frac{a+b}{a-b}}, \tag{4.45}$$

$$K^L = K_1^L - iK_2^L = \frac{P - iQ}{\sqrt{\pi a}}\sqrt{\frac{a-b}{a+b}}, \tag{4.46}$$

where the superscripts R and L denote the right and the left tips, respectively.

It should be emphasized that the concentrated tractions P and Q in equations (4.45) and (4.46) are not the real tractions acting on the sub-interface crack faces (see Figures 4.3(a) and 4.3(b)). These tractions are applied on the crack faces in an infinite homogeneous body first, and then the crack is put beneath the interface. Thus, the tractions acting on the sub-interface crack faces are no longer P and Q. They should be calculated by using superimposing technique, *i.e.*, by using equation (4.43) although the complex stress intensity factor could still be formulated by equations (4.45) and (4.46). In fact, the tractions P and Q should be determined by using the real traction conditions on the sub-interface crack, so that they will be influenced implicitly by the distance h, the crack oriented angle α^*, the bimaterial constants such as Π and Λ. The unknown tractions P and Q are sometimes called the pseudo-tractions (Horii and Nemat Nasser, 1987), but the basic idea was introduced in (Chen, 1984a,b;1986).

2. Pseudo-traction methods and Fredholm integral equations

Consider N arbitrarily located sub-interface microcracks near the tip of an interface macrocrack and along an arbitrarily oriented angle α_k (see Figure 4.4(a)). The prescribed complex stress intensity factor $K^\infty = K_1^\infty + iK_2^\infty$ is taken to specify the applied remote stress field. Using the 'pseudo-traction' method proposed by Horri and Nemat Nasser (1987) and discussed in Chapter 3, the problem shown in Figure 4.4(a) is then decomposed into $N+2$ subproblems (see Figures 4.4(b,c,d)), each of which contains one single crack. The unknown normal tractions $P_0(s)$, $P_k(t_k)$ $(k = 1, 2, \ldots, N)$ and the shear tractions $Q_0(s)$, $Q_k(t_k)$ $(k = 1, 2, \ldots, N)$ on the interface crack faces and the sub-interface crack faces, respectively, are the so called pseudo-tractions which need to be determined. Using the fundamental solutions presented in the above section and the well known superimposing technique, the following Fredholm integral equations with infinite integral intervals from $-\infty$ to 0 are deduced as Chen did in isotropic homogeneous cases (1984a,b) and Chen and Hasebe did in orthotropic homogeneous cases (1994c, 1995a,b,c):

$$P_0(s) + \sum_{k=1}^{N} [\int_{-a_k}^{a_k} P_k(t_k) f_{nn,k0}^3(t_k, s) dt_k + \int_{-a_k}^{a_k} Q_k(t_k) f_{tn,k0}^3(t_k, s) dt_k] = 0, (-\infty < s < 0),$$

$$Q_0(s) + \sum_{k=1}^{N} [\int_{-a_k}^{a_k} P_k(t_k) f_{nt,k0}^3(t_k, s) dt_k + \int_{-a_k}^{a_k} Q_k(t_k) f_{tt,k0}^3(t_k, s) dt_k] = 0, (-\infty < s < 0),$$

$$\int_{-a_k}^{a_k} P_k(s_k) f_{nn,kk}^1(s_k, t_k) ds_k + \int_{-a_k}^{a_k} Q_k(s_k) f_{tn,kk}^1(s_k, t_k) ds_k$$

$$+\sum_{\substack{i=1\\ i\neq k}}^{N}\left[\int_{-a_i}^{a_i} P_i(t_i)f^2_{nn,ik}(t_i,t_k)dt_i+\int_{-a_i}^{a_i} Q_i(t_i)f^2_{tn,ik}(t_i,t_k)dt_i\right]$$

$$+\int_{-\infty}^{0}[P_0(s_k)f^*_{nn,0k}(s,t_k)+Q_0(s_k)f^*_{tn,0k}(s,t_k)]ds_k=p_k(t_k),$$

$$(-a_k<t_k<a_k),$$

$$\int_{-a_k}^{a_k} P_k(s_k)f^1_{nt,kk}(s_k,t_k)ds_k+\int_{-a_k}^{a_k} Q_k(s_k)f^1_{tt,kk}(s_k,t_k)ds_k$$

$$+\sum_{\substack{i=1\\ i\neq k}}^{N}\left[\int_{-a_i}^{a_i} P_i(t_i)f^2_{nt,ik}(t_i,t_k)dt_i+\int_{-a_i}^{a_i} Q_i(t_i)f^2_{tt,ik}(t_i,t_k)dt_i\right]$$

$$+\int_{-\infty}^{0}[P_0(s_k)f^*_{nt,0k}(s_k,t_k)+Q_0(s_k)f^*_{tt,0k}(s_k,t_k)]ds_k=q_k(t_k),$$

$$(-a_k<t_k<a_k), \tag{4.47}$$

where the 16 kernel functions $(f^*_{nn,0k})$, $(f^*_{tn,0k})$, $(f^*_{nt,0k})$, $(f^*_{tt,0k})$, $(f^1_{nn,kk})$, $(f^1_{nt,kk})$, $(f^1_{tn,kk})$, $(f^1_{tt,kk})$, $(f^2_{nn,ik})$, $(f^2_{nt,ik})$, $(f^2_{tn,ik})$, $(f^2_{tt,ik})$, $(f^3_{nn,ik})$, $(f^3_{nt,ik})$, $(f^3_{tn,ik})$ and $(f^3_{tt,ik})$ are given by equations (4.26), (4.27), (4.43), respectively; the subscripts, *n* , *t, o, k, i* indicate the normal quantities, the tangential quantities, the interface crack, the kth or ith microcrack, respectively; $p_k(t_k)$ and $q_k(t_k)$ are the known residual stresses to be released at the location of the kth microcrack, which could be expressed by the near-tip stress field at the macrocrack tip, *i.e.,* equation (4.7) as

$$\begin{aligned} p_k(t_k) &= -[\sigma^t_{xx}\cdot\sin^2(\alpha_k)+\sigma^t_{yy}\cdot\cos^2(\alpha_k)-\sigma^t_{xy}\cdot\sin(2\alpha_k)],\\ q_k(t_k) &= -[(\sigma^t_{yy}-\sigma^t_{xx})\cdot\cos\alpha_k\cdot\sin\alpha_k+\sigma^t_{xy}\cdot\cos(2\alpha_k)]. \end{aligned} \tag{4.48}$$

Of course, the first term in the general solutions proposed by Rice (1988) as given in equations (4.10),(4.11),(4.12),(4.13), should be used.

However, equation (4.47) can not be solved directly by using a traditional numerical integral technique since the infinite integral intervals ($-\infty$,0) are concerned. In order to reduced them into finite intervals, the following simple changes of variables are taken (Fleck *et al.,* 1991):

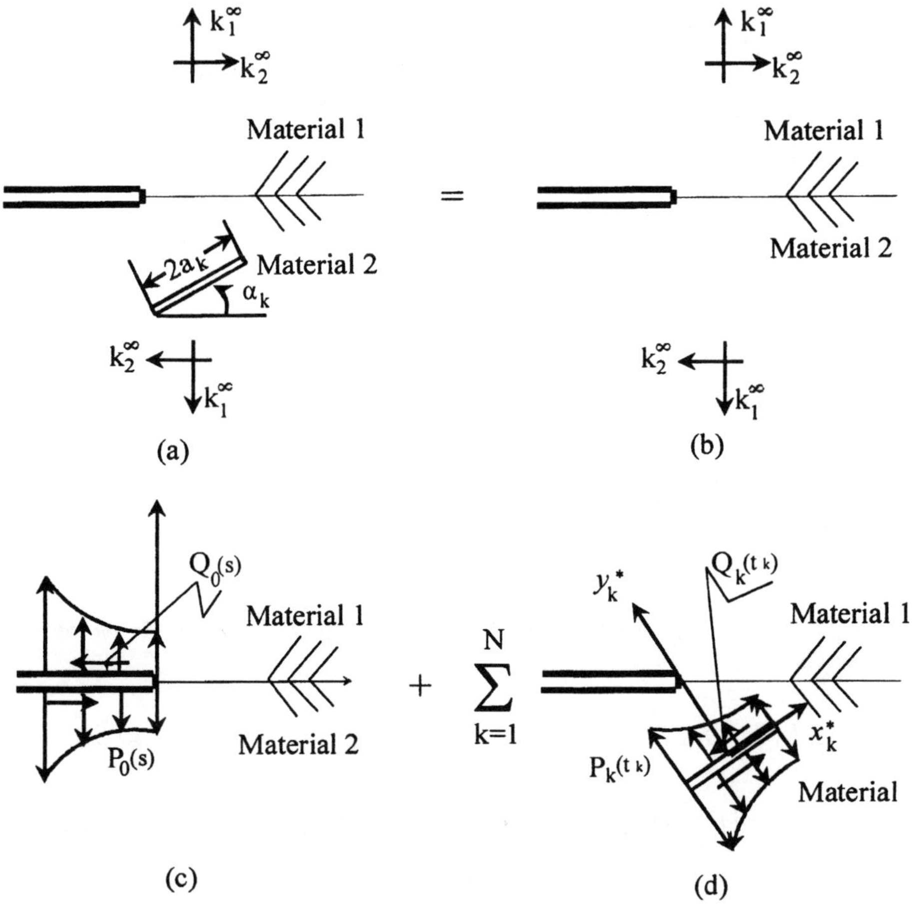

Figure 4.4. Superimposing method: (a) the original problem; (b) the single interface macro-crack problem under the remote stress field specified by the applied stress intensity factor $K^\infty = K_1^\infty + iK_2^\infty$; (c) pseudo-traction applied on the single interface crack faces; (d) superimposing of N sub-interface crack problems with applied pseudo-tractions.

$$
\begin{aligned}
s &= \frac{(u-1)}{(u+1)}, \\
ds &= \frac{2du}{(u+1)^2}.
\end{aligned}
\tag{4.49}
$$

After doing so, a new system of Fredholm integral equations is reduced:

$$P_0(u) + \sum_{k=1}^{N} [\int_{-a_k}^{a_k} P_k(t_k) f^3_{nn,k0}(t_k, u) dt_k + \int_{-a_k}^{a_k} Q_k(t_k) f^3_{tn,k0}(t_k, u) dt_k]$$
$$= 0, (-1 < u < 1),$$

$$Q_0(u) + \sum_{k=1}^{N} [\int_{-a_k}^{a_k} P_k(t_k) f^3_{nt,k0}(t_k, u) dt_k + \int_{-a_k}^{a_k} Q_k(t_k) f^3_{nt,k0}(t_k, u) dt_k]$$
$$= 0, (-1 < u < 1),$$

$$\int_{-a_k}^{a_k} P_k(t_k) f^1_{nn,kk}(s_k, t_k) ds_k + \int_{-a_k}^{a_k} Q_k(s_k) f^1_{tn,kk}(s_k, t_k) ds_k$$
$$+ \sum_{\substack{i=1 \\ i \neq k}}^{N} [\int_{-a_i}^{a_i} P_i(t_k) f^2_{nn,ik}(t_i, t_k) dt_i + \int_{-a_i}^{a_i} Q_i(t_i) f^2_{tn,ik}(t_i, t_k) dt_i]$$
$$+ \int_{-1}^{1} [P_0(u) f^*_{nn,0k}(u, t_k) + Q_0(u) f^*_{tn,0k}(u, t_k)] \frac{2}{(u+1)^2} du = p_k(t_k),$$
$$(-a_k < s_k < a_k),$$

$$\int_{-a_k}^{a_k} P_k(s_k) f^1_{nt,kk}(s_k, t_k) ds_k + \int_{-a_k}^{a_k} Q_k(s_k) f^1_{tt,kk}(s_k, t_k) ds_k$$
$$+ \sum_{\substack{i=1 \\ i \neq k}}^{N} [\int_{-a_i}^{a_i} P_i(t_k) f^2_{nt,ik}(t_i, t_k) dt_i + \int_{-a_i}^{a_i} Q_i(t_i) f^2_{tt,ik}(t_i, t_k) dt_i]$$
$$+ \int_{-1}^{1} [P_0(u) f^*_{nt,0k}(u, t_k) + Q_0(u) f^*_{tt,0k}(u, t_k)] \frac{2}{(u+1)^2} du = q_k(t_k),$$
$$(-a_k < s_k < a_k),$$

(4.50)

where, for the sake of convenience, the following notations are taken (Chen, 1996; Zhao and Chen, 1997, 1998):

$$f^3_{nn,k0}(t_k, s) = f^3_{nn,k0}\left(t_k, \frac{u-1}{u+1}\right) \equiv f^3_{nn,k0}(t_k, u)$$
$$f^3_{nt,k0}(t_k, s) \equiv f^3_{nt,k0}(t_k, u)$$
$$f^3_{tn,k0}(t_k, s) \equiv f^3_{tn,k0}(t_k, u)$$
$$f^3_{tt,k0}(t_k, s) \equiv f^3_{tt,k0}(t_k, u)$$

(4.51)

and

$$
\begin{aligned}
& f^{*}_{nn,0k}\left(s,t_k\right)=f^{*}_{nn,0k}\left(\tfrac{u-1}{u+1},t_k\right)\equiv f^{3}_{nn,0k}\left(u,t_k\right)\\
& f^{*}_{nt,0k}\left(s,t_k\right)\equiv f^{*}_{nt,0k}\left(u,t_k\right)\\
& f^{*}_{tn,0k}\left(s,t_k\right)\equiv f^{3}_{tn,0k}\left(u,t_k\right)\\
& f^{*}_{tt,0k}\left(s,t_k\right)\equiv f^{3}_{tt,0k}\left(u,t_k\right)
\end{aligned}
\tag{4.52}
$$

and

$$
\begin{aligned}
& P_0\left(s\right)=P_0\left(\tfrac{u-1}{u+1}\right)=P_0\left(u\right)\\
& Q_0\left(s\right)=Q_0\left(u\right)
\end{aligned}
\tag{4.53}
$$

The factor of $(u+1)^2$ in the last two equations of equation (4.50) presents a new problem in evaluation of the integrals near $u=-1$.

Calling the asymptotic behaviors of the pseudo-tractions $P_0\left(u\right)$ and $Q_0\left(u\right)$:

$$
\begin{aligned}
& P_0\left(u\right)\sim O\left(\left(u+1\right)^2\right)\ \text{as}\ u\rightarrow\ -1\\
& Q_0\left(u\right)\sim O\left(\left(u+1\right)^2\right)\ \text{as}\ u\rightarrow\ -1
\end{aligned}
\tag{4.54}
$$

and introducing two new functions $F_0\left(u\right)$ and $G_0\left(u\right)$ as follows

$$
\begin{aligned}
& P_0\left(u\right)\sim F_0\left(u\right)\cdot\left(u+1\right)^2/2\\
& Q_0\left(u\right)\sim G_0\left(u\right)\cdot\left(u+1\right)^2/2
\end{aligned}
\tag{4.55}
$$

equation (4.50) can then be reduced to the normal Fredholm integral equations without further problems:

$$F_0(u)\cdot\frac{(u+1)^2}{2}+\sum_{k=1}^{N}\int_{-a_k}^{a_k}P_k(t_k)\,f^3_{nn,k0}(t_k,u)\,dt_k$$
$$+\sum_{k=1}^{N}\int_{-a_k}^{a_k}Q_k(t_k)\,f^3_{tn,k0}(t_k,u)\,dt_k=0,(-1<u<1),$$

$$G_0(u)\cdot\frac{(u+1)^2}{2}+\sum_{k=1}^{N}\int_{-a_k}^{a_k}P_k(t_k)\,f^3_{nt,k0}(t_k,u)\,dt_k$$
$$+\sum_{k=1}^{N}\int_{-a_k}^{a_k}Q_k(t_k)\,f^3_{tt,k0}(t_k,u)\,dt_k=0,(-1<u<1),$$

$$\int_{-a_k}^{a_k}P_k(s_k)\,f^1_{nn,kk}(s_k,t_k)\,ds_k+\int_{-a_k}^{a_k}Q_k(s_k)\,f^1_{tn,kk}(s_k,t_k)\,ds_k$$

$$+\sum_{\substack{i=1\\i\neq k}}^{N}\int_{-a_i}^{a_i}P_i(t_i)\,f^2_{nn,ik}(t_i,t_k)\,dt_i+\sum_{\substack{i=1\\i\neq k}}^{N}\int_{-a_i}^{a_i}Q_i(t_i)\,f^2_{tn,ik}(t_i,t_k)\,dt_i$$

$$+\int_{-1}^{1}F_0(u)\,f^*_{nn,0k}(u,t_k)\,du+\int_{-1}^{1}G_0(u)\,f^*_{tn,0k}(u,t_k)\,du=p_k(t_k),$$

$$(-a_k<s_k<a_k),$$

$$\int_{-a_k}^{a_k}P_k(s_k)\,f^1_{nt,kk}(s_k,t_k)\,ds_k+\int_{-a_k}^{a_k}Q_k(s_k)\,f^1_{tt,kk}(s_k,t_k)\,ds_k$$

$$+\sum_{\substack{i=1\\i\neq k}}^{N}\int_{-a_i}^{a_i}P_i(t_i)\,f^2_{nt,ik}(t_i,t_k)\,dt_i+\sum_{\substack{i=1\\i\neq k}}^{N}\int_{-a_i}^{a_i}Q_i(t_i)\,f^2_{tt,ik}(t_i,t_k)\,dt_i$$

$$+\int_{-1}^{1}F_0(u)\,f^*_{nt,0k}(u,t_k)\,du++\int_{-1}^{1}G_0(u)\,f^*_{tt,0k}(u,t_k)\,du=q_k(t_k),$$

$$(-a_k<s_k<a_k),$$

$$(4.56)$$

Using the Chebyshev numerical integration (Chen, 1984a,b; Chen and Hasebe, 1994c; 1995a,b,c) and taking M as the partitioning number, a linear system with $2\times(N+1)\times M$ unknowns, *i.e.*, $F_0(u_m)$, $G_0(u_m)$,$G_0(u_m)$ and $Q_k\left(t_k^j\right)$ with $(m,j=1,2...,M;K=1,2,...,N)$, is then given as

$$\frac{(u_m+1)^2}{2} F_0(u_m) + \sum_{k=1}^{N}\sum_{j=1}^{M} P_k\left(t_k^j\right) f_{nn,k0}^3\left(t_k^j, u_m\right)\delta_j$$

$$+ \sum_{k=1}^{N}\sum_{j=1}^{M} Q_k\left(t_k^j\right) f_{tn,k0}^3\left(t_k^j, u_m\right)\delta_j = 0,$$

$$\frac{(u_m+1)^2}{2} G_0(u_m) + \sum_{k=1}^{N}\sum_{j=1}^{M} P_k\left(t_k^j\right) f_{nt,k0}^3\left(t_k^j, u_m\right)\delta_j$$

$$+ \sum_{k=1}^{N}\sum_{j=1}^{M} Q_k\left(t_k^j\right) f_{tt,k0}^3\left(t_k^j, u_m\right)\delta_j = 0,$$

$$\sum_{r=1}^{M} P_k\left(s_k^r\right) f_{nn,kk}^1\left(s_k^r, t_k^j\right)\delta_r + \sum_{r=1}^{M} Q_k\left(s_k^r\right) f_{tn,kk}^1\left(s_k^r, t_k^j\right)\delta_r$$

$$+ \sum_{\substack{i=1\\ i\neq k}}^{N}\sum_{p=1}^{M} P_i\left(t_i^p\right) f_{nn,ik}^1\left(t_i^p, t_k^j\right)\delta_p + \sum_{\substack{i=1\\ i\neq k}}^{N}\sum_{p=1}^{M} Q_i\left(t_i^p\right) f_{tn,ik}^1\left(t_i^p, t_k^j\right)\delta_p$$

$$+ \sum_{m=1}^{M} F_0(u_m) f_{nn,0k}^*\left(u_m, t_k^j\right)\delta_m + \sum_{m=1}^{M} G_0(u_m) f_{tn,0k}^*\left(u_m, t_k^j\right)\delta_m$$

$$= p_k\left(t_k^j\right),$$

$$\sum_{r=1}^{M} P_k\left(s_k^r\right) f_{nt,kk}^1\left(s_k^r, t_k^j\right)\delta_r + \sum_{r=1}^{M} Q_k\left(s_k^r\right) f_{tt,kk}^1\left(s_k^r, t_k^j\right)\delta_r$$

$$+ \sum_{\substack{i=1\\ i\neq k}}^{N}\sum_{p=1}^{M} P_i\left(t_i^p\right) f_{nt,ik}^1\left(t_i^p, t_k^j\right)\delta_p + \sum_{\substack{i=1\\ i\neq k}}^{N}\sum_{p=1}^{M} Q_i\left(t_i^p\right) f_{tt,ik}^1\left(t_i^p, t_k^j\right)\delta_p$$

$$+ \sum_{m=1}^{M} F_0(u_m) f_{nt,0k}^*\left(u_m, t_k^j\right)\delta_m + \sum_{m=1}^{M} G_0(u_m) f_{tt,0k}^*\left(u_m, t_k^j\right)\delta_m$$

$$= q_k\left(t_k^j\right),$$

$$(m, j = 1, 2, ..., M) \tag{4.57}$$

where

$$u_m = \cos\frac{(2m-1)\,\pi}{2M},$$

$$t_k^j = a_k \cos\frac{(2j-1)\,\pi}{2M},$$

$$s_k^j = a_k \cos\frac{(2r-1)\,\pi}{2M},$$

$$t_i^p = a_i \cos\frac{(2p-1)\,\pi}{2M}; \tag{4.58}$$

$$\delta_m = \frac{\pi}{M}\sin\frac{(2m-1)\,\pi}{2M},$$

$$\delta_j = \frac{a_k\pi}{M}\sin\frac{(2j-1)\,\pi}{2M},$$

$$\delta_r = \frac{a_k\pi}{M}\sin\frac{(2r-1)\,\pi}{2M},$$

$$\delta_p = \frac{a_i\pi}{M}\sin\frac{(2p-1)\,\pi}{2M}. \tag{4.59}$$

Once the system (4.57) is solved, the incremental values of the stress intensity factors at the interface macrocrack tips and those at the both tips of the N microcracks can be evaluated by using the well known quadrature rule (Erdogan, 1978; Krenk, 1975) in equations (4.45) and (4.46)

$$K_1^t = K_1^\infty + \Delta K_1,$$

$$\Delta K_1 = \tfrac{\sqrt{2\pi}}{M}\cosh(\pi\varepsilon)\cdot\sum_{m=1}^{M} F_0\left(\cos\tfrac{(2m-1)\pi}{2M}\right)\left(1+\cos\tfrac{(2m-1)\pi}{2M}\right)\cos(\varepsilon\delta)$$
$$+\tfrac{\sqrt{2\pi}}{M}\cosh(\pi\varepsilon)\cdot\sum_{m=1}^{M} G_0\left(\cos\tfrac{(2m-1)\pi}{2M}\right)\left(1+\cos\tfrac{(2m-1)\pi}{2M}\right)\sin(\varepsilon\delta), \tag{4.60}$$

$$K_2^t = K_2^\infty + \Delta K_2,$$

$$\Delta K_2 = -\tfrac{\sqrt{2\pi}}{M}\cosh(\pi\varepsilon)\cdot\sum_{m=1}^{M} F_0\left(\cos\tfrac{(2m-1)\pi}{2M}\right)\left(1+\cos\tfrac{(2m-1)\pi}{2M}\right)\sin(\varepsilon\delta)$$

$$+\tfrac{\sqrt{2\pi}}{M}\cosh(\pi\varepsilon)\cdot\sum_{m=1}^{M} G_0\left(\cos\tfrac{(2m-1)\pi}{2M}\right)\left(1+\cos\tfrac{(2m-1)\pi}{2M}\right)\cos(\varepsilon\delta), \tag{4.61}$$

for the interface crack with

$$\delta = \ln\left[\left(1-\cos\frac{(2m-1)\pi}{2M}\right)\Big/\left(1+\cos\frac{(2m-1)\pi}{2M}\right)\right]$$

and

$$\begin{aligned} K_1^R &= \tfrac{\sqrt{\pi a_k}}{M}\cdot\sum_{m=1}^{M} P_k\left(a_k\cdot\cos\tfrac{(2m-1)\pi}{2M}\right)\left(1+\cos\tfrac{(2m-1)\pi}{2M}\right)\\ K_2^R &= \tfrac{\sqrt{\pi a_k}}{M}\cdot\sum_{m=1}^{M} Q_k\left(a_k\cdot\cos\tfrac{(2m-1)\pi}{2M}\right)\left(1+\cos\tfrac{(2m-1)\pi}{2M}\right)\end{aligned} \tag{4.62}$$

$$\begin{aligned} K_1^L &= \tfrac{\sqrt{\pi a_k}}{M}\cdot\sum_{m=1}^{M} P_k\left(a_k\cdot\cos\tfrac{(2m-1)\pi}{2M}\right)\left(1-\cos\tfrac{(2m-1)\pi}{2M}\right)\\ K_2^L &= \tfrac{\sqrt{\pi a_k}}{M}\cdot\sum_{m=1}^{M} Q_k\left(a_k\cdot\cos\tfrac{(2m-1)\pi}{2M}\right)\left(1-\cos\tfrac{(2m-1)\pi}{2M}\right)\end{aligned} \tag{4.63}$$

for the kth sub-interface microcrack.

Equations (4.57)–(4.63) are easy to be programmed, from which numerical results can be derived for any given material pair and relatively located macrocrack microcrack situation.

3. The J integral analysis: the projected conservation law of the J_k vector

As shown in Figure 4.5, an interface macrocrack interacting with a sub-interface microcrack is considered and three closed contours C_∞, C_t and C^* are specially introduced which enclose both macrocrack tip and microcrack,

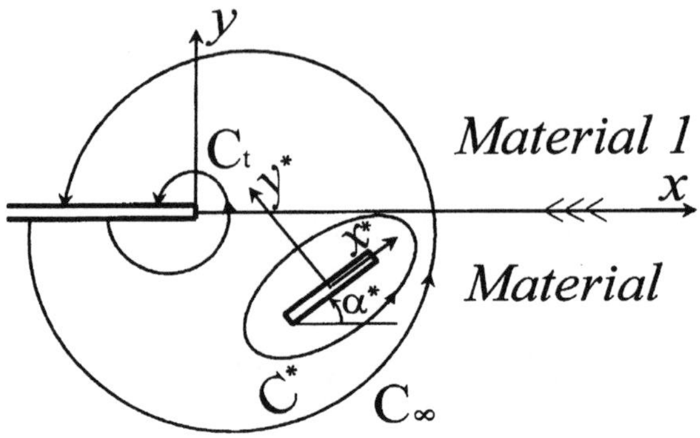

Figure 4.5. An interface macrocrack and a sub-interface microcrack with three contours.

only the macrocrack tip, and only the microcrack, respectively. From the path independent nature of the J integral, as discussed in Chapter 3, it is easy to prove the following relation in the global coordinate system (x, y):

$$J_\infty = J_t + \Delta J, \tag{4.64}$$

where

$$J_\infty = \int_{C_\infty} (W dy - T_i \cdot u_{i,1} ds), \tag{4.65}$$

$$J_t = \int_{C_t} (W dy - T_i \cdot u_{i,1} ds), \tag{4.66}$$

$$\Delta J = \int_{C^*} (W dy - T_i \cdot u_{i,1} ds), \tag{4.67}$$

where W, $T_i = (T_x, T_y)$ and $u_i = (u_x, u_y)$ have the same meanings as defined in Chapter 3, *i.e.,* the strain energy density, the tractions acting on the outside of the contour and the displacement components, respectively, and $i = (1, 2)$ refers to the global coordinate system (x, y).

The values of J_∞ and J_t integrals can be evaluated by using the results of the corresponding stress intensity factors derived by equations (4.60) and (4.61) without any difficulty, whilst the local coordinate system (x^*, y^*) should be specially considered for calculating the value of the ΔJ integral. Therefore the components of the J_k^* vector can be introduced in the local coordinate system (x^*, y^*) shown in Figure 4.5, which are formulated as follows

$$J_1^* = \int_{C^*} \left(W^* dy^* - T_i^* \cdot u_{i,1}^* ds\right),$$

$$J_2^* = \int_{C^*} \left(-W^* dx^* - T_i^* \cdot u_{i,2}^* ds\right), \tag{4.68}$$

where $i = (1, 2)$ refers to (x^*, y^*).

By using the transformation of the two coordinate systems mentioned above the projected conservation relation between the J integral in the global system and the J_k^* vector in the local system can then be obtained as discussed in Chapter 3 or Chen (1996) and Zhao and Chen (1997):

$$\Delta J = J_1^* \cos\alpha^* - J_2^* \sin\alpha^*, \tag{4.69}$$

where α^* is the oriented angle of the microcrack shown in Figure 4.5.

Dividing by J_∞ on the both sides of equation (4.64), the consistent relation of the J integral is given:

$$\frac{J_t}{J_\infty} + \frac{\Delta J}{J_\infty} = 1. \tag{4.70}$$

This important formulation together with equations (4.68) and (4.69) has been called as the projected conservation law of the J_k integral vector (Chen, 1996; Chen and Ma, 1997; Zhao and Chen, 1997, 1998). Although the expression of the law in the present bimaterial case is the same as that discussed in Chapter 3 for brittle solids, the formulations of the J_∞ and J_t integral are quite different from those calculated by using equations (3.44) and (3.46), which are expressed in the following forms (Gao, 1991; Zhang and Li, 1992):

$$J_\infty = \left[\frac{(1-\nu_1)/\mu_1 + (1-\nu_2)/\mu_2}{4\cosh^2(\pi\varepsilon)}\right]\left[(K_1^\infty)^2 + (K_2^\infty)^2\right], \tag{4.71}$$

$$J_t = \left[\frac{(1-\nu_1)/\mu_1 + (1-\nu_2)/\mu_2}{4\cosh^2(\pi\varepsilon)}\right]\left[\left(K_1^t\right)^2 + \left(K_2^t\right)^2\right]. \tag{4.72}$$

As is known, both integrals (4.71) and (4.72) are dependent on the material combination or on the bimaterial constants ν_1, μ_1, ν_2 and μ_2 as well as the oscillatory index ε (Rice, 1988; Gao, 1991; Zhang and Li, 1992).

In the material 2 of Figure 4.5, the components of the J_k vector in the local coordinate system (x^*, y^*) can be evaluated along the closed contour C^* and the following formulations are then given:

$$J_1^* = \frac{1-\upsilon_2}{2\mu_2}(K^R\overline{K^R} - K^L\overline{K^L}), \tag{4.73}$$

$$J_2^* = -\frac{1-\upsilon_2}{\mu_2}(K_1^R K_2^R - K_1^L K_2^L) + \int_{-a}^{a}(W^+ - W^-)dx^*, \tag{4.74}$$

where the superscripts R and L denote the quantities at the right and left tips of the microcrack, respectively; the J_2^* integral consists of two parts Herrmann and Herrmann, 1981), one of which is induced from both tips of the microcrack and the other from the traction-free faces of the microcrack; W^+ and W^- denote the boundary values of the strain energy densities on the upper and lower faces, respectively; W has the following form as we have well discussed in Chapters 1, 2, and 3 (Herrmann and Herrmann, 1981):

$$W = \frac{1-\upsilon_2^2}{2E_2}(\sigma_{x^*x^*}^2 + \sigma_{y^*y^*}^2 - 2\upsilon_2\sigma_{x^*x^*}\sigma_{y^*y^*}) + \frac{(1+\upsilon_2)(1-\upsilon_2)}{E_2}\sigma_{x^*x^*}^2. \tag{4.75}$$

The above results can be easily extended to treat the multiple sub-interface microcrack problems by using the superimposing technique discussed in the Section 4.2 (Zhao and Chen, 1997). Assuming there exist N sub-interface microcracks in the near-tip process zone of the interface crack (See Figure 4.5), the projected theory of the J_k vector becomes:

$$J_\infty = J_t + \sum_{k=1}^{N} \Delta J_k,$$

or

$$\frac{J_t}{J_\infty} + \sum_{k=1}^{N} \frac{\Delta J_k}{J_\infty} = 1, \tag{4.76}$$

where, as we have well addressed,

$$\Delta J_k = J_1^*(k)\cos\alpha_k - J_2^*(k)\sin\alpha_k, \tag{4.77}$$

and α_k is the oriented angle of the kth microcrack with respect to the global system.

Equation (4.76) shows the consistent relation of the J integral between the contributions induced from the multiple sub-interface microcracks and those induced from the macro-interface crack tip. Both kinds of contributions are normalized by the remote stress level J_∞, although the coupled feature of the remote loading should always taken into account (Rice, 1988).

As discussed in Chapter 3, the contribution induced from the sub-interface microcracks is related to both J_1^* and J_2^* components of the well known J_k^* vector. This means that the J_1^* integral and J_2^* integral are of equal importance in the macrocrack microcrack interaction problems in bimaterial solids when evaluating the energy release rate induced from microcracking. It should be noted in equations (4.73) and (4.74) that although the formulations of the J_1^* and J_2^* components in the local coordinate systems are not related directly to the combination of the two phase materials or the material constants of the upper material, the stress intensity factors K_1^R, K_2^R, K_1^L and K_2^L as well as the boundary values of the strain energy densities W^+ and W^- are dependent implicitly on the combination and the material constants of the upper material.

The second term in the right hand side of equation (4.74) should be reconsidered carefully. It is seen that the J_2^* component is related to the upper boundary value W^+ and the lower boundary value W^- of the strain energy densities of the microcrack. Thus it is necessary to determine the two boundary values before calculating the J_2^* component. From the traction-free conditions on the faces of the microcrack as shown in Figure 4.5, the values W^+ and W^- are shown to be only related to the stress component $\sigma_{x^*x^*}$ which is parallel to the crack faces as Herrmann and Herrmann discussed (1981):

$$W^+ = \frac{1-v_2^2}{2E_2}(\sigma_{x^*x^*}^+(x^*))^2,$$

$$W^- = \frac{1-v_2^2}{2E_2}(\sigma_{x^*x^*}^-(x^*))^2. \qquad (4.78)$$

The stress $\sigma_{x^*x^*}^+(x^*)$ and $\sigma_{x^*x^*}^-(x^*)$ defined on the upper and lower faces should be evaluated which, as a result of the principle of superimposing (see Figure 4.4a,b,c,d), consists of three parts:

(i) The first part is the influence from the interface macrocrack near-tip stress field (see Figure 4.4b) and the pseudo-tractions acting on crack faces (see Figure 4.4c).

(ii) The second part is the influence from the considered sub-interface microcrack with the pseudo-tractions (see Figure 4.4d).

(iii) The third part is the influence from the other sub-interface microcrack with the pseudo-tractions (see Figure 4.4d).

Readers can understand that the first and third parts induced from the interface macrocrack and other sub-interface microcracks at the site of the considered sub-interface microcrack can be easily evaluated since the whole stress field has been solved. However, to calculate the second part (ii), we should use the following formulations derived from equations (4.29) and (4.42):

$$\sigma^{+}_{x^*x^*}(t) = 4Re[\Phi^{+}_{2}(t)] - \sigma^{+}_{y^*y^*}(t),$$

$$\sigma^{-}_{x^*x^*}(t) = 4Re[\Phi^{-}_{2}(t)] - \sigma^{-}_{y^*y^*}(t), \tag{4.79}$$

where

$$\Phi^{+}_{2}(t) = \Phi^{+}_{1p}(t) + \Phi^{+}_{2C}(t),$$

$$\Phi^{-}_{2}(t) = \Phi^{-}_{1p}(t) + \Phi^{-}_{2C}(t),$$

$$\Phi^{+}_{2C}(t) = \Phi^{-}_{2C}(t),$$

$$\Phi^{+}_{1p}(t) + \Phi^{-}_{1p}(t) = -\left(P(t) - iQ(t)\right), \tag{4.80}$$

$$\Phi^{+}_{1p}(t) = \frac{1}{2\pi i\sqrt{a^2 - t^2}} \int_{-a}^{a} \frac{P(s) - iQ(s)}{t - s} \sqrt{a^2 - s^2} ds - \frac{P(t) - iQ(t)}{2}, \tag{4.81}$$

and $P(t)$ or $P(s)$ and $Q(t)$ or $Q(s)$ are the pseudo-tractions of the considered sub-interface microcrack, which have been determined numerically by solving equations (4.47), whilst $\sigma^{+}_{x^*x^*}(t)$and $\sigma^{-}_{x^*x^*}(t)$ are the normal stress components induced by the pseudo-tractions, which can be evaluated by using the first formulations of equations (4.43) and (4.44), respectively.

It can be seen from equation (4.81) that the following singular integral must be calculated numerically for determining $\sigma^{+}_{x^*x^*}(t)$ and $\sigma^{-}_{x^*x^*}(t)$:

$$I = \frac{1}{2\pi i\sqrt{a^2 - t^2}} \int_{-a}^{a} \frac{P(s) - iQ(s)}{t - s} \sqrt{a^2 - s^2} ds, \tag{4.82}$$

with $u = s/a$ it follows that

$$I = \frac{1}{2\pi i\sqrt{1-[\frac{t}{a}]^2}} \int_{-1}^{1} \frac{P(u)-iQ(u)}{\frac{t}{a}-u}\sqrt{1-u^2}du. \tag{4.83}$$

Expanding $P(u) - iQ(u)$ into the second kind of the Chebyshev polynomial $U_j(u)$ as follows (Erdogan, 1978; Krenk, 1975)

$$P(u) - iQ(u) = \sum_{j=1}^{M-1} B_j U_j(u), \tag{4.84}$$

where

$$B_j = \frac{\pi}{2}\int_{-1}^{1} \sqrt{1-u^2}(P(u) - iQ(u))U_j(u)du, \tag{4.85}$$

and substituting equation (4.84) into equation (4.83) it follows that

$$\begin{aligned} I &= \frac{1}{2\pi i\sqrt{1-[\frac{t}{a}]^2}} \int_{-1}^{1} \frac{P(u)-iQ(u)}{\frac{t}{a}-u}\sqrt{1-u^2}du \\ &= \frac{1}{2i\sqrt{1-[\frac{t}{a}]^2}} \sum_{j=1}^{M-1} B_j T_{j+1}(\frac{t}{a}), \end{aligned} \tag{4.86}$$

where $T_{j+1}(\frac{t}{a})$ is the first kind of Chebyshev polynomial.

Thus the singular integral (4.82) can be calculated by using equation (4.86) and then $\sigma^{+}_{x^*x^*}(x^*)$ and $\sigma^{-}_{x^*x^*}(x^*)$ can be evaluated by using equation (4.79). Therefore, the values of W^+,W^- and J_2^* can be determined numerically without any difficulty.

4. Numerical examples and discussions

In this section the abovementioned pseudo-traction method and the Fredholm integral equations are programmed and numerical results derived are then discussed in detail. In all of the following computational examples, the number of the Gauss-Chebyshev collocation points is always taken to be $M = 41$ since the desired accuracy is achievable.

In order to prove the effectiveness and the validity of the numerical technique, a special case is considered in which the two materials are the same (homogeneous case) and a parallel sub-interface microcrack of length $2a$ is centered just below the semi-infinite crack (see Figure 4.6, let $\alpha^* = 0°$, β^*= 90 degrees. The solution of this special problem was given by Cai and Faber (1992). Let h be the distance between the macrocrack tip and the microcrack center. With the variation of $h/2a$ and by using the method established in section 4.2, the

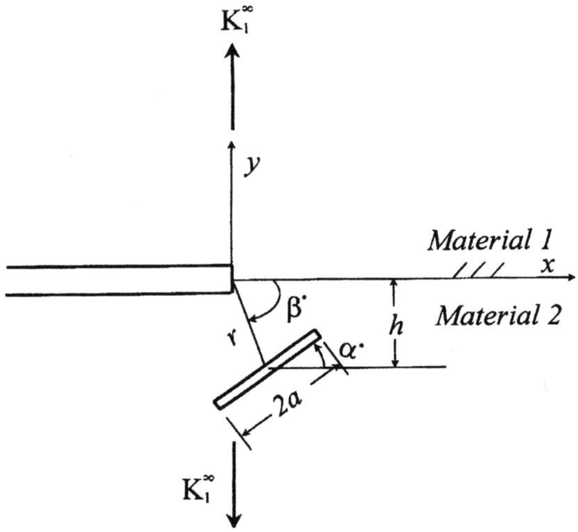

Figure 4.6. An interface macrocrack and a sub-interface microcrack.

results of the normalized stress intensity factor $\Delta k_1/k_1^\infty$ at the macrocrack tip listed in Table 4.1 agree well with those in Cai and Faber (1992), which proves the accuracy of the numerical scheme.

Table 4.1. Numerical results of $\Delta k_1/k_1^\infty$ for the homogeneous case.

$h/2a$	Numerical Results	Results by Cai and Faber (1992)
0.20	−0.5337	−0.5337
0.40	−0.2889	−0.2889
0.60	−0.1747	−0.1747
0.80	−0.1141	−0.1141
1.00	−0.0791	−0.0791

Table 4.2. Material constants.

Material		Cu	Al_2O_3	Ni	MgO
Shear-modulus $(10^{11} N/m^2)$	μ	0.487	1.792	0.808	1.283
Poisson's ratio v		0.345	0.207	0.314	0.175

Table 4.3. Remote complex stress intensity factor.

Material Combination	Cu/Al_2O_3	Ni/MgO
$K_2^\infty/K_1^\infty = 2\varepsilon$	0.056	−0.0098

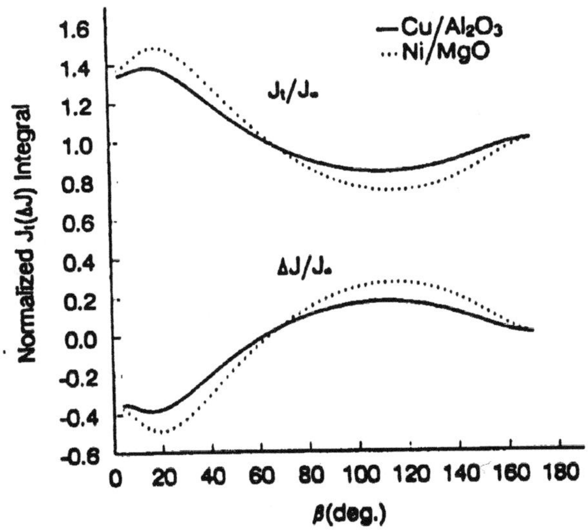

Figure 4.7. The J integral analysis for the parallel sub-interface microcrack in Cu/Al_2O_3 and Ni/MgO bimaterial solids.

Two kinds of material combinations provided by Hutchinson *et al.,* (1997) and used in Chapter 2 are then considered. Their materials constants are shown in Table 4.2 and their Dundur' parameters have already been shown in Table 2.2. It should be noted that the brittle materials, say, $Al_2\ O_3$ or MgO, always occupy the lower half-plane, so it is more possible for microcracks to appear in Material 2, namely, sub-interface microcracks. The condition of plane strain with ($\kappa_j = 3 - 4v_j$) is assumed. It should be noted also in Table 2.2 that the Dundur first parameter α is always negative, whilst the second parameter β is negative for $Cu/Al_2\ O_3$ and positive for Ni/MgO.

Here the remote stress field is specified by the Mode I intensity $K_1^\infty$and keeping the property in mind that for all of practical loading conditions the stress intensity factor of an interface crack is inherently coupled with a mixed mode, *i.e.,* $K^\infty = K_1^\infty + iK_2^\infty$ since, as shown in Table 4.3, the imaginary part of the applied complex stress intensity factor is always much smaller than the real part for the purely tensile loading along the y-axis.

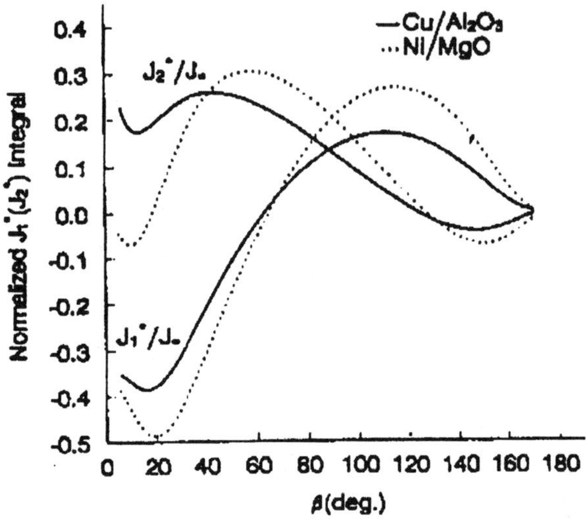

Figure 4.8. The components of the J_k vector induced from the parallel sub-interface microcrack.

Table 4.4. Numerical results for the parallel sub-interface microcrack for Cu/Al_2O_3 and Ni/MgO bimaterial solids.

Bimaterials	Cu/Al_2O_3			Ni/MgO		
Located angle β^* (degrees)	14	94	134	25	75	150
$\Delta K_I^t/K_I^\infty$	0.1759	−0.0785	−0.0687	0.2113	−0.0578	−0.0680
$\Delta K_{II}^t/K_I^\infty$	−0.0291	0.0469	−0.0216	−0.0027	0.1157	−0.0451
$\Delta K_I^R/K_I^\infty$	0.7345	0.7426	0.5097	0.6244	0.7109	0.3367
$\Delta K_{II}^R/K_I^\infty$	−0.1569	−0.0791	0.2296	−0.0991	0.0018	0.2230
$\Delta K_I^L/K_I^\infty$	1.1580	0.4414	0.1388	0.9688	0.5859	−0.0513
$\Delta K_{II}^L/K_I^\infty$	−0.0501	0.2467	0.1612	−0.0569	0.2178	0.1025
J_t/J_∞	1.38357	0.85133	0.86771	1.46733	0.90106	0.87075
$\Delta J/J_\infty$	−0.38349	0.14863	0.13219	−0.46736	0.09891	0.12926
$(J_t + \Delta J)/J_\infty$	1.00008	0.99996	0.99990	0.99997	0.99997	1.00001

Numerical results for an interface crack interacted with a parallel sub-interface microcrack are shown in Figure 4.7 and 4.8, and in Table 4.4, respectively, taking the normalized distance d/a =1.5. It is seen that the projected conservation law of the J_k integral vector, *i.e.,* equations (4.69) and (4.70), is actually and always satisfied as it was done in Chapter 3 for homogeneous brittle elastic solids.The normalized J integral, *i.e.*, $J_t/J_\infty + \Delta J/J_\infty$ is always equal to unity with errors less than about 0.02%.

Table 4.5. Results for an oriented sub-interface microcrack in Cu/Al_2O_3 bimaterial solid.

Located angle β^* (degrees)	60		90		120	
Oriented angle α^* (degrees)	0	155	45	90	60	110
$\Delta K_I^t/K_I^\infty$	0.0064	0.0502	−0.0674	0.0448	0.0144	−0.0547
$\Delta K_{II}^t/K_I^\infty$	0.0901	0.0932	0.0361	0.0289	0.0187	−0.0050
$\Delta K_I^R/K_I^\infty$	0.7769	0.9530	0.5655	0.3530	0.1328	0.4861
$\Delta K_{II}^R/K_I^\infty$	−0.0732	−0.0523	0.2318	−0.3026	−0.2138	−0.0788
$\Delta K_I^L/K_I^\infty$	0.7836	0.6820	0.1950	0.2452	0.1082	0.4689
$\Delta K_{II}^L/K_I^\infty$	0.1927	−0.1989	0.1339	−0.2007	−0.1371	−0.2189
J_t/J_∞	1.0209	1.11163	0.87099	1.09243	1.02937	0.89363
$\Delta J/J_\infty$	−0.0207	−0.11144	0.12899	−0.09234	−0.02935	0.10625
$(J_t+\Delta J)/J_\infty$	1.00013	1.00019	0.99998	1.00009	1.00002	0.99988

Table 4.6. Results for an oriented sub-interface microcrack in Ni/MgO bimaterial solid.

Located angle β^* (degrees)	60		90		120	
Oriented angle α^* (degrees)	38	128	0	135	90	160
$\Delta K_I^t/K_I^\infty$	−0.0309	0.1621	−0.1105	−0.0592	−0.0489	−0.1426
$\Delta K_{II}^t/K_I^\infty$	0.1052	−0.0499	−0.0863	0.0900	−0.0076	−0.0014
$\Delta K_I^R/K_I^\infty$	0.6448	0.6751	0.7020	0.6789	0.3714	0.3392
$\Delta K_{II}^R/K_I^\infty$	0.1318	−0.3820	−0.0634	−0.0707	−0.2347	0.1755
$\Delta K_I^L/K_I^\infty$	0.4003	0.4112	0.4560	0.5559	0.2632	0.6438
$\Delta K_{II}^L/K_I^\infty$	0.2490	−0.2361	0.2351	−0.2238	−0.2378	0.0447
J_t/J_∞	0.95027	1.35290	0.79857	0.89321	0.90469	0.73520
$\Delta J/J_\infty$	0.04970	−0.35293	0.20141	0.10678	0.09533	0.26482
$(J_t+\Delta J)/J_\infty$	0.99997	0.99997	0.99998	0.99999	1.00002	1.00002

Numerical results for an interface crack made to interact with an arbitrarily oriented sub-interface microcrack are shown in Figure 4.9, 4.10 and 4.11, respectively, for the oriented angle α^*= 60, 90 and 120 degrees. Also, the results are shown in Table 4.5 for Cu/Al_2O_3 and Table 4.6 for Ni/MgO. It can also be found that the normalized J integral, *i.e.*, $J_t/J_\infty + \Delta J/J_\infty$ is always equal to unity with error less than about 0.02 %. This implies not only the effectiveness of the technique for evaluating the J_2^* integral (4.78), (4.79), and (4.86), but also the fairness of the projected conservation theory of the J_k vector in the present situation.

It is concluded that the interaction effect between an interface macrocrack tip and sub-interface microcrack, whatever it is shielding corresponding to $J_t/J_\infty < 0$ or anti-shielding corresponding to $J_t/J_\infty > 0$, could be considered as the redistribution of the remote J integral. In other words, the remote J integral, as a measure of the loading level, is divided into two distinct parts. One is induced from the existence of the microcracks and the other is contributed from the interface macrocrack tip.

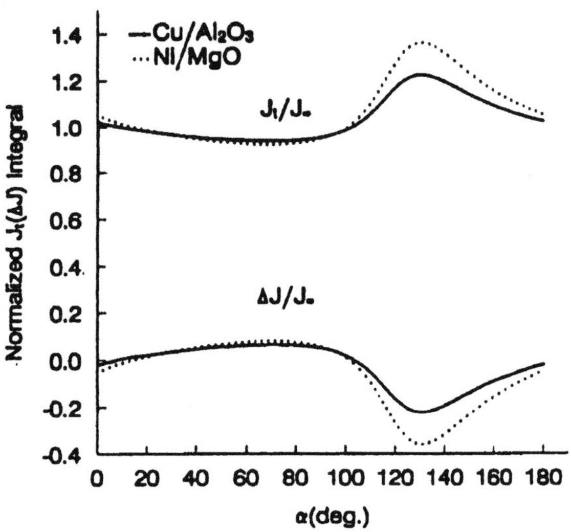

Figure 4.9. Results for an oriented sub-interface microcrack ($\beta^* = 60°$).

Indeed, the simple but universal conservation law amongst the three values of the J integral found in brittle solids and discussed in Chapter 3 is still valid in bimaterial solids although the characteristic oscillating stress singularity as well as the bimaterial mismatch nature disturbs the near-tip stress field significantly.

5. Brief summary

In this chapter we have shown that the pseudo-traction method could be extended to treat the macrocrack microcrack interaction in metal/ceramic bimaterials, providing that the fundamental solutions established above are derived

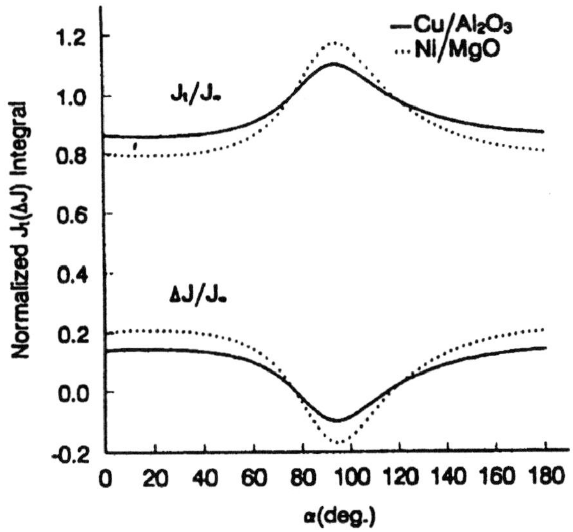

Figure 4.10. Results for an oriented sub-interface microcrack ($\beta^* = 90°$).

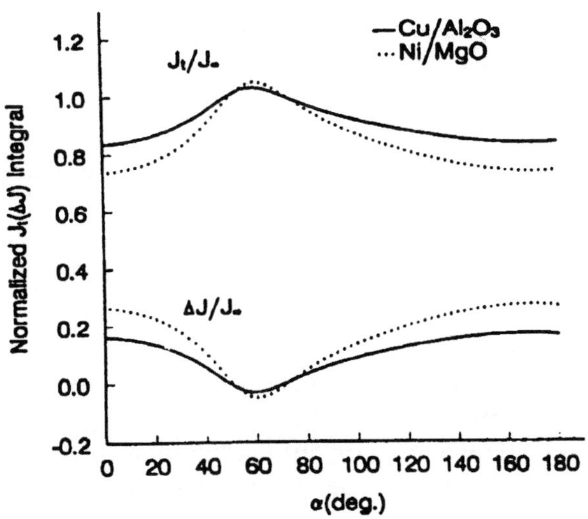

Figure 4.11. Results for an oriented sub-interface microcrack ($\beta^* = 120°$).

beforehand. Moreover, the projected conservation law of the J_k vector established in Chapter 3 for a discrete model of shielding problems in brittle solids is still valid in bimaterial solids although the well known characteristic oscillating stress singularity in the near-tip stress field of a macro-interface crack adds considerable complication in manipulations and calculations. In other words, the summation of the contributions induced from the interface crack tip and those induced from the microcracks should be equal to the remote J integral. This law not only leads to a consistency check for numerical examinations, but also provides, from the physical point of view, a better understanding of the shielding effect in metal/ceramic bimaterials.

It will be seen in the next chapter that the conservation law is valid too for microcrack macrocrack interaction problems in dissimilar anisotropic elastic solids such as an interface crack interacting with near-tip microcracks in 0/90-ply unidirectional fibre reinforced composites.

References for Chapter 4

Atkinson, N., (1977) On stress singularities and interfaces in linear elastic fracture mechanics , *International Journal of Fracture*, **13**, 807 – 820.

Cai, H., and Faber, T.K., (1992) On the use of approximation methods for microcrack shielding problems , *ASME Journal of Applied Mechanics*, **59**, 497 – 501.

Cai, H. and Evans, A.G., (1989) An experimental study of the fracture resistance of bimaterial interfaces , *Mechanics of Materials*, **7**, 295 – 304.

Chen, Y.H., and Hasebe, N., (1994a) Further investigation of Comninou's EEF for an interface crack with completely closed faces , *International Journal of Engineering Science*, **32**, 1037 – 1046.

Chen, Y. Z., and Hasebe, N., (1994b) Eigenfunction expansion and higher order weight functions of interface cracks , *ASME Journal of Applied Mechanics*, **50**, 843 – 849.

Chen, Y.H., and Hasebe, N., (1994c) Interaction between an main-crack and a parallel microcrack in an orthotropic plane elastic solid , *International Journal of Solids and Structures*, **31**, 1877 – 1890.

Chen, Y.H., and Hasebe, N., (1995a) Investigation of main-micro crack interaction problems in a unidirectional fibre reinforced composite , *Engineering Fracture Mechanics.*, **50**, 283 – 293.

Chen, Y.H., and Hasebe, N., (1995b) On the influence of orthotropic material constants on the neutral-shielding angle in main-micro crack interaction problems , *Engineering Fracture Mechanics*, **51**, 349 – 360.

Chen, Y. H., and Hasebe, N., (1995c) Interaction of two-off-axis cracks exhibiting orthotropic and general anisotropic behavior under arbitrary extension , *Theor. Appl. Fract. Mech.*, **22**, 249 – 260.

Chen, Y., (1996) On the contribution of discontinuities in a near-tip stress field to the J-integral , *International Journal of Engineering Science*, **34**, 819 – –829.

Chen, Y.H., and Ma, H., (1997) Explicit formulations of the J_2 integral in anisotropic bodies and its application, *Science in China, E.*, **27**, 310 – 317 (in Chinese).

Chen, Y.Z., (1984a) General case of multiple crack problems in an infinite plate , *Engineering Fracture Mechanics,* **20**, 591 – 597.

Chen, Y.Z., (1984b) Multiple crack problems of antiplane elasticity in an infinite body , *Engineering Fracture Mechanics*, **20**, 767 – 776.

Chen, Y.Z., (1986) Multiple crack problem crack problems for two bounded half planes in plane antiplane elasticity , *Engineering Fracture Mechanics*, **25**, 1 – 9.

Cherepanov, G.P., (1979) *Mechanics of Brittle Fracture* , McGraw-Hill, New York.

Clausen, N., (1976) Fracture toughness of Al_2O_3 with a unstabilized ZrO_2 dispersed phase , *J. Am. Ceram. Soc.*, **59**, 49 – 51.

Comninou, M., (1977a) The interface crack , *ASME Journal of Applied Mechanics*, **44**, 631 – 636.

Comninou, M., (1977b) Interfacial crack with friction in the contact zone , *ASME Journal of Applied Mechanics*, **44**, 780 – 781.

Comninou, M., (1978) The interface crack in a shear field , *ASME Journal of Applied Mechanics*, **45**, 287 – 290.

Comninou, M., and Schmueser, D. (1979) The interface crack in a combined tension-compression and shear field , *ASME Journal of Applied Mechanics*, **46**, 345 – 348.

England, A.H., (1965) A crack between dissimilar media , *ASME Journal of Applied Mechanics*, **32**, 400 – 402.

Erdogan, F., (1978) *Mixed boundary value problem* , *Mechanics Today*, **4**, Pergamen Press, Oxford.

Erdogan, F., (1965) Stress distributions in bonded dissimilar materials with cracks , *ASME Journal of Applied Mechanics*, **32**, 403 – 410.

Evans, A.G., and Faber, K.T., (1981) Toughening of ceramics by circumferential microcracking , *J. Am. Ceram. Soc.*, **64**, 394 – 398.

Evans, A.G., and Faber, K.T., (1984) Crack-growth resistance of microcracking brittle materials , *J. Am Ceram. Soc.*, **67**, 255 – 260.

Faber, A.G., and Fu, Y., (1985) Some effects of microcracks on the mechanical properties of brittle solids-II. Microcrack toughening , *Acta. Metall.*, **33**, 1525 – 1531.

Fleck, N.A., Hutchinson, J.W., and Suo, Z., (1991) Crack path selection in a brittle adhensive layer , *International Journal of Solids and Structures*, **27**, 1683 – 1703.

Fu, Y., and Evans, A.G., (1985) Some effects of microcracks on the mechanical properties of brittle solids-I. Stress, Strain relations , *Acta. Metall.*, **33**, 1515 – 1523.

Gao, H., (1991) Weight function analysis of interface cracks: Mismatch versus oscillation , *ASME Journal of Applied Mechanics*, **58**, 931 – 938.

He, M.Y., and Hutchison, J.W., (1989) King of a crack out of an interface , *ASME Journal of Applied Mechanics*, **56**, 270 – 278.

Herrmann, A.G., and Herrmann, G. (1981) On energy-release rates for a plane crack , *ASME Journal of Applied Mechanics*, **48**, 525 – 528.

Hoagland, R.G., Hahn, G.T., and Rosenfield, A.R., (1973) Influence of microstructure on the fracture propagation in rocks , *Rock Mech.*, **5**, 77 – 106.

Hoagland, R.G., and Embury, J.D., (1980) A treatment of inelastic deformation around a crack tip Owing to microcracking , *J. Am. Ceram. Soc.*, **63**, 404 – 410.

Hong, C., and Stern, M., (1978) The computation of stress intensity factors in dissimilar materials , *J. Elasticity,* **8**, 21 – 34.

Horii, M., and Nemat Nasser, N., (1987) Interacting microcracks near the tip in the process zone of a macrocrack , *Journal of Mechanics and Physics of Solids*, **35**, 601 – 629.

Hutchinson, J.W., Mear, M.E., and Rice, J.R.,(1987) Crack paralleling an interface between dissimilar materials, *ASME Journal of Applied Mechanics*, **54**, 828 – 832.

Hutchinson, J.W., (1983) Constitutive behavior and crack tip fields for materials undergoing creep-constrained grain boundary cavitation , *Acta Metall.*, **31**, 1079 – 1088.

Hutchinson, J. W. (1987) Crack Tip Shielding by Microcracking in Brittle Solids , *Acta Metall*, **35**, 1605 – 1619.

Knowles, T.K., and Sternberg, E., (1983) Large deformation near a tip of an interface crack between two Neo-Hooken Sheets , *J. Elasticity*, **13**, 257 – 293.

Krenk, S., (1975) On the use of the interpolation polynomial for solutions of singular integral equations , *Quarterly of Applied Mathematics*, **32**, 479 – 484.

Muskhelishvili, N.I., (1953) *Some Basic Problems of Mathematical Theory of Elasticity* , Noordhoff Publishers, Groningen, The Netherlands.

Park, J.W., and Earmme, Y.Y., (1986) A application of conservation integrals to interface crack problems , *Mechanics of Materials*, **5**, 261 – –276.

Rice, J.R., and Sih, G.C.,(1965) Plane problems of cracks in dissimilar media , *ASME Journal of Applied Mechanics*, **32**, 418 – 423.

Rice, J.R., (1988) Elastic fracture mechanics concepts for interfacial cracks , *ASME Journal of Applied Mechanics*, **55**, 98 – 103.

Suo, Z., and Hutchison, J.W., (1989a) Steady-state cracking in brittle substrate between adherent films , *International Journal of Solids and Structures*, **25**, 1337 – 1353.

Suo, Z., and Hutchison, J.W., (1989b) Sandwich specimens for measuring interface crack toughness, *J. Mat. Sci. Eng.*, **A107**, 135 – 143.

Suo, Z., (1989) Singularities interacting with interface cracks , *International Journal of Solids and Structures*, **25**, 1133 – 1142.

Symington, M.F., (1987) Eigenvalues for interface cracks in linear elasticity , *ASME Journal of Applied Mechanics*, **54**, 973 – 974.

Wang, J. S. and Suo, Z. (1990) Experimental determination of interfacial toughness curves using Brazial Nut sandwiches , *Acta Metall. Mater.*, **38**, 1279 – 1290.

Williams, M.L., (1959) The stress around a fault or crack in dissimilar material , *Bull. Soc. Am.*, **49**, 199 – 204.

Wu, C., Freiman, S.W., Rice, J.R., and Mecholsky, J.J., (1978) Microstructural aspects of crack propagation in ceramics, *J. Mater. Sci.*, **13**, 2659 – 2670.

Yau, J. F. and Wang, S.S., (1984) Analysis of interface cracks between dissimilar isotropic materials using conservation integral in elasticity , *Engineering Fracture Mechanics*, **20**, 423 – 432.

Zhao, L. G. and Chen, Y. H. (1997) On the contribution of sub-interface microcracks near the tip of an interface crack to the J-integral , *International Journal of Engineering Science*, **35**, 387 – 407.

Zhao L.G. and Chen Y. H., (1998) Effect of the T-stress in microcrack shielding problems, *ASME Journal of Applied Mechanics*, **65**, 71 – 75.

Zhang, T. Y. and Li, J. C. M. (1992) Interaction of an edge dislocation with an interfacial crack , *J. Appl. Phys.*, **72**, 2215 – 2226.

Chapter 5

MACROCRACK MICROCRACK INTERACTION IN DISSIMILAR ANISOTROPIC MATERIALS

Because of their potential for reducing weight and the capacities of optimizing the structural strength and stiffness, laminated fiber reinforced composite materials are widely used in light structures, and are applied to replace many metallic components usually used before. However, the mechanical behaviors of such materials influenced by the fiber and matrix interaction, the matrix microcracks and the multiple ply configurations show very complicated features, which have challenged designers with a new class of problems. One particular area which has received considerable attention in the past two decades, has been their low tolerance to interfacial damage and the other is concerned with damage of micro-defects such as micro-cracks, micro-voids, and micro-inclusions. The occurrences of these types of damage, which are frequently caused by impact or other sources, are common and unavoidable during manufacturing, maintenance, and service of the light structures (see Figure 5.1).

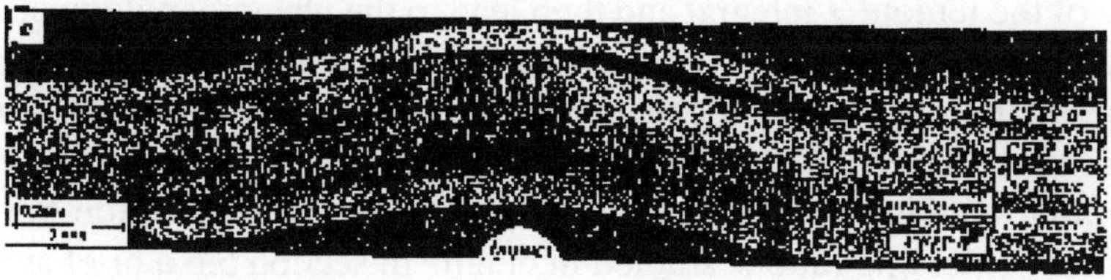

Figure 5.1. A macrocrack formed on the interface of 0/90 ply-carbon fiber *CFRP* provided by Blimenauer *et al.*,(1999).

To date, although many significant progresses have been made in solving the interface crack problems in both dissimilar isotropic materials and dissimilar anisotropic materials [see, *e.g.*, Wang and Choi (1983a, 1983b), Hutchinson *et al.,* (1987b), Rice (1988), Miller (1989a, 1989b), Suo (1990), Ting (1986, 1990), Gao (1991), Lu and Lardner (1992), Chen and Hasebe (1994a, 1994b)], the influence of microcracks in the near-tip process zone of an interface macrocrack in dissimilar anisotropic materials on the macrocrack tip parameters remains inadequately treated [Chen and Tian, (2000); Tian and Chen, 2000(a, b)].

This chapter addresses the interaction problem between a macro-interface crack and sub-interface matrix microcracks in a dissimilar anisotropic material. Following those cited in the above chapters, the conservation law of the J integral is extended to treat this kind of problem taking dissimilar material anisotropic nature into account. In Section 5.1 the elementary solutions for an interface crack subjected to different loads and those for an edge dislocation beneath the interface in dissimilar anisotropic materials are presented by using the Lekhenitskii anisotropic theorem (1963). In Section 5.2 the interaction problem between a semi-infinite interface crack and multiple sub-interface matrix microcracks in the near tip process zone in dissimilar anisotropic materials is deduced to a system of singular integral equations with the aid of superimposing technique. As discussed in the above chapters, the singular integral equations can be solved numerically by using the Chebyshev integration technique (Erdogan, 1978). In Section 5.3 the conservation law found in the above chapters among three kinds of the J integral values is re-examined in the present much more complicated cases. The three kinds of the values are induced from the macro-interface crack tip, the sub-interface microcracks in the near-tip process zone and the remote stress field, respectively. As pointed out in the above chapters, the conservation law not only provides a necessary condition with which the numerical results derived here should meet, but also reveals that the microcrack shielding effect in these dissimilar materials could also be considered as a redistribution of the remote J integral. It is this redistribution that leads to the wastage of the remote J integral and then lead to the phenomenological shielding effect when the remote J integral transmits across a microcracking zone from the infinity to the macrocrack tip. In Section 5.4 numerical results and discussions are given for $0/90$ ply fibre reinforced composites. Three kinds of dissimilar configurations are used for comparisons. In Section 5.5 the T effect on the shielding ratio is studied in detail. In section 5.6 a brief summary is given.

1. Fundamental formulations in dissimilar anisotropic materials

The interface crack problem in dissimilar anisotropic materials has been studied in quite details by Wang and Choi (1983a,b), Miller (1989a,b), Suo (1990). It is well known (Lekhnitskii, 1963, Suo, 1990) that, under plane stress or plane stain conditions, the elastic field in an anisotropic material could be represented in terms of two complex functions $f_1(z_1)$ and $f_2(z_2)$, each of which is holomorphic in its argument $z_j = x + \mu_j y\,(j = 1, 2)$. Here μ_j denotes two distinct complex numbers with positive imaginary parts. They could be determined as the roots of a fourth-order characteristic equation. Using these holomorphic functions, the representations of displacements u_i, stresses σ_{ji}, and resultant forces T_i could be put in the following forms:

$$\begin{aligned} u_i &= 2Re[\sum_{j=1}^{2} A_{ij} f_j(z_j)], \\ T_i &= -2Re[\sum_{j=1}^{2} L_{ij} f_j(z_j)], \\ \sigma_{2i} &= 2Re[\sum_{j=1}^{2} L_{ij} f_j'(z_j)], \\ \sigma_{1i} &= -2Re[\sum_{j=1}^{2} L_{ij} \mu_j f_j'(z_j)], \end{aligned} \tag{5.1}$$

where the over prime is designated as the derivative with respect to the associated arguments, and **L** and **A** are two 2×2 complex matrices depending on elastic constants (Suo, 1990):

$$\mathbf{L} = \begin{bmatrix} -\mu_1 & -\mu_2 \\ 1 & 1 \end{bmatrix}, \tag{5.2}$$

$$\begin{aligned} A_{1j} &= s_{11}\mu_j^2 + s_{12} - s_{16}\mu_j, \\ A_{2j} &= s_{21}\mu_j + s_{22}/\mu_j - s_{26}, \end{aligned} \quad (j = 1, 2) \tag{5.3}$$

where, μ_j satisfies the following fourth-order characteristic equation

$$s_{11}\mu^4 - 2s_{16}\mu^3 + (2s_{12} + s_{66})\mu^2 - 2s_{26}\mu + s_{22} = 0. \tag{5.4}$$

Equations (5.2)–(5.4) are valid for plane stress deformation, whilst the corresponding ones for plane strain deformation could be given by using the following change of compliances

$$\tilde{s_{ij}} = s_{ij} - \frac{s_{i3}s_{j3}}{s_{33}}. \tag{5.5}$$

For the sake of convenience, Suo (1990) has introduced a positive definite Hermitian matrix **B** and a related matrix **Y**:

$$\mathbf{B} = i\mathbf{A}\mathbf{L}^{-1}, \tag{5.6}$$

$$\mathbf{Y} = \mathbf{B} + \bar{\mathbf{B}}, \tag{5.7}$$

where $i = \sqrt{-1}$, the over bar denotes the complex conjugate.

For interface problems in dissimilar materials in which material 1 and 2 occupy the upper and lower half planes, respectively (see, *e.g.*, Figure 5.2), another positive definite Hermitian matrix **H** involving bimaterial elastic constants is defined as follows (Suo, 1990):

$$\mathbf{H} = \mathbf{B}_1 + \bar{\mathbf{B}}_2, \tag{5.8}$$

where subscripts 1 and 2 are attached to the upper and lower materials, respectively, and the over bar denotes the corresponding complex conjugate.

Furthermore, Suo (1990) has also defined the following function vector $\mathbf{f}(z)$:

$$\mathbf{f}(z) = [f_1(z), f_2(z)]^T, \tag{5.9}$$

where $f_1(z)$ and $f_2(z)$ describe the functions for the upper and lower materials, respectively.

Once the solution of $\mathbf{f}(z)$ is derived for a given boundary value problem, a replacement of variable z_1 or z_2 from variable z should be made for each component function to calculate field quantities from equation (5.1). Moreover, the set of vectors along the interface is given as follows:

$$\begin{aligned} \mathbf{u}_x &= \{u_j(x,0)\} = \mathbf{A}\mathbf{f}(x) + \bar{\mathbf{A}}\bar{\mathbf{f}}(x), \\ \mathbf{t}(x) &= \{\sigma_{2j}(x,0)\} = \mathbf{L}\mathbf{f}(x) + \bar{\mathbf{L}}\bar{\mathbf{f}}(x), \end{aligned} \tag{5.10}$$

where $\mathbf{u}_x = [u_{1x}, u_{2x}]^T$, $\mathbf{t}(x) = [t_1(x), t_2(x)]^T$, and the x-axis is along the interface.

1.1 Fundamental solution for an interface crack

The formulations presented above are used to derive the fundamental solution for a semi-infinite interface crack under concentrated tractions as shown in Figure 5.2. Without going into detail (Appendix of this chapter), the following Hilbert problem is deduced:

$$\mathbf{h}^+(x) + \bar{\mathbf{H}}^{-1}\mathbf{H}\mathbf{h}^-(x) = \mathbf{t}_0(x), \qquad x \in C. \tag{5.11}$$

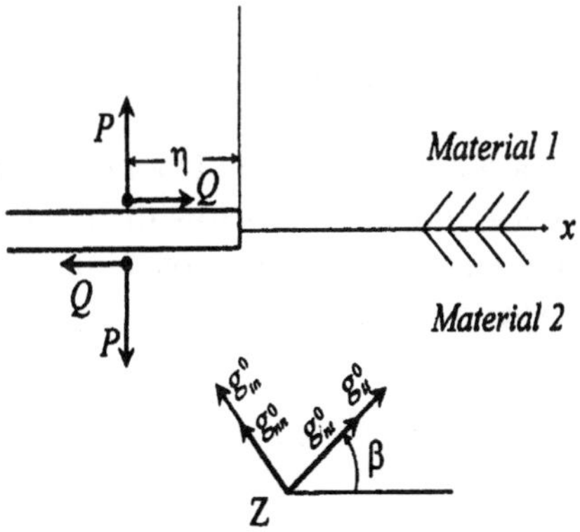

Figure 5.2. A pair of normal and tangential concentrated tractions acting on an interface macrocrack.

where C is the crack surface, the complete solution could easily be given as follows:

$$\mathbf{h}(z) = h_1(z)\mathbf{w} + h_2(z)\bar{\mathbf{w}}, \tag{5.12}$$

where

$$\begin{aligned} h_1(z) &= \frac{\chi(z)}{2\pi i}\int_C \frac{t_{01}(x)dx}{\chi^+(x)(x-z)}, \\ h_2(z) &= \frac{\bar{\chi}(z)}{2\pi i}\int_C \frac{\bar{t}_{01}(x)dx}{\bar{\chi}^+(x)(x-z)}, \end{aligned} \tag{5.13}$$

$$t_{01} = \frac{\bar{\mathbf{w}}\mathbf{H}\mathbf{t}_0(x)}{\bar{\mathbf{w}}\mathbf{H}\mathbf{w}}, \tag{5.14}$$

$$\chi(z) = z^{-\frac{1}{2}-i\varepsilon}, \tag{5.15}$$

and the vector $\boldsymbol{w}$ and the complex index ε satisfy the following equation

$$\bar{\mathbf{H}}\mathbf{w} = e^{2\pi\varepsilon}\mathbf{H}\mathbf{w}. \tag{5.16}$$

It should be pointed out that for the case in Figure 5.2,

$$\mathbf{t}_0(x) = [0, P\delta(x-\eta)]^T, \quad \text{for the traction P,}$$

$$\mathbf{t}_0(x) = [Q\delta(x-\eta), 0]^T, \quad \text{for the traction Q,} \tag{5.17}$$

where η is the distance of the traction point from the origin (Figure 5.2).

Therefore, the stresses at any point z below the interface in the $x-y$ plane induced from the normal concentrated traction P or from the tangential concentrated traction Q, respectively, are:

$$\begin{aligned} g^0_{nn} &= 2Re[f_1'(z_1) + f_2'(z_2)], \\ g^0_{nt} &= -2Re[\mu_1 f_1'(z_1) + \mu_2 f_2'(z_2)], \quad \text{for the traction P,} \\ g^0_{tn} &= 2Re[f_1(z_1) + f_2(z_2)], \\ g^0_{tt} &= -2Re[\mu_1 f_1(z_1) + \mu_2 f_2(z_2)], \quad \text{for the traction Q,} \end{aligned} \tag{5.18}$$

Equation (5.18) could be evaluated by using (5.12) without any difficulty. As well known (Crouch and Starfield, 1983), the stresses could be considered as the influence functions or the kernel functions in the boundary element method, which will be used in the forthcoming manipulations.

1.2 Fundamental solution for an edge dislocation

As regards a sub-interface crack, it could be modelled by using continuously distributed dislocations at its location beneath the interface and then the second elementary solution is concerned with a dislocation in the lower half plane. It is well known that such a dislocation in a homogeneous anisotropic elastic body has been studied by Obata, *et. al.*, (1989) and the complex potentials corresponding to the basic singular solution have been given by them as

$$f_k'(z) = \frac{B_k}{z - z_{0k}}, \quad (k = 1, 2), \tag{5.19}$$

in which $z_{0k} = \xi_0 + \mu_k \eta_0$ with (ξ_0, η_0) denoting the location of the singularity and

$$\begin{aligned} B_1 &= \frac{1}{2\pi i}\frac{1}{s_{11}}\frac{\mu_1 B_x - B_y}{(\mu_1-\mu_2)(\mu_1-\overline{\mu_1})(\mu_1-\overline{\mu_2})}, \\ B_2 &= \frac{1}{2\pi i}\frac{1}{s_{11}}\frac{\mu_2 B_x - B_y}{(\mu_2-\mu_1)(\mu_2-\overline{\mu_1})(\mu_2-\overline{\mu_2})}, \end{aligned} \tag{5.20}$$

where B_x and B_y are the displacement discontinuities across the dislocation line in the x- and y-directions, respectively.

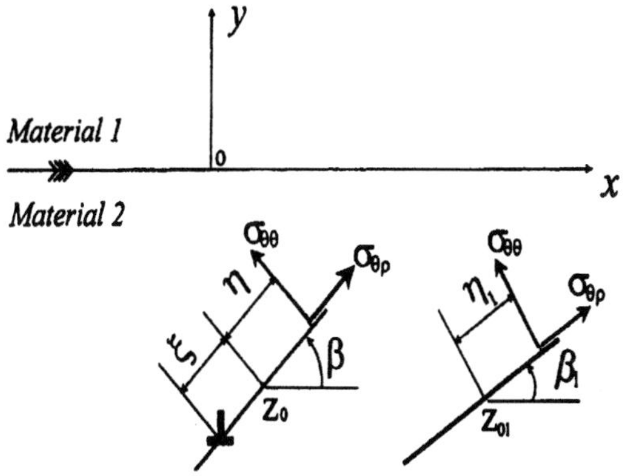

Figure 5.3. An edge dislocation near the interface.

However, an edge dislocation near and beneath the interface in a dissimilar anisotropic material (see Figure 5.3) is much more complicated than that in a homogeneous anisotropic material. Miller (1989a,b) proposed the complex potentials f_k modeling an edge dislocation at point Z_0 beneath the interface, which are formulated as follows

$$f_k'(z) = \frac{B_k}{z - z_{0k}} - \sum_{j=1}^{4} b_{kj} \sum_{i=1}^{2} \overline{m_{ji}} \frac{\overline{B_i}}{z - \overline{z_{0i}}}, \tag{5.21}$$

where

$$\begin{bmatrix} b_{1j} \\ b_{2j} \\ b_{3j} \\ b_{4j} \end{bmatrix} = \begin{bmatrix} \begin{pmatrix} A_{11} & A_{12} \\ A_{21} & A_{22} \end{pmatrix}_2 & \begin{pmatrix} -\overline{A_{11}} & -\overline{A_{12}} \\ -\overline{A_{21}} & -\overline{A_{22}} \end{pmatrix}_1 \\ \begin{pmatrix} L_{21} & L_{22} \\ L_{11} & L_{12} \end{pmatrix}_2 & \begin{pmatrix} -\overline{L_{21}} & -\overline{L_{22}} \\ -\overline{L_{11}} & -\overline{L_{12}} \end{pmatrix}_1 \end{bmatrix}^{-1}, \tag{5.22}$$

and

$$m_{1k} = (A_{1k})_2, \quad m_{2k} = (A_{2k})_2, \quad m_{3k} = (L_{2k})_2, \quad m_{4k} = (L_{1k})_2, \tag{5.23}$$

where the notations $(\cdot)_1$ and $(\cdot)_2$ correspond to the upper and lower materials, respectively.

Therefore the stress components at any point $z = z_0 + \xi e^{i\beta}$ induced from the edge dislocation can be formulated as:

$$\sigma_{\theta\theta} + i\sigma_{\rho\theta} = H_1 + H_3 + H_2 + H_4, \tag{5.24}$$

where $z = z_0 + \eta e^{i\beta}$ on the line at which the dislocation is located as shown in Figure 5.3; or

$$\sigma_{\theta\theta} + i\sigma_{\rho\theta} = H_5 + H_7 + H_6 + H_8, \tag{5.25}$$

where $z = z_{01} + \eta_1 e^{i\beta_1}$ on the line at which the dislocation is not located as shown in Figure 5.3.

Here

$$\begin{aligned}
H_1 &= \frac{1}{\eta - \xi}\sum_{k=1}^{2}\frac{[1+\mu_k^2+e^{2i\beta}(1-\mu_k^2-2i\mu_k)]B_k}{(1-i\mu_k)e^{i\beta}+(1+i\mu_k)e^{-i\beta}},\\
H_2 &= \frac{1}{\eta - \xi}\sum_{k=1}^{2}\frac{[1+\overline{\mu_k}^2+e^{2i\beta}(1-\overline{\mu_k}^2-2i\overline{\mu}_k)]\overline{B_k}}{(1+i\overline{\mu}_k)e^{-i\beta}+(1-i\overline{\mu}_k)e^{i\beta}},\\
H_3 &= -\frac{1}{2}\sum_{k=1}^{2}\sum_{j=1}^{4}\sum_{i=1}^{2}\frac{\overline{b_{kj}}m_{ji}[1+\overline{\mu_k}^2+e^{2i\beta}(1-\overline{\mu_k}^2-2i\overline{\mu_k})]B_i}{(\eta-\xi)\cos\beta+(y_0+\eta\sin\beta)\overline{\mu_k}-(y_0+\xi\sin\beta)\mu_i},\\
H_4 &= -\frac{1}{2}\sum_{k=1}^{2}\sum_{j=1}^{4}\sum_{i=1}^{2}\frac{b_{kj}\overline{m_{ji}}[1+\mu_k^2+e^{2i\beta}(1-\mu_k^2-2i\mu_k)]\overline{B_i}}{(\eta-\xi)\cos\beta+(y_0+\eta\sin\beta)\mu_k-(y_0+\xi\sin\beta)\overline{\mu_i}},\\
H_5 &= \frac{1}{2}\sum_{k=1}^{2}\frac{[1+\mu_k^2+e^{2i\beta_1}(1-\mu_k^2-2i\mu_k)]B_k}{V_1},\\
H_6 &= \frac{1}{2}\sum_{k=1}^{2}\frac{[1+\overline{\mu_k}^2+e^{2i\beta_1}(1-\overline{\mu}_k^2-2i\overline{\mu_k})]\overline{B_k}}{V_2},\\
H_7 &= -\frac{1}{2}\sum_{k=1}^{2}\sum_{j=1}^{4}\sum_{i=1}^{2}\frac{\overline{b_{kj}}m_{ji}[1+\overline{\mu_k}^2+e^{2i\beta_1}(1-\overline{\mu_k}^2-2i\overline{\mu_k})]B_i}{V_3},\\
H_8 &= -\frac{1}{2}\sum_{k=1}^{2}\sum_{j=1}^{4}\sum_{i=1}^{2}\frac{b_{kj}\overline{m_{ji}}[1+\mu_k^2+e^{2i\beta_1}(1-\mu_k^2-2i\mu_k)]\overline{B_i}}{V_4},
\end{aligned} \tag{5.26}$$

where V_1, V_2, V_3, and V_4 are defined by

$$V_1 = (x_{01} + \eta_1 \cos\beta_1 - x_0 - \xi\cos\beta) + (y_{01} + \eta_1 \sin\beta_1 - y_0 - \xi\sin\beta)\mu_k,$$

$$V_2 = (x_{01} + \eta_1 \cos\beta_1 - x_0 - \xi\cos\beta) + (y_{01} + \eta_1 \sin\beta_1 - y_0 - \xi\sin\beta)\overline{\mu_k},$$

$$V_3 = (x_{01} + \eta_1 \cos\beta_1 - x_0 - \xi\cos\beta) + (y_{01} + \eta_1 \sin\beta_1)\overline{\mu_k} - (y_0 + \xi\sin\beta)\mu_k,$$

$$V_4 = (x_{01} + \eta_1 \cos\beta_1 - x_0 - \xi\cos\beta) + (y_{01} + \eta_1 \sin\beta_1)\mu_k - (y_0 + \xi\sin\beta)\overline{\mu_k}.$$

1.3 Remote loading conditions

As regards the remote loading conditions for a semi-infinite crack in an infinite dissimilar anisotropic material loaded by the remote stress intensity factors K_1^∞ and K_2^∞, the complex potentials for the two half planes are formulated as

$$\mathbf{L}_1\mathbf{f}_1'(z) = \frac{e^{\pi\varepsilon}Kz^{i\varepsilon}w + e^{-\pi\varepsilon}\bar{K}z^{-i\varepsilon}\bar{w}}{2(2\pi z)^{\frac{1}{2}}\cosh\pi\varepsilon}, \quad \text{in upper half plane 1,}$$

$$\mathbf{L}_2\mathbf{f}_2'(z) = \frac{e^{-\pi\varepsilon}Kz^{i\varepsilon}w + e^{\pi\varepsilon}\bar{K}z^{-i\varepsilon}\bar{w}}{2(2\pi z)^{\frac{1}{2}}\cosh\pi\varepsilon}, \quad \text{in lower half plane 2,} \tag{5.27}$$

where $K = K_1^\infty + iK_2^\infty$.

It should be emphasized that all the manipulations performed above are based on Suo' work (1990), which are given in a compact form following Lekhenitskii's theorem(1963). When calculating the field quantities via (5.1), it is necessary to replace z by $z_j = x + \mu_j y$, respectively for each component of $f(z)$ in (5.27).

Consequently the residual stresses to be released on the location of sub-interface crack are given as:

$$p = 2Re[f_1'(z_1) + f_2'(z_2)],$$
$$q = -2Re[\mu_1 f_1'(z_1) + \mu_2 f_2'(z_2)], \tag{5.28}$$

where $f_1'(z_1)$ and $f_2'(z_2)$ are given by equation (5.27).

2. Superimposing technique and singular integral equations

2.1 Decomposition of the original problem

Consider N sub-interface microcracks near the tip of a semi-infinite interface macrocrack and take the prescribed complex interface intensity factors K_1^∞ and

K_2^{∞} to specify the applied remote stress field as shown in Figure 5.4. Using the pseudo-traction method proposed by Chen (1984) and Horii and Nemat Nasser(1985), the problem shown in Figure 5.4(a) is decomposed into $N+2$ subproblem [see Figures 5.4(b)(c)(d)], each of which contains one single crack.

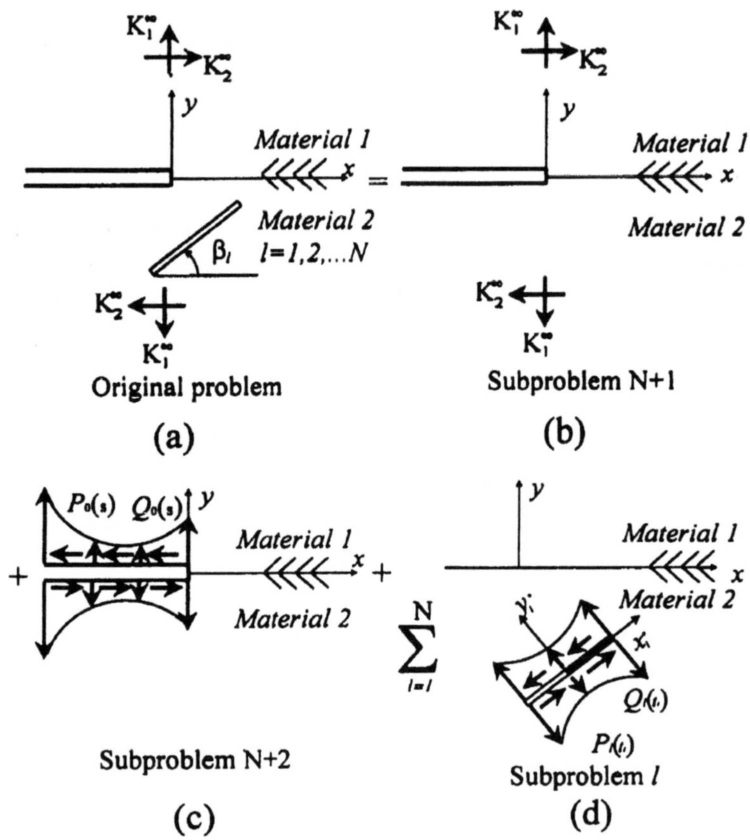

Figure 5.4. N sub-interface cracks in the near-tip process zone.

The unknown normal tractions $P_0(t), P_k(s_k)$ and the tangential tractions $Q_0(t), Q_k(s_k)$ on the crack faces (k=*1,2, ..., N*) are the so called pseudo-tractions to be determined. Moreover, each of the N sub-interface cracks could be considered or modeled by using the edge dislocations distributed continuously on the crack location. In order to satisfy the traction condition on the crack surfaces, the stresses on the crack surface of the $N+2$ subproblems foregoing mentioned should be superimposed. As done by Chen (1994), the following singular integral equations with $P_l(x_l)$ and $Q_l(x_l)$ being expressed in term of continuously distributed dislocations (Lu and Lardner,1992) could be deduced

$$P_0(\eta) + Re\left\{\sum_{l=1}^{N} \int_{-a_l}^{a_l} [H_5^\star + H_6^\star + H_7^\star + H_8^\star] d\xi_l\right\} = 0, \quad (-\infty < \eta < 0) ,$$

$$Q_0(\eta) + Im\left\{\sum_{l=1}^{N} \int_{-a_l}^{a_l} [H_5^\star + H_6^\star + H_7^\star + H_8^\star] d\xi_l\right\} = 0, \quad (-\infty < \eta < 0) ,$$

$$\int_{-1}^{1} P_0(\eta) f_{nn,0l}^0(\eta,\eta_l) d\eta + \int_{-1}^{1} Q_0(\eta) f_{tn,0l}^0(\eta,\eta_l) d\eta$$

$$+Re\left\{\int_{-a_l}^{a_l} [H_1^\star + H_2^\star + H_3^\star + H_4^\star] d\xi_l\right\}$$

$$+Re \sum_{\substack{i=1 \\ i \neq l}}^{N} \int_{-a_i}^{a_i} [H_5^{\star\star} + H_6^{\star\star} + H_7^{\star\star} + H_8^{\star\star}] d\xi_i = p_l(\eta_l),$$

$$\int_{-1}^{1} P_0(\eta) f_{tn,0l}^0(\eta,\eta_l) d\eta + \int_{-1}^{1} Q_0(\eta) f_{tt,0l}^0(\eta,\eta_l) d\eta$$

$$+Im \int_{-a_l}^{a_l} [H_1^\star + H_2^\star + H_3^\star + H_4^\star] d\xi_l$$

$$+Im \sum_{\substack{i=1 \\ i \neq l}}^{N} \int_{-a_i}^{a_i} [H_5^{\star\star} + H_6^{\star\star} + H_7^{\star\star} + H_8^{\star\star}] d\xi_i = q_l(\eta_l),$$

$$(-a_l < \xi_l, \eta_l < a_l), \tag{5.29}$$

where the quantities with over stars are related to equation (5.26) by

$$\begin{aligned} H_5^\star &= H_5(\xi_l, \beta_l, x_{0l}, y_{0l}, \eta, 0, 0, 0), \\ H_6^\star &= H_6(\xi_l, \beta_l, x_{0l}, y_{0l}, \eta, 0, 0, 0), \\ H_7^\star &= H_7(\xi_l, \beta_l, x_{0l}, y_{0l}, \eta, 0, 0, 0), \\ H_8^\star &= H_8(\xi_l, \beta_l, x_{0l}, y_{0l}, \eta, 0, 0, 0), \end{aligned}$$

$$\begin{aligned} H_1^\star &= H_1(\xi_l, \beta_l, \eta_l), \\ H_2^\star &= H_2(\xi_l, \beta_l, \eta_l), \\ H_3^\star &= H_3(\xi_l, \beta_l, \eta_l, y_{0l}), \\ H_4^\star &= H_4(\xi_l, \beta_l, \eta_l, y_{0l}), \end{aligned}$$

$$\begin{aligned} H_5^{\star\star} &= H_5(\xi_i,\beta_i,x_{0i},y_{0i},\eta_l,x_{0l},y_{0l},\beta_l) \\ H_6^{\star\star} &= H_6(\xi_i,\beta_i,x_{0i},y_{0i},\eta_l,x_{0l},y_{0l},\beta_l) \\ H_7^{\star\star} &= H_7(\xi_i,\beta_i,x_{0i},y_{0i},\eta_l,x_{0l},y_{0l},\beta_l) \\ H_8^{\star\star} &= H_8(\xi_i,\beta_i,x_{0i},y_{0i},\eta_l,x_{0l},y_{0l},\beta_l) \end{aligned}$$

In addition, the unique-value conditions of displacements require that

$$\int_{-a_l}^{a_l} B_x(\xi_l) \quad d\xi_l = 0 \quad , \int_{-a_l}^{a_l} B_y(\xi_l) \quad d\xi_l = 0 \quad , \tag{5.30}$$

where the kernel functions $f^0_{nn,0k}$, $f^0_{tn,0k}$, $f^0_{nt,0k}$ and $f^0_{tt,0k}$ are given by equation (5.18), respectively, the subscripts n, t, 0, l indicate the normal quantities, the tangential quantities, the macrocrack and the lth microcrack, respectively; H_1 , H_2, H_3, H_4 , H_5 , H_6, H_7 , and H_8 are given by equation (5.26), $B_x(\xi)$ and $B_y(\xi)$ are dislocation density functions; $p_l(\eta_l)$ and $q_l(\eta_l)$ are given by equation (5.28).

2.2 Solution of the integral equations

Obviously equations (5.30) cannot be solved directly by using a normal numerical integration technique since infinite integral intervals are concerned. In order to reduce the range of the singular integral equations (5.30) to finite intervals the following variable changes are taken (Hutchinson *et al.*, 1987):

$$\begin{aligned} \eta &= \frac{u-1}{u+1}, \\ d\eta &= \frac{2du}{(u+1)^2}, \end{aligned} \tag{5.31}$$

After doing so, the above singular integral equation (5.30) become:

$$P_0(u) + Re\left\{ \sum_{l=1}^{N} \int_{-a_l}^{a_l} [H_5^\star + H_7^\star + H_6^\star + H_8^\star] d\xi_l \right\} = 0, \quad (-1 < u < 1),$$

$$Q_0(u) + Im\left\{ \sum_{l=1}^{N} \int_{-a_l}^{a_l} [H_5^\star + H_7^\star + H_6^\star + H_8^\star] d\xi_l \right\} = 0, \quad (-1 < u < 1),$$

$$\int_{-1}^{1} P_0(u) f^0_{nn,0l}(u,\eta_l)\frac{2}{(u+1)^2}du + \int_{-1}^{1} Q_0(u) f^0_{tn,0l}(u,\eta_l)\frac{2}{(u+1)^2}du$$

$$+Re\left\{ \int_{-a_l}^{a_l} [H_1^\star + H_2^\star + H_3^\star + H_4^\star] d\xi_l \right\}$$

$$+Re\left\{ \sum_{i=1, i\neq l}^{N} \int_{-a_i}^{a_i} [H_5^{\star\star} + H_6^{\star\star} + H_7^{\star\star} + H_8^{\star\star}] d\xi_i \right\} = p_l(\eta_l),$$

$$\int_{-1}^{1} P_0(u) f^0_{tn,0l}(u,\eta_l)\frac{2}{(u+1)^2}du + \int_{-1}^{1} Q_0(u) f^0_{tt,0l}(u,\eta_l)\frac{2}{(u+1)^2}du$$

$$+Im\left\{ \int_{-a_l}^{a_l} [H_1^\star + H_2^\star + H_3^\star + H_4^\star] d\xi_l \right\}$$

$$+Im\left\{ \sum_{i=1, i\neq l}^{N} \int_{-a_i}^{a_i} [H_5^{\star\star} + H_6^{\star\star} + H_7^{\star\star} + H_8^{\star\star}] d\xi_i \right\} = q_l(\eta_l). \tag{5.32}$$

Using the Chebyshev numerical integration, equation (5.32) can be solved and $B(\xi)$ can be expressed by the first kind of the Chebyshev polynomial as follows (see Chen and Tien, 2000)

$$B_x(\xi_l) = \frac{1}{\sqrt{1-\eta_l^2}} \sum_{k=0}^{N} d_k T_k(\eta_l), \tag{5.33}$$

$$B_y(\xi_l) = \frac{1}{\sqrt{1-\eta_l^2}} \sum_{k=0}^{N} b_k T_k(\eta_l), \tag{5.34}$$

where

$$-1 < \eta_l = \frac{\xi_l}{a_l} < 1, \qquad -a_l \leq \xi_l \leq a_l,$$

and d_k, b_k are coefficients of the Chebyshev polynomial needed to be determined.

Once equation (5.32) is solved the incremental values of the *SIF*'s at the macrocrack tip as influenced by microcracking and the *SIF*'s at both tips of the microcrack can be evaluated.

The stress intensity factor defined by Suo(1990) for the semi-infinite interface crack under arbitrarily distributed loads $P_0(x)$ and $Q_0(x)$ is given as

$$K_1 + iK_2 \;=\; -\left(\frac{2}{\pi}\right)^{\frac{1}{2}} \cosh \pi\varepsilon \int_{-\infty}^{0} (-x)^{-\frac{1}{2}-i\varepsilon} t_{01}(x)dx,$$

$$t_{01}(x) = \frac{\bar{\mathbf{w}}^T \mathbf{H}[Q_0(x), P_0(x)]}{\bar{\mathbf{w}}^T \mathbf{H} \mathbf{w}}, \tag{5.35}$$

where the matrices **W** and **H** are given in (5.16), with the solutions of those in (5.32), K can be calculated by using the Chebyshev numerical integration method.

The stress intensity factor for the lth finite sub-interface microcrack is given as

$$K = K_1 + iK_2 = \frac{(2\pi)^{\frac{3}{2}}}{\sqrt{a}} \sum_{k=1}^{2} [(1 - i\mu_k) B_k(l) + (1 - i\bar{\mu}_k) \bar{B}_k(l)], \tag{5.36}$$

where

$$B_1(l) = \frac{1}{2\pi i} \frac{1}{s_{11}} \frac{\mu_1 \sum_{j=1}^{N} d_j - \sum_{j=1}^{N} b_j}{(\mu_1 - \mu_2)(\mu_1 - \bar{\mu}_1)(\mu_1 - \bar{\mu}_2)},$$

$$B_2(l) = \frac{1}{2\pi i} \frac{1}{s_{11}} \frac{\mu_2 \sum_{j=1}^{N} d_j - \sum_{j=1}^{N} b_j}{(\mu_2 - \mu_1)(\mu_2 - \bar{\mu}_1)(\mu_2 - \bar{\mu}_2)}, \tag{5.37}$$

for the right crack tip with the numerical solutions of the coefficients d_j, b_j, $(j = 0, N)$ given in equations (5.32), (5.33), and (5.34).

3. Analysis of the J integral

3.1 Conservation law of the J integral

The conservation law of the J integral discussed in the above chapters are re-examined in this section for the macrocrack microcrack interaction problem in dissimilar anisotropic materials. Consider the original problem with three integral contours as shown in Figure 5.5 [see Tian and Chen, 2000(a)], it is easy to obtain the following relation between the three values of the J integral along the three integral contours in the global (x, y) coordinate system:

$$J_t + \Delta J = J_\infty,$$

$$\frac{J_t}{J_\infty} + \frac{\Delta J}{J_\infty} = 1. \tag{5.38}$$

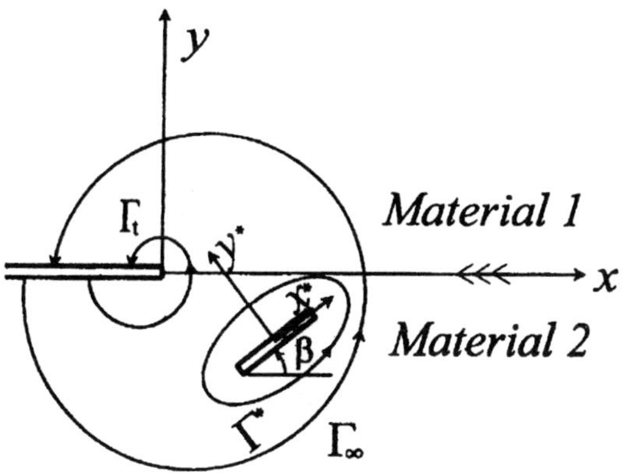

Figure 5.5. Three closed contours for the J integral analysis.

Although the contours Γ_∞ and Γ_t cut the interface at quite different points as shown in Figure 5.5, it has been proved that the segments cut by the two contours have no contributions to the J integral (see the proof in Chapter 2 or Zhao and Chen, 1997). Here J_∞ is the contribution from the remote stress field parameterized by K_1^∞ and K_2^∞, J_t is the contribution from the macrocrack tip, and ΔJ is the contribution from the existing microcracks, and

$$\begin{aligned} J_\infty &= \int_{\Gamma_\infty} (wdy - T_i u_{i,1} dl), \\ J_t &= \int_{\Gamma_t} (wdy - T_i u_{i,1} dl), \\ \Delta J &= \oint_{\Gamma^*} (wdy - T_i u_{i,1} dl), \end{aligned} \tag{5.39}$$

and

$$\Delta J = J_1^* \cos\beta - J_2^* \sin\beta, \tag{5.40}$$

where β is the orientation angle measured from x-axis (see Figure 5.5) to x^*-axis, the star denotes the local coordinate system, and the two components of the J_k^* vector, denoted by J_1^* and J_2^* along the closed integral contour Γ^* are given as

$$J_1^* = \oint_{\Gamma^*} (w^* dy^* - T_i^* u_{i,1}^* dl),$$

$$J_2^* = \oint_{\Gamma^*} (-w^* dx^* - T_i^* u_{i,1}^* dl), \tag{5.41}$$

When the Hermitian matrix **H** is a purely real matrix corresponding to the homogeneous cases or the dissimilar orthotropic cases (see Figure 5.1), the calculations of J_∞, J_t and J_1^* in the above equations can be expressed in the following forms (Suo, 1990):

$$J_\infty = \bar{\mathbf{w}}^T(\mathbf{H}+\bar{\mathbf{H}})\mathbf{w}\,|K_1^\infty + iK_2^\infty|^2 \big/ (4\cosh^2 \pi\varepsilon),$$

$$J_t = \bar{\mathbf{w}}^T(\mathbf{H}+\bar{\mathbf{H}})\mathbf{w}\,|K_1^\infty + \Delta K_1 + i(K_2^\infty + \Delta K_2)|^2 \big/ (4\cosh^2 \pi\varepsilon), \tag{5.42}$$

$$J_1^* = \frac{1}{4}\mathbf{K}^{R^T}\mathbf{Y}^*\mathbf{K}^R - \frac{1}{4}\mathbf{K}^{L^T}\mathbf{Y}^*\mathbf{K}^L, \tag{5.43}$$

where $\mathbf{Y}^* = 2Re(i\mathbf{A}^*(\mathbf{L}^*)^{-1})$, $\mathbf{L}^*$ and $\mathbf{A}^*$ are given in equations (5.2) and (5.3), $\mathbf{K}^R = [K_2^R, K_1^R]^T$ and $\mathbf{K}^L = [K_2^L, K_1^L]^T$ denote the stress intensity factors, and the superscripts R and L refer to the right and left tips of the microcrack, respectively.

3.2 Calculation of the J_2 integral

However, the calculation of J_2^* should be specially considered since it consists of two parts, one of which is induced from the *SIF's* at both microcracks tips and the other from the microcrack traction free faces (Herrmann and Herrmann, 1981). Chen and Ma (1997) have proved that the contribution of the right crack tip to the J_2^* integral, namely J_2^{tR}, should be equal to the energy release rate specially proposed as follows

$$J_2^{tR} = \lim_{\Delta\to 0} \frac{-1}{2\Delta} \int_0^\Delta \sigma_{i1}^*(\Delta - r)u_i^*(r)dr, \tag{5.44}$$

where $\sigma_{i1}(\Delta - r)$ and $u_i(r)$ is the stress at a distance Δ ahead the right crack tip and the displacement jump at a distance r behind the right crack tip, respectively:

$$\sigma_{i1}^*(r) = \mathbf{X}^*\mathbf{K}^R(2\pi r)^{-\frac{1}{2}},$$
$$u_i^*(r) = \mathbf{Y}^*\mathbf{K}^R(2r/\pi)^{\frac{1}{2}}, \tag{5.45}$$

in which the superscript R refers to the right tip of the microcrack, and the matrices

$$\mathbf{X}^* = -Re(\mathbf{D}^*\mathbf{L}^{*-1}),$$
$$\mathbf{D}^* = \begin{bmatrix} L_{11}^*\mu_1^* & L_{12}^*\mu_2^* \\ L_{21}^*\mu_1^* & L_{22}^*\mu_2^* \end{bmatrix}. \tag{5.46}$$

Substituting equation (5.45) into equation (5.44), the contribution of the right crack tip to the J_2^* integral is obtained explicitly as:

$$J_2^{tR} = \frac{1}{4}(\mathbf{K}^R)^T\mathbf{X}^*\mathbf{Y}^*\mathbf{K}^R, \tag{5.47}$$

Similarly the contribution of the left tip to the J_2^* integral, namely J_2^{tL}, can be given as

$$J_2^{tL} = -\frac{1}{4}(\mathbf{K}^L)^T\mathbf{X}^*\mathbf{Y}^*\mathbf{K}^L. \tag{5.48}$$

The remaining task is to account the contribution of the crack faces to the J_2^* integral as Herrmann and Herrmann (1981) mentioned, which can be derived as:

$$J_2^f = \int_{-a}^{a}(w^+ - w^-)dx, \tag{5.49}$$

where w^+ and w^- are the boundary values of the strain energy densities on the upper and lower crack faces, respectively, which can be evaluated after the interaction problem is solved.

Thus the J_2^* integral calculating along the contour Γ surrounding the complete crack are:

$$J_2^* = \frac{1}{4}\mathbf{K}^{R^T}\mathbf{X}^*\mathbf{Y}^*\mathbf{K}^R - \frac{1}{4}\mathbf{K}^{L^T}\mathbf{X}^*\mathbf{Y}^*\mathbf{K}^L + \int_{-a}^{a}(w^+ - w^-)dx^*. \tag{5.50}$$

Using the superimposing technique, the above results can be extended to the multiple sub-interface microcracks situation without any difficulty.

Indeed, equation (5.38) provides a necessary condition with which the numerical results should meet, no matter how they are derived by using what kind of technique.

3.3 Multiple microcracks situation

In addition, using the superimposing technique the above results can be extended to the multiple sub-interface microcracks situation. Assuming that there exist N sub-interface microcracks enclosed by the closed contour Γ, the conservation law or the consistent relation (5.38) becomes (Chen, 1996)

$$\frac{J_t}{J_\infty} + \sum_{l=1}^{N} \frac{\Delta J_l}{J_\infty} = 1, \tag{5.51}$$

Equation (5.51) shows the inherent law of the J integral whose values are induced from the sub-interface microcracks ($\sum \Delta J/J_\infty$) and the tip of the interface macrocrack (J_t/J_∞), respectively. This has already been proved to be valid in brittle solids and in bimaterial isotropic solids as discussed in chapters 3 and 4 (Chen, 1996, Zhao and Chen 1997). Moreover, it provides a necessary condition as well as a powerful tool to examine the numerical results.

4. Numerical results and consistency check

4.1 Composite material properties

In this section a unidirectional fiber reinforced composite with the fibre direction along the interface is taken to be the upper half material (0 ply). The material constants are listed below:$E_1 = 5.50 \times 10^6$ psi, $E_2 = E_3 = 1.33 \times 10^6$ psi, $G_{21} = 0.50 \times 10^6$ psi, $\nu_{12} = 0.28$, $\nu_{23} = 0.36$. Quite contrary, the lower half material is taken to be the same material but its fibre direction is perpendicular to the plane under consideration (the degenerated isotropic case, *i.e.*, $E_1 = E_2 = 1.33 \times 10^6$ psi). Therefore, matrix microcracks are more likely to be formed in the lower half material (90 ply) near the tip of a macro-interface crack (see, *e.g.*, Figure 5.2).

4.2 Crack interaction configuration and numerical results

Assume a semi-infinite interface macrocrack interacted with a sub-interface matrix microcrack of length $2a$ in the near-tip process zone as shown in Figure 5.6. Here r is the distance between the macrocrack tip and the center of the sub-interface microcrack, α is the angle between the segment r and the x-axis, and β is the oriented angle of the sub-interface microcrack with respect to the x-axis.

The remote stress field is specified by the intensity K_1^∞ only because in this material combination K_2^∞ is extremely small. Let $\beta = 30$ degrees and 45 degrees, respectively, and $r/a = 2.0$, the computed values of J_t/J_∞ and $\Delta J/J_\infty$ against the location angle α are shown in Figure 5.7.

From Figure 5.6 it is found that the consistent relation of the J integral (5.38) is really satisfied, *i.e.*, the computed values of J_t/J_∞ plus the computed

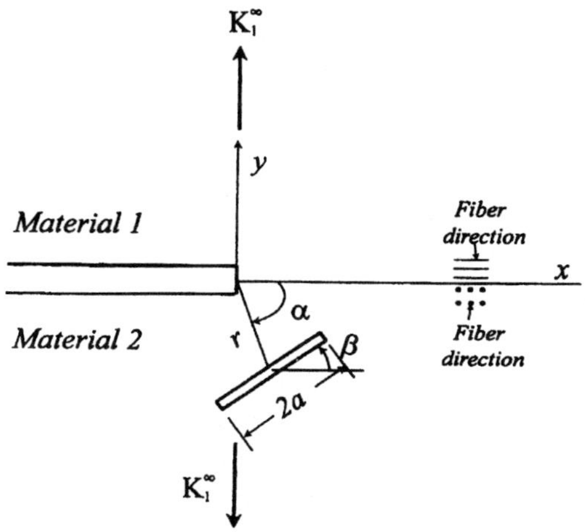

Figure 5.6. Crack configuration for a dissimilar anisotropic material.

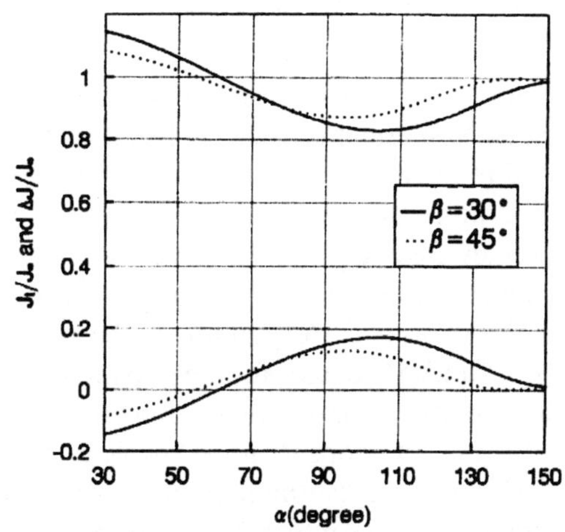

Figure 5.7. Results of J_t/J_∞ and $\Delta J/J_\infty$ vs. the angle α.

values of $\Delta J/J_\infty$ always equal to unity. This confirms that the superimposing method proposed in this chapter is effective for solving the interaction problem in dissimilar anisotropic materials, although the different mismatch nature from those by Zhao and Chen (1997) is taken into account. Moreover, this relation actually provides a powerful tool to check the numerical results no matter how they are derived by whatever kind of technique.

It is also shown that the microcrack has the anti-shielding effect on J_t for smaller values of the location angle ($\alpha < 60^0$ for $\beta = 30^0$, and $\alpha < 54^0$ for $\beta = 45^0$) corresponding to the normalized value of $J_t/J_\infty > 1$, whilst the microcrack has the shielding effect on J_t for larger values of the location angle corresponding to the value of $J_t/J_\infty < 1$. Moreover, there is only single neutral shielding angle for each curve at which a transform from the anti-shielding effect to shielding effect occurs.

Again, this consistent relation reveals that the remote J integral has been redistributed owing to the existence of microcracks in the near-tip stress field. From the physical point of view, this redistribution nature of the remote J integral can be used to explain the microcrack shielding effect in dissimilar anisotropic materials. As pointed out by Hutchinson (1987), there are two sources of the redistribution of stress in the near-tip stress field of a macrocrack induced from microcracks. One is owed to the reduction in the effective elastic moduli and the other is the strain arising from the release of residual stresses. Both sources are energy dissipation processes. Obviously it is this redistribution of stress in the near-tip stress field that does lead to the redistribution of the remote J integral and in turn yields the phenomenological shielding effect.

4.3 In homogeneous anisotropic cases

Setting the upper and lower materials to be the same composite with the same fiber direction along the semi-infinite crack surfaces (the x-axis) yields the homogeneous cases. Numerical results are listed in Table 5.1, where two microcracks are located in the near-tip process zone. Two examples are given. One of them is corresponding to the two microcracks case in the near-tip process zone. Their location angles are chosen arbitrarily to be, for example, 45 and 90 degrees, respectively, and their oriented angles are arbitrarily, say, 135 and 120 degrees, respectively. The other is corresponding to the case where the location angles are chosen to be 45 and 90 degrees for the two microcracks, respectively, and the oriented angles are 30 and 45 degrees, respectively (see Figure 5.8). It is shown from Table 5.1 that the conservation law or consistent check (5.51) is really confirmed. Moreover, these numerical results confirm the explicit formulations for calculating the J_2 integral at each microcrack tip as well as that on the traction-free surfaces. It is seen that in all the macrocrack microcrack configurations the normalized contribution induced from the traction-free surfaces, *i.e.*, F_2^*, is always very small comparing with those induced from microcrack tips. Therefore, for the sake of convenience, the contribution could be neglected.

4.4 Different dissimilar materials combinations

In order to give a fundamental understanding of the interaction behaviors, numerical results for different dissimilar material combinations are studied and

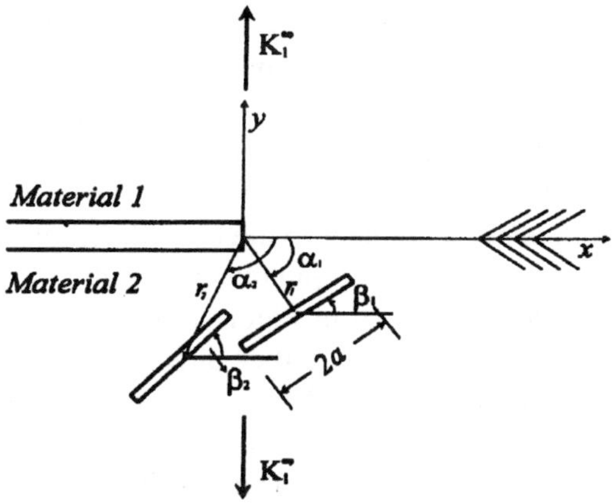

Figure 5.8. Two microcracks in the near-tip process zone of a macrocrack.

Table 5.1. Two microcracks interacting with a macrocrack tip in homogeneous anisotropic case.

Location angles $\left\{\begin{array}{c}\alpha_1 = 45^0 \\ \alpha_2 = 90^0\end{array}\right\}$	Orientation angles $\left\{\begin{array}{c}\beta_1 = 135^0 \\ \beta_2 = 120^0\end{array}\right\}$	Orientation angles $\left\{\begin{array}{c}\beta_1 = 30^0 \\ \beta_2 = 45^0\end{array}\right\}$
K_I^t/K_I^∞ Macrocrack	0.8161	0.4074
K_{II}^t/K_I^∞ Macrocrack	−0.2704	−0.2986
K_I^R/K_I^∞ first microcrack	0.1516	0.7566
K_{II}^R/K_I^∞ first microcrack	0.0106	0.1180
K_I^L/K_I^∞ first microcrack	1.2426	0.5612
K_{II}^L/K_I^∞ first microcrack	0.4363	−0.1202
F_2^*/J_∞ first microcrack	−0.0078	0.0164
K_I^R/K_I^∞ second microcrack	0.4048	0.7070
K_{II}^R/K_I^∞ second microcrack	0.0578	0.0854
K_I^L/K_I^∞ second microcrack	0.6217	1.2931
K_{II}^L/K_I^∞ second microcrack	0.0824	−0.1742
F_2^*/J_∞ second microcrack	0.0056	0.0887
$J_t/J_\infty =$	0.8391	0.1660
$\Delta J/J_\infty =$	0.1608	0.8340
$(J_t + \Delta J)/J_\infty =$	0.9999	1.0000

discussed in detail in this section. Here, three kinds of material combinations are considered. The first is associated with the dissimilar material combination of $0/90$ type as just mentioned in Figure 5.6 (labelled as Case I in Figure 5.9(a)). The second is associated with another kind of material combination (labelled as Case II in Figure 5.9(b)) whose upper material is also the unidirectional fiber reinforced composite mentioned above, but its fiber direction is perpendicular to the interface and whose lower material is the same composite and its fiber direction is perpendicular to the considered plane. The third is associated with a homogeneous anisotropic material with fiber direction parallel to the interface (labelled as Case III in Figure 5.9(c)) whose upper material is a fiber reinforced composite with the fiber direction parallel to the interface corresponding to an orthotropic case. Numerical results of the *shielding ratio* $(K_I^\infty + \Delta K_I^t)/K_I^\infty$ against the local angle of a parallel microcrack $(\beta = 0)$ are shown in Figures 5.10(a,b,c), where the normalized distance r/a is, respectively, taken to be 2.0, 2.2, and 2.5. It is found that the variable tendencies of the dotted curves, the imaginary curves, and the real curves corresponding to the three kinds of material combinations labelled as Cases I, II, and III, are similar, however, the changes in magnitude among them are remarkable. Particularly, the dissimilar composite materials shift the maximum anti-shielding angle, the shielding angle, and the neutral shielding angle, respectively. In other words, the values of the maximum anti-shielding angles and the maximum shielding angles in Cases I and II are always larger than those in Case III corresponding to a homogeneous material. However, the neutral shielding angles in Cases I and II are always little less than those in Case III.

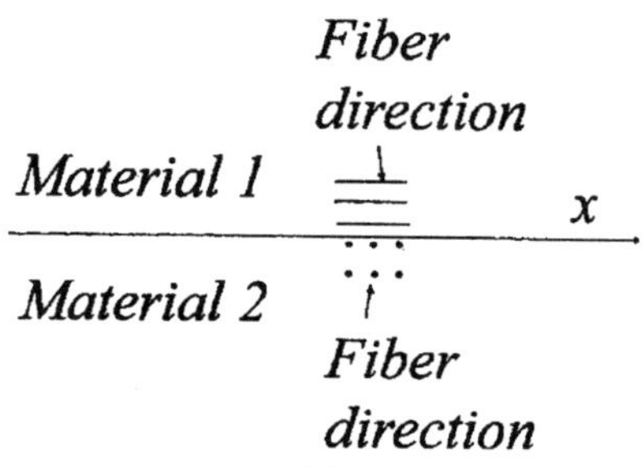

(a)

Numerical results of the *shielding ratio* against the orientation angle β of the microcrack are shown in Figures 5.11(a,b,c), where the location angle α is taken to be 60, 90, and 120 degrees, respectively. It is found that the variable tendencies of the dotted curves (Case I), the imaginary curves (Case II), and the real curves (Case III), are also similar. However, in most range of β, the divergences in magnitude of the *ratio* in Cases I and II from those in Case

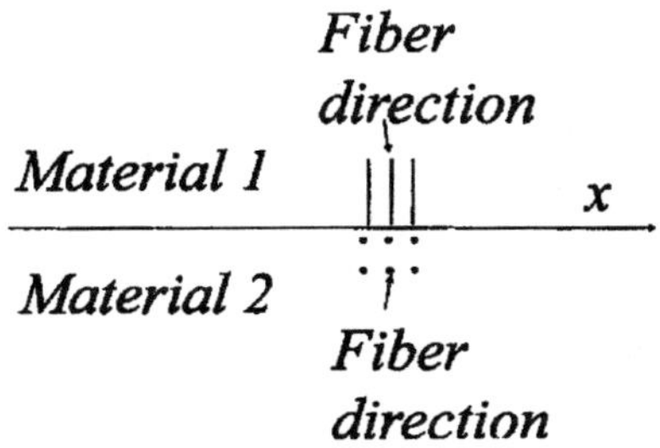

(b)

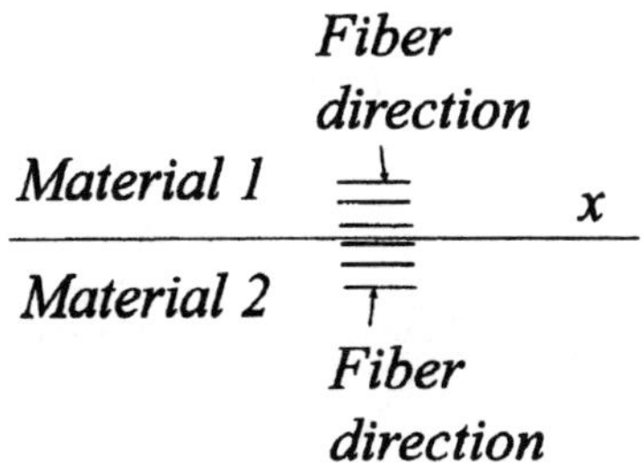

(c)

Figure 5.9. Three kinds of material combinations.

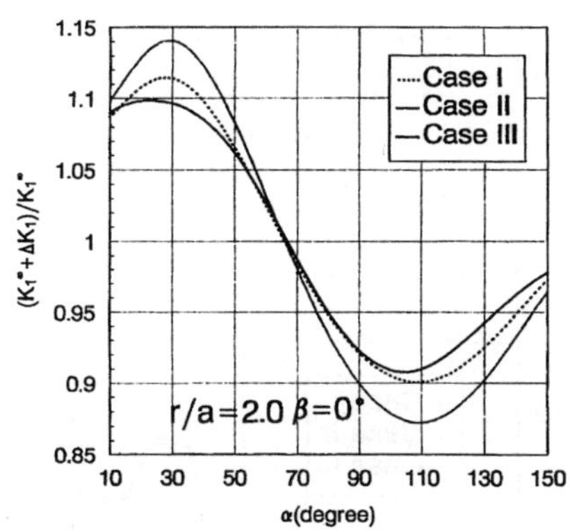

(a) $r/a = 2.0$.

III are remarkable. It is seen from Figure 5.11(a) or (b) that, the homogeneous orthotropic material corresponding to Case III does always lead to smaller values of the *ratio* than those for Cases I and II when the maximum anti-shielding effect occurs. For example, in Figure 5.11(a), the values of the *ratio* at the maximum anti-shielding angles are 1.185 in Case I, 1.235 in Case II, and 1.180 in Case III.

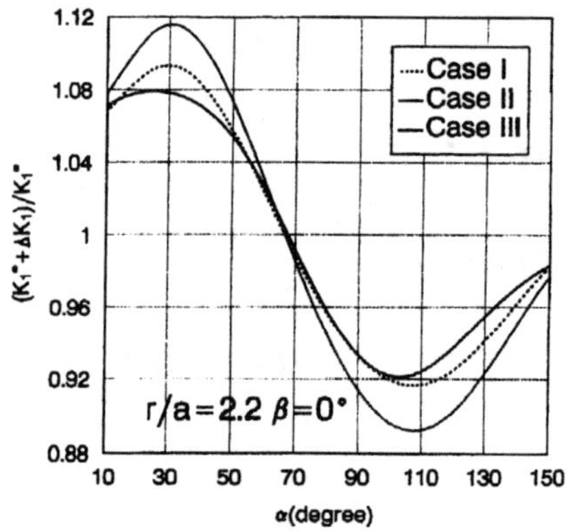

(b) $r/a = 2.2$.

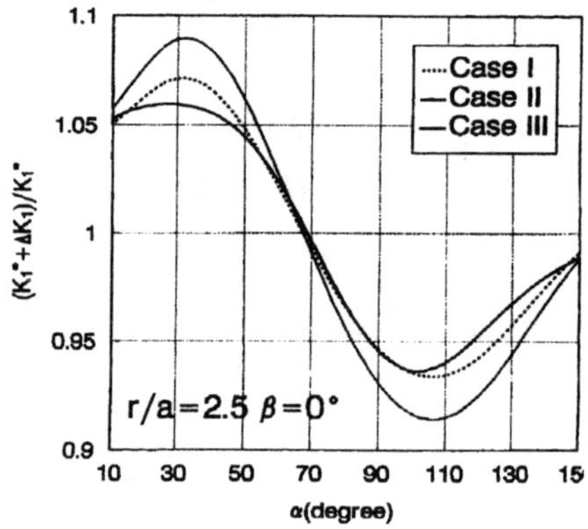

(c) $r/a = 2.5$.

Figure 5.10. Shielding ratio vs. α for a parallel microcrack.

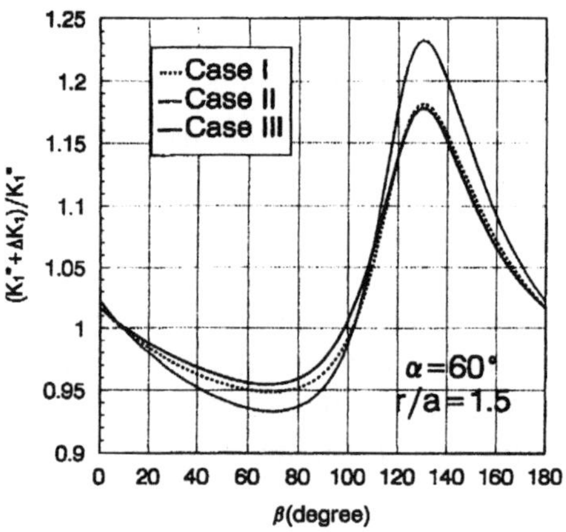

(a) $\alpha = 60^0$.

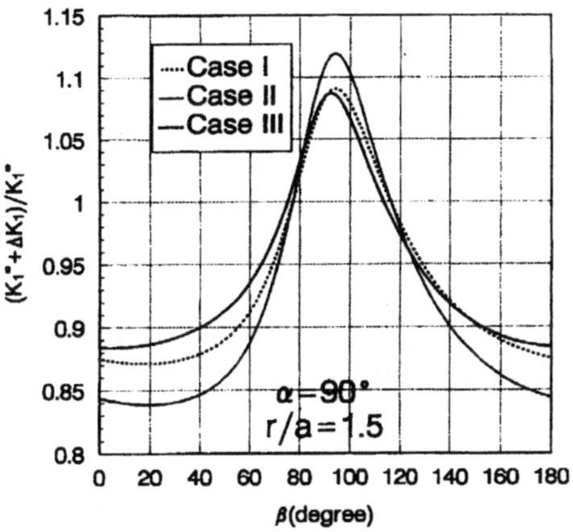

(b) $\alpha = 90^0$.

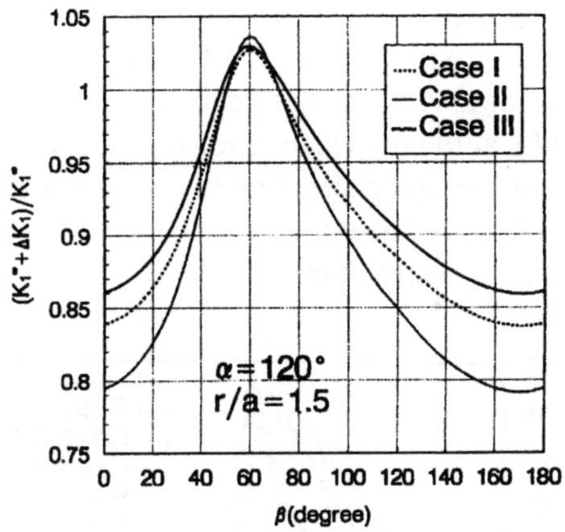

(c) $\alpha = 120^0$.

Figure 5.11. Shielding ratio vs. the angle β for a microcrack.

However, such a behavior is not a general phenomenon for every case. For example, the above phenomenon could not be seen in Figure 5.11(c), where the values of the maximum anti-shielding effect corresponding to Case III is obviously greater than that corresponding to Case I, although it is smaller that that corresponding to Case II. In addition, it could be seen that the maximum anti-shielding angles in Figure 5.11(b) are 95 degrees for Cases I and II, and 92 degrees for Case III. Thus, as concluded above, the material mismatch nature on the interface induced from dissimilar materials shown in Figures 5.9(a) and (b) shifts the maximum anti-shielding angle and in turn shifts the anti-shielding region below the interface.

5. The T stress effect

The T stress effect on the *shielding ratio* in dissimilar anisotropic materials is discussed in this section. The biaxial parameter B as defined by Leevers and Radon (1983) and as formulated in equation (3.56) of Chapter 3 is used. The crack interaction configuration under biaxial loading conditions is no longer repeated, keeping in mind that the remote T stress only influences the right loading terms in the integral equations (5.32), whilst the left side of (5.32) remains unaltered.

Computed values of the *shielding ratio* vs. the oriented angle β as influenced by the T stress are plotted in Figure 5.12(a,b) for the three material combinations labelled as Cases I, II, and III. In Figure 5.12(a), the T stress is taken to be negative values, *i.e.*, $B = -0.5$ and -1.0, respectively. Whereas in Figure 5.12(b), the T stress is positive, *i.e.*, $B = 0.5$ and 1.0, respectively.

It is found that for the parallel microcrack corresponding to $\beta = 0$ and $\beta = 180$ degrees the remote T stress has little influence on the *shielding ratio* (*i.e.*, the normalized *SIF* at the interface crack tip) although the Modes I and II fracture are coupled in the near-tip stress field owing to the mismatch nature of dissimilar materials. It is found also that the T stress has significant influence on the *ratio* for an oriented microcrack.

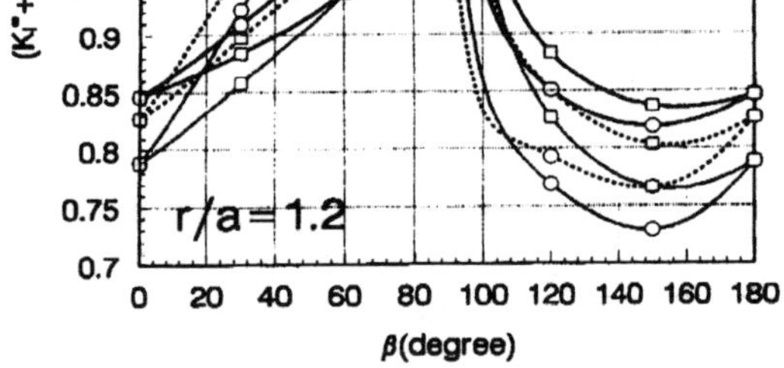

(a) Remote negative T stress.

What is more, the influence is sensitive to the material combinations although the variable tendencies of the *shielding ratio* as influenced in the three

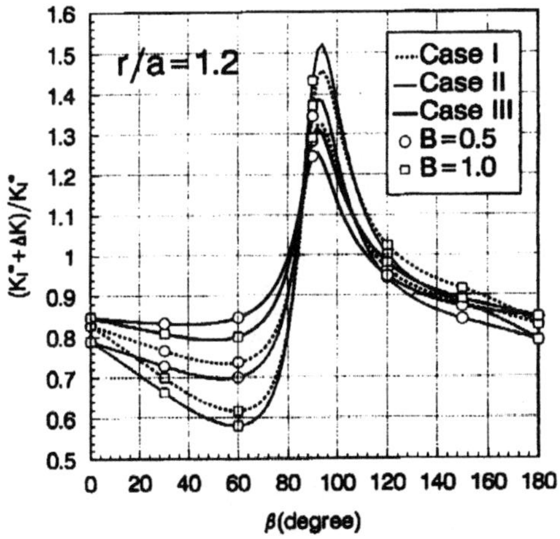

(b) Remote positive T stress.

Figure 5.12. Shielding ratio vs. the oriented angle β as influenced by the remote T stress.

cases are similar. The transformation angle β_T from which the T stress yields opposite influences on the *ratio* is found to be about 87 degrees no matter how the material combinations are considered and whatever the remote T stress is negative corresponding to $B < 0$ or positive corresponding to $B > 0$. It is also seen that the influence of the material combinations on the *shielding ratio* in magnitude for a certain value of T^∞ is remarkable. For example, it is seen in Figure 5.12(a) that the maximum value of the *ratio* is about 1.16 for Case I at β= 80 degrees, 1.17 for Case II at β= 84 degrees, 1.125 for Case III at β= 88 degrees, when taking $B = -1.0$, the minimum value of *shielding ratio* is about 0.765 for Case I at β= 150 degrees, 0.73 for Case II at β= 150 degrees, and 0.82 for Case III at β= 150 degrees.

It is concluded that the divergences of the influence of T^∞ on the *shielding ratio* for the two dissimilar cases from those for the orthotropic case are remarkable, but the divergences of Case I from Case II are smaller. Similar conclusions could also be seen in Figure 5.12(b) although T^∞ is positive.

The influence of T^∞ on the redistribution of the J integral is plotted in Figure 5.13 for Case I only. It is found that although T^∞ influences the redistribution of the integral significantly the conservation law (5.38) is still valid whatever the values of T^∞ are positive or negative. This conclusion is identical to those discussed in Chapter 3 for homogeneous isotropic materials and in Chapter 4 for dissimilar isotropic materials (also see, Zhao and Chen, 1998).

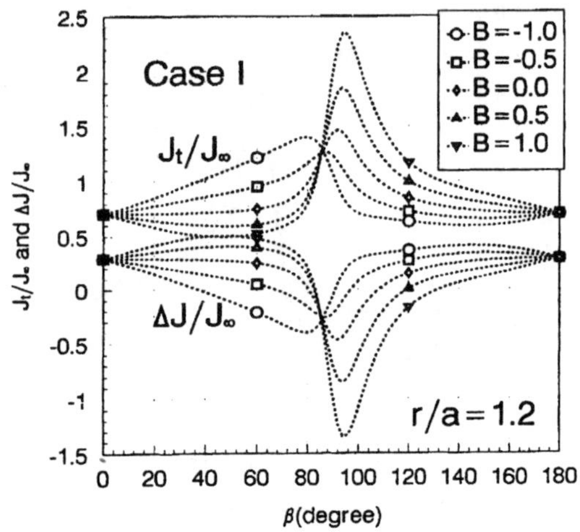

Figure 5.13. Influence of the T stress on the redistribution of the J integral.

6. Brief Summary

From the forgoing discussions, one can summarize as follows:

(A) The conservation law of the J integral found in macrocrack microcrack interaction problems in brittle solids and in dissimilar isotropic materials is still valid in the present dissimilar anisotropic materials, although the material mismatch nature influences the near-tip stress field significantly. This law provides a useful consistency check to examine the numerical results.

(B) The microcrack shielding effect in dissimilar anisotropic materials could also be considered, from the phenomenological point of view, as the result of the J integral redistribution since the redistribution of the stresses induced from the two sources owing to microcracking in the near-tip field does lead to the redistribution of the J integral.

(C) The variable tendencies of the interaction effect in dissimilar anisotropic materials are similar to those in homogeneous orthotropic materials with the stiffer axis parallel to the macrocrack. This means that the material mismatch nature at the interface crack tip has little influence on the variable tendencies when the remote loading is preferred by the nearly Mode I K_I^∞. However, the divergences of the effect in magnitude from those in the orthotropic materials are remarkable.

(D) The material mismatch nature on the interface shifts the maximum anti-shielding effect angle, the maximum shielding angle, and the neutral shielding angle, at which the maximum anti-shielding effect occurs, the maximum shielding effect occurs, and a transformation from the former effect to the latter effect occurs.

(E) The T stress does bring great influences on the interaction behaviors in present problems. However, the T stress only alters the values of the J integral at the interface crack tip and the J_k vector on the microcracks, whilst the conservation law of the J integral (5.38) or (5.51) is always unchanged.

References for Chapter 5

Blimenauer, H., Dehne, F., Michel, G., Mook, G., and Pohl, J., (1999) Nondestructive damage evaluation of adaptive materials, *International Journal of Materials and Product Technology (Geneva)*, **14**, 291 – 303.

Chen, Y. H. and Hasebe, N. (1994a) Further investigation of Comninou's EEF for an interface crack with completely closed faces, *International Journal of Engineering Science*, **32**, 1037 – 1046.

Chen, Y. Z. and Hasebe, N. (1994b) Eigenfunction expansion and higher order weight function of interface cracks, *ASME Journal of Applied Mechanics*, **61**, 843 – 849.

Chen, Y.H. and Tian, W.Y. (2000) A semi-infinite interface crack interacting with sub-interface matrix cracks in dissimilar anisotropic materials, Part II. Numerical results and discussions, *International Journal of Solids and Structures*, **37**, 7731 – 7742.

Chen, Y.Z. (1984) General case of multiple crack problems in an infinite plate, *Engineering Fracture Mechanics*, **20**, 591 – 597.

Chen, Y. H. (1996) On the contribution of discontinuities in a near-tip stress field to J integral, *International Journal of Engineering Science*, **34**, 819 – 829.

Chen, Y. H. and Ma, H. (1997) Explicit function expression of J_2 in anisotropic body, *Science in China, E*, **27**, 310 – 317.

Crouch, S.L. and Starfield, A.M. (1983) *Boundary Element Methods in Solids Mechanics,* George Allen & Unwin, London.

Erdogen, F. (1978) *Mixed Boundary Value Problems* , Mechanics Today, **4**, Pergamen Press, Oxford.

Gao, H. (1991) Weight function analysis of interface crack, *ASME Journal of Applied Mechanics*, **58**, 931 – 938.

Herrmann, A. G. and Herrmann, G. (1981) On energy release rates for a plane cracks, *ASME Journal of Applied Mechanics*, **48**, 525 – 528.

Horii, H. and Nemat Nasser, S. (1985) Elastic field of interacting inhomogeneities, *Journal of Mechanics and Physics of Solids,* **33,** 731 – 745.

Hutchinson, J. W. (1987a) Crack tip shielding by microcracking in brittle solids, *Acta Metall.*, **35**, 1605 – 1619.

Hutchinson, J. W., Mear, M. E., and Rice, J. R. (1987b) Crack paralleling an interface between dissimilar materials, *ASME Journal of Applied Mechanics,* **54**, 828 – 832.

Leevens, P.S. and Radon, J.C. (1983) Inherent stress biaxiality in various fracture specimen geometries, *International Journal of Fracture*, **19**, 311 – 325.

Lekhenitskii, S.G. (1963) *Theory of Elasticity of an Anisotropic Body*, Holden-Day, San Francisco.

Lu, H. and Lardner, T. J. (1992) Mechanics of sub-interface crack in layered material, *International Journal of Solids and Structures*, **29**, 669 – 688.

Miller, G. R. (1989a) Analysis of branched interface cracks between dissimilar anisotropic madia, *ASME Journal of Applied Mechanics*, **56**, 844 – 849.

Miller, G. R. (1989b) Analysis of cracks near interfaces between dissimilar anisotropic materials, *International Journal of Engineering Science*, **27**, 667 – 678.

Obata, M., Nemat Nasser, S., And Goto, Y. (1989) Branched cracks in anisotropic elastic solids, *ASME Journal of Applied Mechanics* **56**, 858 – 864.

Rice, J. R. (1988) Elastic fracture mechanics concepts for interface cracks, *ASME Journal of Applied Mechanics*, **55**, 98 – 103.

Suo, Z. (1990) Singularities, interfaces and cracks in dissimilar anisotropic media, *Proc. Roy. Soc. London.* **A 427**, 331 – 358.

Tian, W.Y. and Chen, Y.H. (2000a) A semi-inf inite interface crack interacting with sub-interface matrix cracks in dissimilar anisotropic materials., Part I. Fundamental formulations and the J analysis, *International Journal of Solids and Structures*, **37**, 7717 – 7730.

Tian W.Y. and Chen Y.H. (2000b) Interaction between an interface crack and a parallel sub-interface crack in dissimilar anisotripic materials, *International Journal Fracture*, **102**, 305 – 322.

Ting, T.C.T. (1986) Explicit solution and invariance of the singularities at an interface crack in anisotropic composites, *International Journal of Solids and Structures*, **22**, 965 – 983.

Ting, T.C.T. (1990) Interface crack in anisotropic bimaterials, *Journal of Mechanics and Physics of Solids*, **38**, 505 – 513.

Wang, S. S. and Choi, I. (1983a) The interface crack between dissimilar anisotropic composite materials, *ASME Journal of Applied Mechanics*, **50**, 169 – 178.

Wang, S. S. and Choi, I. (1983b) The interface crack behavior in dissimilar anisotropic composites under mixed mode loading, *ASME Journal of Applied Mechanics*, **50**, 179 – 183.

Zhao, L.G. and Chen, Y.H. (1997) Further investigation of sub-interface cracks, *Archive of Applied Mechanics*, **67**, 393 – 406.

Zhao, L.G. and Chen, Y.H. (1998) Effect of the T stress in microcrack shielding problems, *ASME Journal of Applied Mechanics*, **65**, 71 – 75.

Appendix of Chapter 5

According to the continuous condition of the traction across the whole x-axis, there is

$$\mathbf{L}_1\mathbf{f}_1'(x) + \bar{\mathbf{L}}_1\bar{\mathbf{f}}_1'(x) = \mathbf{L}_2\mathbf{f}_2'(x) + \bar{\mathbf{L}}_2\bar{\mathbf{f}}_2'(x), \tag{A5.1}$$

which, to facilitate the analytic continuation, is rearranged as

$$\mathbf{L}_1\mathbf{f}_1'(x) - \bar{\mathbf{L}}_2\bar{\mathbf{f}}_2'(x) = \mathbf{L}_2\mathbf{f}_2'(x) - \bar{\mathbf{L}}_1\bar{\mathbf{f}}_1'(x). \tag{A5.2}$$

By using the standard analytic continuity argument, equation (A5.2) leads to

$$\mathbf{L}_1\mathbf{f}_1(z) = \bar{\mathbf{L}}_2\bar{\mathbf{f}}_2(z) \qquad z \in 1 \ , \tag{A5.3}$$

Def ine the displacement jump across the interface as

$$\mathbf{d}(x) = \mathbf{u}(x, 0+) - \mathbf{u}(x, 0-), \tag{A5.4}$$

with the aid of equation (A5.3), a direct calculation is given by

$$\mathbf{t}(x) = \mathbf{L}_1\mathbf{f}_1(x) + \mathbf{L}_2\mathbf{f}_2(x), \tag{A5.5}$$

and

$$i\mathbf{d}(x) = \mathbf{H}\mathbf{L}_1\mathbf{f}_1(x) - \bar{\mathbf{H}}\mathbf{L}_2\mathbf{f}_2(x), \tag{A5.6}$$

On the other hand, according to the continuity of the displacement across the bonded interface as inferred from (A5.6), it implies that $\mathbf{L}_1\mathbf{f}_1(x)$ and $\mathbf{L}_2\mathbf{f}_2(x)$ can be analytically extended to the whole plane except the crack line and they satisfy

$$\mathbf{h}(z) = \mathbf{L}_1\mathbf{f}_1(z) = \mathbf{H}^{-1}\bar{\mathbf{H}}\mathbf{L}_2\mathbf{f}_2(z) \quad z \notin C, \tag{A5.7}$$

where C is denoted as the crack line.

Hence, attention could be focused on the complex function $\mathbf{h}(z)$. Once $\mathbf{h}(z)$ is obtained, the fully-field solution could be given by (A5.7). In terms of $\mathbf{h}(z)$, the traction (A5.5) can be expressed as:

$$\mathbf{t}(x) = \mathbf{h}+(x) + \bar{\mathbf{H}}^{-1}\mathbf{H}\mathbf{h}-(x) \quad z \in C. \tag{A5.8}$$

For a semi-infinite interface crack in a dissimilar anisotropic material, the prescribed traction $\mathbf{t}_0(x)$ on the crack lines results in the Hilbert problem.

Chapter 6

MACROCRACK MICROCRACK INTERACTION IN PIEZOELECTRIC MATERIALS

Owing to the well known performance of the mechanical electric coupling in piezoelectric ceramics, these kinds of materials are found to have wide technological applications such as transducers, sensors, and actuators. Generally speaking, piezoelectric ceramics show a brittle nature, from the mechanical point of view, and are susceptible to fracture when a macrocrack is formed. The understanding of the fracture behavior of piezoelectric ceramics is of great importance, consequently it received considerable attention in the past decade. Sosa and Pak (1990), Sosa (1991, 1992), Pak (1990, 1992), Park and Sun (1995a, b), and Pak and Carman (1997) studied the two-dimensional problems for an elliptic hole or crack in piezoelectric materials. They defined two kinds of intensity factors, *i.e.,* the traditional mode I and mode II stress intensity factors (*SIF's*) and the electric displacement intensity factor (*EDIF*). As regards the conservation integrals in piezoelectric materials, Pak (1990), Suo *et. al.* (1992) proposed the J integral which also has a clear physical significance as the total potential energy release rate (*TPERR*). However, Park and Sun (1995a, b) pointed out that neither the *SIF's* nor the *TPERR* are suitable for describing the fracture behavior of piezoelectric ceramics. They showed that a new fracture criterion should be proposed, which is based on the concept of Mechanical Strain energy release rate (*MSERR*).

On the other hand, the fracture toughness of brittle materials in engineering is significantly influenced by the microstructure of the materials, such as microcracks, microvoids, and micro-inclusions. The interaction between a macrocrack tip and the near-tip material microstructure has been studied by a number of researchers as discussed in the above chapters. It could be imagined that the fracture toughness of piezoelectric materials would also be influenced by the microstructure of the near-tip process zone. However, the shielding problem in such materials has never yet been studied, neither analytically nor experimentally. In particular, how the electric loading influences the shielding effect is not clear and thus is worthy of investigation.

The goal of this chapter is to supply the lack of analytical estimations for the shielding problem in piezoelectric materials. Attention is focussed on the macrocrack microcrack interaction in a transversely isotropic piezoelectric material. In Section 6.1 elementary solutions with two kinds of crack configurations and subjected to three kinds of loads are given (Chen and Han, 1999a,b). In Section 6.2 a pseudo-traction electric displacement method (*PTED*) is proposed (Chen and Han, 1999a,b), which is motivated by the well known pseudo-traction method (Chen, 1984; Horii and Nemat Nasser, 1985, 1987). In Section 6.3 a consistency check is introduced which is based on the conservation law of the J integral. In Section 6.4 the influence of the electric loading on the mechanical strain energy release rate is studied in details. It is concluded that, besides the two sources of shielding effect defined by Hutchinson (1987), the disturbed near-tip electric field owed to microcracking provides another source of shielding (Chen and Han, 1999a,b). Finally, a summary is given.

1. Elementary solutions

Consider a two-dimensional problem for a transversely isotropic piezoelectric material under plane-strain conditions. Assume that the poling direction of the piezoelectric material is parallel to the y-axis (see Figures 6.1 and 6.2). The constitutive equations of the material can be written as (Sosa, 1992):

$$\begin{pmatrix} \varepsilon_{11} \\ \varepsilon_{22} \\ 2\varepsilon_{12} \end{pmatrix} = \begin{bmatrix} a_{11} & a_{12} & 0 \\ a_{12} & a_{22} & 0 \\ 0 & 0 & a_{33} \end{bmatrix} \begin{pmatrix} \sigma_{11} \\ \sigma_{22} \\ \sigma_{12} \end{pmatrix} + \begin{bmatrix} 0 & b_{21} \\ 0 & b_{22} \\ b_{13} & 0 \end{bmatrix} \begin{pmatrix} D_1 \\ D_2 \end{pmatrix},$$

$$\begin{pmatrix} E_1 \\ E_2 \end{pmatrix} = -\begin{bmatrix} 0 & 0 & b_{13} \\ b_{21} & b_{22} & 0 \end{bmatrix} \begin{pmatrix} \sigma_{11} \\ \sigma_{22} \\ \sigma_{12} \end{pmatrix} + \begin{bmatrix} \delta_{11} & 0 \\ 0 & \delta_{22} \end{bmatrix} \begin{pmatrix} D_1 \\ D_2 \end{pmatrix}, \tag{6.1}$$

whereσ_{ij}, ε_{ij}, D_i, E_i are the stresses, the strains, the electric displacements, the electric field components, respectively. And *(i,j)=1,2,* a_{ij}, b_{ij},and δ_{ij} are reduced material constants which were discussed by Sosa (1991).

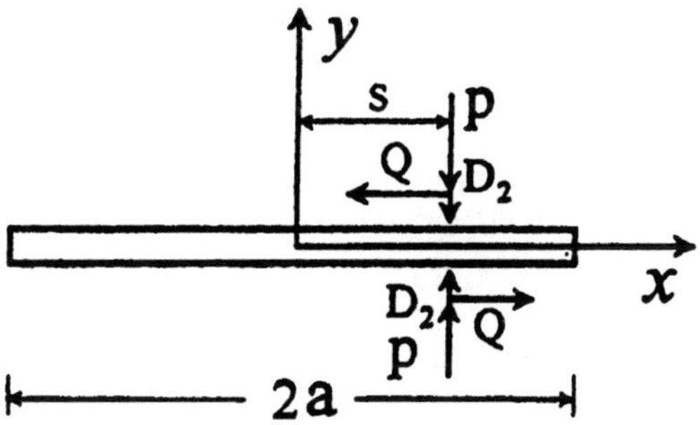

Figure 6.1. A finite crack loaded by concentrated tractions and concentrated electric displacement.

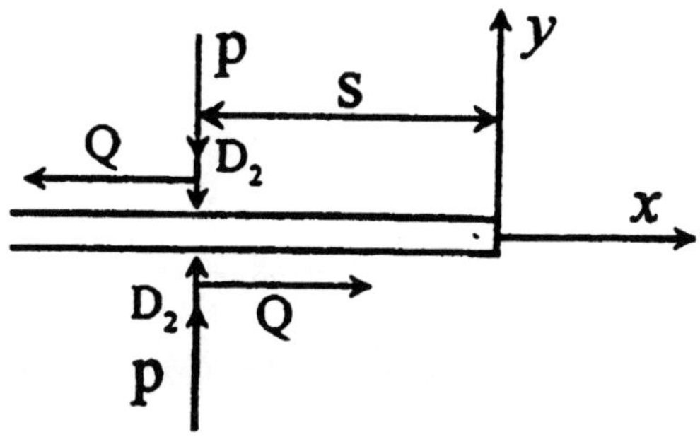

Figure 6.2. A semi-infinite crack loaded by concentrated tractions and electric displacement.

The stress components and the electric displacement components could be expressed by three complex potentials $\Phi_k(Z_k)$ *(k=1,2,3)* as follows (Sosa, 1992):

$$\sigma_{11} = G_4(\Phi_k(z_k), z_k) = \sum_{k=1}^{3}[\mu_k^2\Phi_k(z_k) + \overline{\mu_k^2\Phi(z_k)}], \quad (6.2)$$

$$\sigma_{22} = G_1(\Phi_k(z_k), z_k) = \sum_{k=1}^{3}[\Phi_k(z_k) + \overline{\Phi(z_k)}], \quad (6.3)$$

$$\sigma_{12} = G_2(\Phi_k(z_k), z_k) = -\sum_{k=1}^{3}[\mu_k\Phi_k(z_k) + \overline{\mu_k}\overline{\Phi(z_k)}], \quad (6.4)$$

$$D_1 = G_5(\Phi_k(z_k), z_k) = \sum_{k=1}^{3}[\lambda_k\mu_k\Phi_k(z_k) + \overline{\lambda_k\mu_k}\overline{\Phi(z_k)}], \quad (6.5)$$

$$D_2 = G_3(\Phi_k(z_k), z_k) = -\sum_{k=1}^{3}[\lambda_k\Phi_k(z_k) + \overline{\lambda_k}\overline{\Phi(z_k)}], \quad (6.6)$$

where the over bar denotes the complex conjugate, and

$$z_k = x + \mu_k y, \tag{6.7}$$

$$\lambda_k = -\frac{(b_{21}+b_{13})\mu_k^2 + b_{22}}{\delta_{11}\mu_k^2 + \delta_{22}}, \qquad (k = 1, 2, 3), \tag{6.8}$$

in which μ_k are three roots with positive imaginary parts of the following governing equations:

$$A\mu^6 + B\mu^4 + C\mu^2 + D = 0 \tag{6.9}$$

where

$$\begin{aligned} A &= a_{11}\delta_{11} \\ B &= a_{11}\delta_{22} + 2a_{12}\delta_{11} + a_{33}\delta_{11} + b_{21}^2 + b_{13}^2 + 2b_{21}b_{13}, \\ C &= a_{22}\delta_{11} + 2a_{12}\delta_{22} + a_{33}\delta_{22} + 2b_{21}b_{22} + 2b_{13}b_{22}, \\ D &= a_{22}\delta_{22} + b_{22}^2. \end{aligned} \tag{6.10}$$

1.1 Elementary solutions for a finite crack

As shown in Figure 6.1, a finite crack in an infinite piezoelectric material is considered. Its surfaces are loaded by the concentrated tractions P and Q as well as the concentrated electric displacement D_2. From the conditions with superscripts $+$ and $-$ denoting the upper surface boundary values and the lower surface boundary values, respectively:

$$\begin{aligned} &\sigma_{22}^+(x) = \sigma_{22}^-(x), \\ &\sigma_{12}^+(x) = \sigma_{12}^-(x), \qquad y = 0^{\pm} \;, -\infty \prec x \prec \infty, \\ &D_2^+(x) = D_2^-(x), \end{aligned} \tag{6.11}$$

the following relations can be obtained:

$$\sum_{k=1}^{3} [\Phi_k(x) - \bar{\Phi}_k(x)]^+ = \sum_{k=1}^{3} [\Phi_k(x) - \bar{\Phi}_k(x)]^-,$$

$$\sum_{k=1}^{3}[\mu_k\Phi_k(x)-\bar{\mu}_k\bar{\Phi}_k(x)]^+ = \sum_{k=1}^{3}[\mu_k\Phi_k(x)-\bar{\mu}_k\bar{\Phi}_k(x)]^-,$$
$$\sum_{k=1}^{3}[\lambda_k\Phi_k(x)-\bar{\lambda}_k\bar{\Phi}_k(x)]^+ = \sum_{k=1}^{3}[\lambda_k\Phi_k(x)-\bar{\lambda}_k\bar{\Phi}_k(x)]^-, \quad (6.12)$$

which are the simplest Riemann–Hilbert problems whose solutions are holomorphic functions. Considering the remote conditions, the holomorphic functions should be zero and the following relations are derived (Chen and Han, 1999a,b):

$$\sum_{k=1}^{3}\Phi_k(z_k)=\sum_{k=1}^{3}\bar{\Phi}_k(z_k),$$
$$\sum_{k=1}^{3}\mu_k\Phi_k(z_k)=\sum_{k=1}^{3}\bar{\mu}_k\bar{\Phi}_k(z_k), \quad (6.13)$$
$$\sum_{k=1}^{3}\lambda_k\Phi_k(z_k)=\sum_{k=1}^{3}\bar{\lambda}_k\bar{\Phi}_k(z_k).$$

The boundary conditions on the crack surfaces in Figure 6.1 are well known as

$$\sigma_{22}^{+}(x)=\sigma_{22}^{-}(x)=P\delta(x-s),$$
$$\sigma_{12}^{+}(x)=\sigma_{12}^{-}(x)=Q\delta(x-s), \quad (6.14)$$
$$D_2^{+}(x)=D_2^{-}(x)=D_2\delta(x-s), \qquad (y=0^{\pm}, |x|\prec a),$$

where a is the half-length of the crack, s is the distance shown in Figure 6.1, and $\delta(x-s)$ is the Derac function which has a different meaning from δ_{11}andδ_{22}in equation (6.1).

Substituting equation (6.2) into (6.14), it follows that (Chen and Han, 1999a,b)

$$\sum_{k=1}^{3}[\Phi_k^{+}(x)+\bar{\Phi}_k^{-}(x)]=P\delta(x-s),$$
$$\sum_{k=1}^{3}[\mu_k\Phi_k^{+}(x)+\bar{\mu}_k\bar{\Phi}_k^{-}(x)]=-Q\delta(x-s), \quad (6.15)$$
$$\sum_{k=1}^{3}[\lambda_k\Phi_k^{+}(x)+\bar{\lambda}_k\bar{\Phi}_k^{-}(x)]=-D_2\delta(x-s),$$

where the superscripts + and − refer to the upper and lower boundary values of the corresponding complex potentials.

Furthermore, substituting equation (6.13) into (6.15) leads to the following equations:

$$\Phi_k^+(x) + \Phi_k^-(x) = (A_{k1}P + A_{k2}Q + A_{k3}D_2)\delta(x - s), \qquad (k = 1, 2, 3), \tag{6.16}$$

where the A_{ij} are elements of the following matrix:

$$A = [A_{ij}] = \frac{1}{\Delta}\begin{bmatrix} \mu_2\lambda_3 - \mu_3\lambda_2 & \lambda_3 - \lambda_2 & \mu_2 - \mu_3 \\ \mu_3\lambda_1 - \mu_1\lambda_3 & \lambda_1 - \lambda_3 & \mu_3 - \mu_1 \\ \mu_1\lambda_2 - \mu_2\lambda_1 & \lambda_2 - \lambda_1 & \mu_1 - \mu_2 \end{bmatrix}, \tag{6.17}$$

$$\Delta = \mu_1(\lambda_2 - \lambda_3) + \mu_2(\lambda_3 - \lambda_1) + \mu_3(\lambda_1 - \lambda_2). \tag{6.18}$$

Equation (6.16) is a typical Riemann–Hilbert problem whose solution is (Chen and Han, 1999a,b)

$$\Phi_k^a(z_k) = \frac{(A_{k1}P + A_{k2}Q + A_{k3}D_2)}{2\pi(s - z_k)}\left(\frac{a^2 - s^2}{z_k^2 - a^2}\right)^{1/2}, \qquad (k = 1, 2, 3), \tag{6.19}$$

where the superscript a refers to the first elemental solution corresponding to the case of Figure 6.1.

The stresses at any point in the (x, y) plane induced by P, Q, and D_2 could be given by substituting equation (6.19) into (6.2) which will be discussed below as the so called influence functions in integral equations.

1.2 Elementary solutions for a semi-infinite crack

Obviously, the same procedure from (6.11) to (6.19) could be repeated for a semi-infinite crack in an infinite piezoelectric material as shown in Figure 6.2, which is loaded by the concentrated tractions P and Q as well as the concentrated electric displacement D_2 on both surfaces. Without going into the details, the solution is obtained (Chen and Han, 1999a,b):

$$\Phi_k^b(z_k) = \frac{(A_{k1}P + A_{k2}Q + A_{k3}D_2)\sqrt{|s|}}{2\pi(s - z_k)z_k^{1/2}}, \qquad (k = 1, 2, 3), \tag{6.20}$$

where the superscript b refers to the second elemental solution shown in Figure 6.2 and s refers to the distance from the traction acting point to the origin.

Similarly, the stresses at any point in the xy-plane induced by P, Q, and D_2 can be given by substituting equation (6.20) into (6.2) which will be discussed below as the so called influence functions in integral equations.

1.3 Remote loading conditions

As mentioned earlier, generally speaking, the remote loading conditions should be preferred as combined mechanical and electric ones. These kinds of conditions should be considered for a semi-infinite crack in an infinite piezoelectric material loaded by the remote stress intensity factors K_I^{∞} and K_{II}^{∞}, and also by the remote electric displacement intensity factor K_e^{∞}. Without going into the details, the solution is given as

$$\Phi_k^{\infty}(z_k) = f_k(z_k) \cdot z_k^{-1/2} + P_k(z_k), \qquad (k = 1, 2, 3), \qquad (6.21)$$

where $f_k(Z_k)$ and $P_k(Z_k)$ can be determined by using the remote conditions, which lead to the following formulation:

$$\Phi_k^{\infty}(z_k) = (A_{k1}K_I^{\infty} + A_{k2}K_{II}^{\infty} + A_{k3}K_e^{\infty})\ z_k^{-1/2}, \qquad (k = 1, 2, 3), \quad (6.22)$$

where A_{k1} , A_{k2} , and A_{k3} are the elements of the matrix A_{ij} in (6.17).

2. Pseudo-traction electric displacement method ($PTED$)

Consider a semi-infinite macrocrack interacting with N parallel microcracks arbitrarily located in the near-tip process zone in a transversely isotropic piezoelectric material (see Figure 6.3). Here the so called small scale approach has been adopted. At infinity, the loading conditions are preferred by K_I^{∞}, K_{II}^{∞}, and K_e^{∞}. The global coordinate system (x, y) and N local coordinate systems (x_i, y_i) $(i = 1, 2, \ldots, N)$ are preferred with the origins taken to be the macrocrack tip and the microcrack centers, respectively. As mentioned above, the poling direction of the material coincides well with the y-direction. The geometric parameters φ_i, d_i and a_i denote the location angle, the location distance, and the half length of the ith microcrack, respectively. All the crack surfaces are traction-free and charge-free.

By using the superimposing technique, the original problem shown in Figure 6.3 has been decomposed into $N+2$ subproblems as Chen (1984) and Horii and

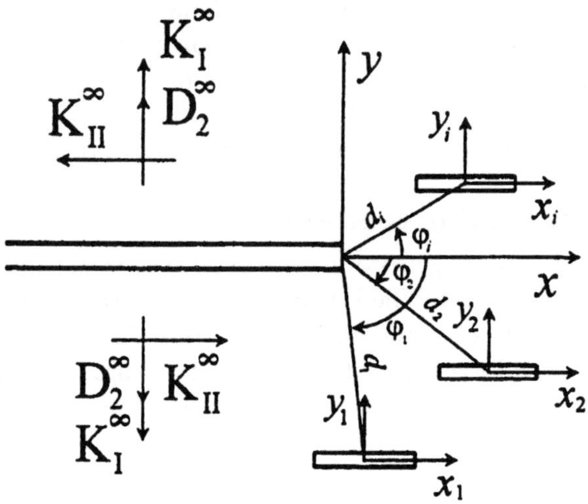

Figure 6.3. Macrocrack-microcrack interaction.

Nemat Nasser (1985,1987) treated in microcracking brittle solids (see Figure 6.4).

In the subproblem 1 of Figure 6.4, the semi-infinite crack is loaded by the remote *SIF's* K_I^∞ and K_{II}^∞, and the remote *EDIF* K_e^∞. According to the discussion presented in the above section, the normal stress $f_{ip0}(s)$, the tangential stress $f_{iq0}(s)$, and the electric displacement $f_{iD0}(s)$ at any point $(s, 0)$ on the ith microcrack location induced from the remote loading conditions are given by equations (6.2) and (6.22):

$$f_{ip0}(s) = G_1(\Phi_k^\infty(z_k), Z_k),$$
$$f_{iq0}(s) = G_2(\Phi_k^\infty(z_k), Z_k), \tag{6.23}$$
$$f_{iD0}(s) = G_3(\Phi_k^\infty(z_k), Z_k),$$

where $(i = 1, 2, ..., N)$ and

$$z_k = Re(d_i e^{i\varphi_i} + s) + \mu_k \cdot Im(d_i e^{i\varphi_i} + s), \tag{6.24}$$

and G_1, G_2, and G_3 are known functions given explicitly in equation (6.2).

In the subproblem 2 of Figure 6.4, the macrocrack is subjected to the unknown pseudo-tractions $P^0(x)$, and $Q^0(x)$, and the unknown pseudo-electric

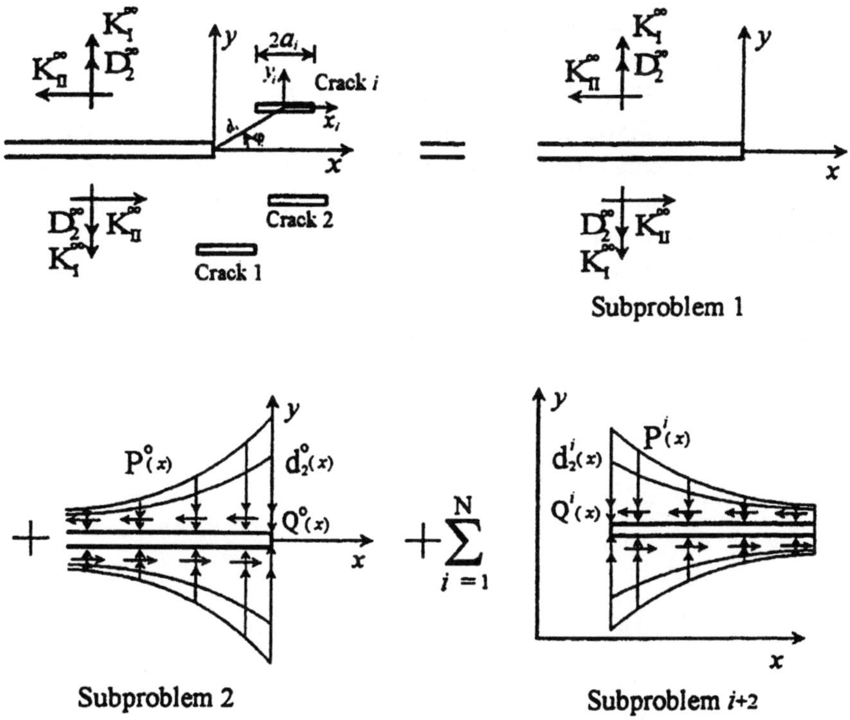

Figure 6.4. Decomposition of the original interaction problem.

displacement $d_2^0(x)$. The solutions can be obtained by using the elementary solution in the above section and the integration technique. The normal stress $\sigma_{22}^{i0}(s)$, the tangential stress $\sigma_{12}^{i0}(s)$, and the electric displacement $D_2^{i0}(s)$, at any point $(s, 0)$ on the ith microcrack location can be expressed as

$$\sigma_{22}^{io}(s)$$

$$= \int_{-\infty}^{0} [f_{ipop}(t,x,y)P^0(t) + f_{ipoq}(t,x,y)Q^0(t) + f_{ipod}(t,x,y)d_2^0(t)]dt,$$

$$\sigma_{12}^{io}(s)$$

$$= \int_{-\infty}^{0} [f_{iqop}(t,x,y)P^0(t) + f_{iqoq}(t,x,y)Q^0(t) + f_{iqod}(t,x,y)d_2^0(t)]dt.$$

$$D_2^{io}(s)$$

$$= \int_{-\infty}^{0} [f_{iDop}(t,x,y)P^0(t) + f_{iDoq}(t,x,y)Q^0(t) + f_{iDod}(t,x,y)d_2^0(t)]dt, \tag{6.25}$$

where $x+iy = d_i e^{i\phi i}+s$, the nine kernel functions $f_{ip0p}(t,x,y),\ldots, f_{iD0d}(t,x,y)$ are known real functions, which can be expressed by using (6.2) and (6.20) without any difficulty. Here, the subscripts *i, 0, p, q, D, d* have definite meanings which refer to the microcrack, the macrocrack, the normal traction, the tangential traction, the poling direction, and the electric displacement, respectively.

In the subproblem $i+2$ of Figure 6.4, the ith microcrack is subjected to the unknown pseudo-tractions $P^0(x)$ and $Q^0(x)$, and the unknown pseudo-electric displacement $d_2^i(x_i)$. Without going into the details, the normal stress $\sigma_{22}^{0i}(s)$, the tangential stress $\sigma_{12}^{0i}(s)$, and the electric displacement $D_2^{0i}(s)$ at any point $(s,0)$ on the macrocrack location are obtained as follows

$$
\begin{aligned}
&\sigma_{22}^{0i}(s)\\
&= \int_{-a_i}^{a_i} [f_{0pip}(t,x_i,y_i)P^i(t) + f_{0piq}(t,x_i,y_i)Q^i(t) + f_{0pid}(t,x_i,y_i)d_2^i(t)]dt,\\
&\sigma_{12}^{0i}(s)\\
&= \int_{-a_i}^{a_i} [f_{0qip}(t,x_i,y_i)P^i(t) + f_{0qiq}(t,x_i,y_i)Q^i(t) + f_{0qid}(t,x_i,y_i)d_2^i(t)]dt,\\
&D_2^{0i}(s)\\
&= \int_{-a_i}^{a_i} [f_{0Dip}(t,x_i,y_i)P^i(t) + f_{0Diq}(t,x_i,y_i)Q^i(t) + f_{0Did}(t,x_i,y_i)d_2^i(t)]dt,
\end{aligned}
\tag{6.26}
$$

where the nine kernel functions $f_{0pip}(t,x,y),\ldots f_{opid}(t,x,y)$ are known real functions, which could be expressed by using (6.2) and (6.19) with $x_i+iy_i = -d_i e^{i\phi_i}+s$.

Similarly, the normal stress $\sigma_{22}^{ji}(s)$, the tangential stress $\sigma_{12}^{ji}(s)$, and the electric displacement $D_2^{ji}(s)$ induced from the ith microcrack at any point on the jth $(j \neq i)$ microcrack are obtained as follows

$$\sigma_{22}^{ji}(s)$$
$$= \int_{-a_i}^{a_i} [f_{jpip}(t, x_i, y_i)P^i(t) + f_{jpiq}(t, x_i, y_i)Q^i(t) + f_{jpid}(t, x_i, y_i)d_2^i(t)]dt,$$
$$\sigma_{12}^{ji}(s)$$
$$= \int_{-a_i}^{a_i} [f_{jqip}(t, x_i, y_i)P^i(t) + f_{jqiq}(t, x_i, y_i)Q^i(t) + f_{jqid}(t, x_i, y_i)d_2^i(t)]dt,$$
$$D_2^{ji}(s)$$
$$= \int_{-a_i}^{a_i} [f_{jDip}(t, x_i, y_i)P^i(t) + f_{jDiq}(t, x_i, y_i)Q^i(t) + f_{jDid}(t, x_i, y_i)d_2^i(t)]dt. \tag{6.27}$$

where the nine kernel functions in equation (6.27) are known real ones, which can be expressed by using (6.2) and (2.19) and $x_i + iy_i = d_j e^{i\phi_j} - d_i e^{i\phi_i} + s$ $(j \neq i)$.

The original problem is now reduced to the following integral equations after performing the superimposing technique (Chen and Han, 1999a,b):

$$\begin{aligned}
&P^0(s) + \sum_{l=1}^{N} \sigma_{22}^{0l}(s) = 0, \\
&Q^0(s) + \sum_{l=1}^{N} \sigma_{12}^{0l}(s) = 0, \qquad (-\infty < s < 0), \\
&d_2^0(s) + \sum_{l=1}^{N} D_2^{0l}(s) = 0, \\
&P^i(s) + \sum_{j=0}^{N} \sigma_{22}^{ij}(s) = -f_{ip0}(s), \qquad (j \neq i), \\
&Q^i(s) + \sum_{j=0}^{N} \sigma_{12}^{ij}(s) = -f_{iq0}(s), \qquad (j \neq i), \quad (-a_i < s < a_i), \\
&d_2^i(s) + \sum_{j=0}^{N} D_2^{ij}(s) = -f_{iD0}(s), \qquad (j \neq i),
\end{aligned} \tag{6.28}$$

where $\sigma_{22}^{0i}(s)$, $\sigma_{12}^{0i}(s)$, $D_2^{0i}(s)$, $\sigma_{22}^{ji}(s)$, $\sigma_{12}^{ji}(s)$, and $D_2^{ji}(s)$ have been defined by (6.25), (6.26), (6.27), whilst f_{ip0}, f_{iq0}, and f_{iD0} are known as expressed by (6.23).

Using the Chebyshev numerical integration technique, the system of integral equations (6.28) could be solved numerically for the unknown pseudo-tractions and pseudo-electric displacement $P^i(s)$, $Q^i(s)$, and $d_2^i(s)$, $(i = 1, 2, \ldots, N)$. After doing so, the intensity factors of the macrocrack tip and those at the ith microcrack tips are given

$$
\begin{aligned}
K_I^t &= K_I^\infty - \int_{-\infty}^0 \sqrt{2}P^0(s) \Big/ \sqrt{-\pi s} ds, \\
K_{II}^t &= K_{II}^\infty - \int_{-\infty}^0 \sqrt{2}Q^0(s) \Big/ \sqrt{-\pi s} ds, \qquad (6.29) \\
K_e^t &= K_e^\infty - \int_{-\infty}^0 \sqrt{2}d_2^0(s) \Big/ \sqrt{-\pi s} ds,
\end{aligned}
$$

$$
\begin{aligned}
K_I^{Ri} &= -\int_{-a_i}^{a_i} P^i(s)(a_i+s)^{1/2}(a_i-s)^{-1/2} \Big/ \sqrt{\pi a_i} ds, \\
K_I^{Li} &= -\int_{-a_i}^{a_i} P^i(s)(a_i-s)^{1/2}(a_i+s)^{-1/2} \Big/ \sqrt{\pi a_i} ds, \\
K_{II}^{Ri} &= -\int_{-a_i}^{a_i} Q^i(s)(a_i+s)^{1/2}(a_i-s)^{-1/2} \Big/ \sqrt{\pi a_i} ds, \\
K_I^{Li} &= -\int_{-a_i}^{a_i} Q^i(s)(a_i-s)^{1/2}(a_i+s)^{-1/2} \Big/ \sqrt{\pi a_i} ds, \qquad (6.30) \\
K_e^{Ri} &= -\int_{-a_i}^{a_i} d_2^i(s)(a_i+s)^{1/2}(a_i-s)^{-1/2} \Big/ \sqrt{\pi a_i} ds, \\
K_e^{Li} &= -\int_{-a_i}^{a_i} d_2^i(s)(a_i-s)^{1/2}(a_i+s)^{-1/2} \Big/ \sqrt{\pi a_i} ds,
\end{aligned}
$$

where the superscripts R and L refer to the right tip and the left tip of the ith microcrack, respectively.

3. Conservation law and consistency check

From the abovementioned manipulations, numerical results for the *SIF's* and the *EDIF* at the macrocrack tip and the microcrack tips could be given, which could be seen in Chen and Han (1999a, b). However, it is not easy to confirm the results surely and steadily since no previously known results could be used for comparisons. A possible way in which the present results could be demonstrated is to make some comparisons in degenerate cases. However, it is not clear, at least not fully clear, from a physical point of view, that if the present results are really correct in general cases, after performing such comparisons. Specially, if there were a tiny mistake in the abovementioned manipulations which has no influence on the results in the degenerate cases and does influence the final results in general cases, the conclusions derived could be incorrect or misleading. Such a mistake could not be recognized by making the abovementioned comparisons. Therefore a question arises as to what the present authors should do to examine their numerical results more surely and steadily (Chen, 1996, Chen and Hasebe, 1998).

In this section a consistency check is proposed which is based on the conservation law of the J integral in piezoelectric materials. As defined by Pak (1990) and Suo *et al.* (1992), the J integral or the total potential energy release rate (*TPERR*) in piezoelectric materials could be formulated as:

$$J = \int_\Gamma \frac{1}{2}(\sigma_{ij}\varepsilon_{ij} - D_i E_i)dy - n_i \sigma_{ip}\frac{\partial u_p}{\partial x_1}ds - n_i D_i \frac{\partial \varphi}{\partial x_1}ds, \tag{6.31}$$

where u_p $(p = 1, 2)$ is the displacement, ϕ is the electric potential, and is a closed contour to be chosen. Customarily, in single crack problems, is generally taken as a smooth curve that starts from one point on the lower surface and ends at another point on the upper surface of a crack.

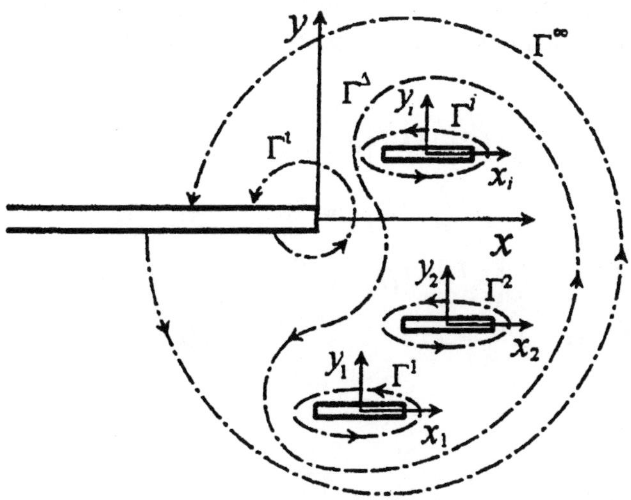

Figure 6.5. Special closed contours for evaluating the J integral.

For making the consistency check, several closed contours are specially introduced (see Figure 6.5) along which the J integral is evaluated. In Figure 6.5

Γ^{∞} is corresponding to the remote mechanical electric field, Γ^{t} is corresponding to the macrocrack tip, whilst Γ^{Δ} is corresponding to the N microcracks and Γ^{i} encloses the ith microcrack only. According to the path independent nature of the J integral (Chen and Han, 1999a,b), it is obvious that

$$\Delta J = \sum_{l=1}^{N} J^{(l)}, \tag{6.32}$$

where ΔJ is evaluated along Γ^{Δ} and $J^{(i)}$ is evaluated along Γ^{l}, and

$$J^{\infty} = J^{t} + \Delta J, \tag{6.33}$$

$$J^{t}/J^{\infty} + \Delta J/J^{\infty} = 1, \tag{6.34}$$

where J^{∞} is evaluated along Γ^{∞} and J^{t} is evaluated along Γ^{t}.

Equation (6.34) is similar to the conclusions derived in the above chapters in brittle solids, in bimaterial solids, in anisotropic solids, and in dissimilar anisotropic solids, respectively. It reveals that the remote *TPERR* (or the remote J integral) is divided into two parts. One of them is contributed from the macrocrack tip and the other is induced from the existence of the microcracks. The detailed formulations of the two parts are given as follows:

$$J^{t} = \frac{1}{2}\left[K^{t}\right]^{T}[H]\left[K^{t}\right], \tag{6.35}$$

$$J^{Rl} = \frac{1}{2}\left[K^{Rl}\right]^{T}[H]\left[K^{Rl}\right], \quad (l = 1, 2, ..., N), \tag{6.36}$$

$$J^{Ll} = \frac{1}{2}\left[K^{Ll}\right]^{T}[H]\left[K^{Ll}\right], \quad (l = 1, 2, ..., N), \tag{6.37}$$

$$J^{(l)} = J^{Rl} - J^{Ll}, \quad (l = 1, 2, ..., N), \tag{6.38}$$

where

$$\left[\mathbf{K}^{t}\right] = \left[K_{II}^{t}\ K_{I}^{t}\ K_{e}^{t}\right]^{T}, \tag{6.39}$$

$$\left[\mathbf{K}^{Rl}\right] = \left[K_{II}^{Rl}\ K_{I}^{Rl}\ K_{e}^{Rl}\right]^{T}, \ (l = 1, 2, 3, ..., N), \tag{6.40}$$

$$\left[\mathbf{K}^{Ll}\right] = \left[K_{II}^{Ll}\ K_{I}^{Ll}\ K_{e}^{Ll}\right]^{T}, \ (l = 1, 2, 3, ..., N), \tag{6.41}$$

and $\mathbf{H}$ is a 3×3 matrix related to the piezoelectric material constants as

$$[H] = 2Re(i\,[A]\,[B]^{-1}), \tag{6.42}$$

in which

$$[\mathbf{A}] = \begin{bmatrix} p_1\, p_2\, p_3 \\ q_1\, q_2\, q_3 \\ r_1\, r_2\, r_3 \end{bmatrix}, \tag{6.43}$$

$$[\mathbf{B}] = \begin{bmatrix} -\mu_1 \; - \mu_2 \; - \mu_3 \\ 1\;1\;1 \\ -\lambda_1 \; - \lambda_2 - \lambda_3 \end{bmatrix}, \tag{6.44}$$

and

$$\begin{aligned} p_k &= a_{22}\mu_k^2 + a_{12} - b_{21}\lambda_k, \\ q &= (a_{12}\mu_k^2 + a_{22} - b_{22}\lambda_k)/\mu_k, \\ r &= -(b_{13} + \delta_{11}\lambda_k)\mu_k. \end{aligned} \tag{6.45}$$

What is more, the relation (6.34) provides a necessary condition as well as a powerful tool for the consistency check to examine the numerical results derived by the abovementioned technique $PTED$. Taking the PZT-4 piezoelectric ceramic as an example whose material constants are presented in Table 6.1, numerical examinations derived by the $PTED$ method are shown in Table 6.2 and Figure 6.6, respectively. Here the normalized distance $d_n = (d - a)/a$ =0.3, 0.5, and 1.0, respectively, and the location angle φ is taken to be a variable from 0 to 160 degrees. For the sake of convenience only one microcrack is considered in the near-tip process zone of a macrocrack which is under combined mechanical electric loading conditions, *i.e.*, $K_I^\infty \neq 0$, $K_{II}^\infty = 0$, $K_e^\infty = 10^{-8} K_I^\infty C N^{-1}$.

Table 6.1. The reduced material constants of the PZT-4 piezoelectric ceramic.

a_{11}	a_{12}	a_{22}	a_{33}
8.205×10^{-12}	-3.144×10^{-12}	7.495×10^{-12}	$19.3\times10^{-12} m^2 N^{-1}$
b_{21}	b_{22}	b_{13}	
-16.62×10^{-3}	23.96×10^{-3}	39.4×10^{-3}	$(m^2 C^{-1})$
δ_{11}	δ_{22}		
7.66×10^{7}	9.82×10^{7}	$(V^2 N^{-1})$	

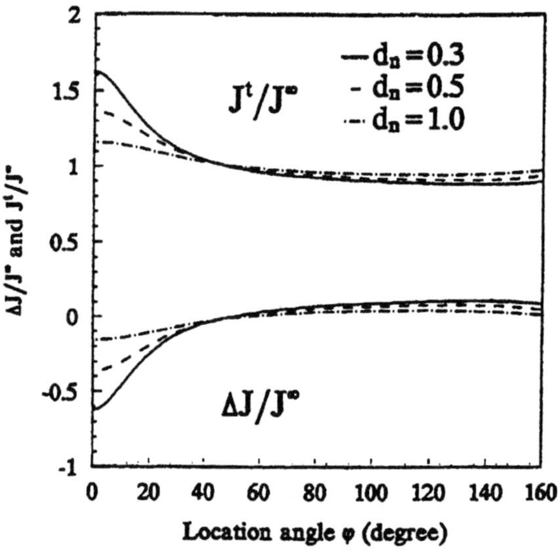

Figure 6.6. Numerical examination for the consistency check taking d_n=*(d-a)/a=0.3, 0.5,* and 1.0, respectively.

It is shown that the consistency check or the conservation law of the J integral (6.34) is really satisfied by the present numerical results. The examinations presented in Table 6.2 and Figure 6.6 confirm the results derived without taking any comparisons with some known results in degeneralized cases since any mistake in the manipulations mentioned in the above sections, if exists, will certainly lead to unsatisfied conclusions contrary to the law (6.34).

It is noted that the relation (6.34), as well as the results shown in Figure 6.6, reveals that there is a wastage when the remote $TPERR$ (J) transmits across the microcracking zone from the infinity to the macrocrack tip. The wastage is owed just to the redistribution of the near-tip stress field and the near-tip electric field induced from microcracking. The major difference between the present discussion and the previous ones done by other researchers is the influence of the disturbed near-tip electric field on the wastage which represents as another consequence induced from microcracking. Thus there are three kinds of energy dissipative processes owed to microcracking in piezoelectric materials which lead to the phenomenon of shielding effect and material toughness enhancement. As discussed by Hutchinson (1987), the first two are well known, *i.e.,* the reduction on the effective moduli and the release of residual stresses at the microcrack location. The last one is just the abovementioned disturbed near-tip

Table 6.2. Numerical examination by using a consistency check under combined mechanical electric loading $K_I^\infty \neq 0, K_{II}^\infty = 0, K_e^\infty = 10^{-8} K_I^\infty CN^{-1}$ with $d_n = 0.3$.

φ (degrees) =	8	20	60	90	120	160
K_I^R/K_I^∞	0.5854	0.6233	0.7721	0.8265	0.7447	0.2250
K_{II}^R/K_I^∞	0.0086	0.0371	0.0691	0.0079	−0.0984	−0.1007
K_e^R/K_I^∞ $10^{-4}CN^{-1}$	0.5735	0.5546	0.4911	0.4546	0.4217	0.3300
K_I^L/K_I^∞	1.0308	1.1659	0.7967	0.4661	0.2116	−0.0004
K_{II}^L/K_I^∞	0.0766	0.1004	−0.1603	−0.1811	−0.1206	−0.0161
K_e^L/K_I^∞ $10^{-4}CN^{-1}$	0.9154	0.7548	0.4530	0.3345	0.2419	0.1171
K_I^t/K_I^∞	1.3127	1.3877	0.9990	0.7928	0.7589	0.9807
K_{II}^t/K_I^∞	0.0534	0.0290	−0.1928	−0.1323	−0.0109	0.0113
K_e^t/K_I^∞ $10^{-4}CN^{-1}$	1.2286	1.1240	0.9818	0.9508	0.9378	0.9518
ΔJ	−0.5033	−0.2458	0.0372	0.0973	0.1107	0.0964
J^t/J^∞	1.5033	1.2458	0.9628	0.9127	0.8892	0.9042
$(J^t + \Delta J)/J^\infty$	1.0000	1.0000	1.0000	1.0000	0.9999	1.0008

electric field (Chen and Han, 1999a,b). In the forthcoming sections, the disturbance of the near-tip electric field owed to microcracking will be discussed in some detail.

4. The mechanical strain energy release rate (*MSERR*)

As concluded by Park and Sun (1995a,b), for a charge-free crack neither the *SIF's* nor the J integral are suitable for describing the fracture behavior of piezoelectric ceramics. They proposed a new criterion, *i.e.*, the Mechanical Strain Energy Release Rate($MSERR$) for this purpose, which is defined as

$$G_I^M = \lim_{\delta \to 0} \frac{1}{2\delta} \int_0^\delta \sigma_{22}(x) \Delta u_2(\delta - x) dx, \tag{6.46}$$

for the mode I, and

$$G_{II}^M = \lim_{\delta \to 0} \frac{1}{2\delta} \int_0^\delta \sigma_{12}(x) \Delta u_1(\delta - x) dx, \tag{6.47}$$

for the mode II. Here the superscript M refers to the $MSERR$ and Δu_1 and Δu_2 are the jumps of the displacement components measured from the lower surface to the upper surface of a macrocrack.

For the Mode I the $MSERR$ is related to the crack tip *SIF's* and the *EDIF* as follows

$$G_I^M = \frac{1}{2}(H_{21}K_IK_{II} + H_{22}K_I^2 + H_{23}K_IK_e), \tag{6.48}$$

where H_{ij} *(i, j=1,2,3)* is the elements of the 3×3 matrix **H** given by equation (6.42).

Unlike the results given by Park and Sun (1995a,b) for single crack problems, the mechanical loading and the electric loading are coupled in the present interaction problem since the microcrack not only releases the near-tip residual stresses and reduces the effective elastic moduli in the process zone, but also disturbs the near-tip electric field. The latter presents another source of shielding as mentioned above. Therefore, the influence of the electric loading on the *MSERR* in the present interaction problem should bring much attention. Of the most interest is the divergence of the variable tendencies of *MSERR* from those of the mode I *SIF*.

4.1 Variable tendencies of the SIF owing to microcracking

For the sake of convenience, the piezoelectric ceramics used above, *i.e.*, the PZT-4, is taken again as an example with the reduced material constants listed in Table 6.1. Only one single microcrack is considered in the near-tip process zone. In the forthcoming numerical results, the plane strain conditions are always assumed and the poling direction of the material is always taken to be perpendicular to both macrocrack and microcrack, *i.e.*, parallel to the y-axis. All the numerical results have been confirmed by using the conservation law or consistency check (6.34).

Numerical results for the interacting effect represented by the traditional mode I *SIF* are plotted in Figure 6.7 against the location angle φ and in Figures 6.8(a,b,c,d) against the normalized distance d_n, respectively.

Here three kinds of remote loading conditions are preferred for making comparisons, *i.e.*, (i) The combined mode I and positive electric loading: $K_I^\infty \neq 0, K_{II}^\infty = 0, K_e^\infty = 10^{-8}K_I^\infty CN^{-1}$ corresponding to the real curves; (ii) The purely mechanical loading $K_I^\infty \neq 0, K_{II}^\infty = 0, K_e^\infty = 0$ corresponding to the imaginary curves; (iii) The combined mode I and negative electric loading $K_I^\infty \neq 0, K_{II}^\infty = 0, K_e^\infty = -10^{-8}K_I^\infty CN^{-1}$ corresponding to the dotted curves, respectively (Chen and Han, 1999a,b).

It is seen that the corresponding three curves show quite different and very complicated variable tendencies, which are dissimilar to those found by Park and Sun (1995a,b) in single crack problems. The mechanical loading and the

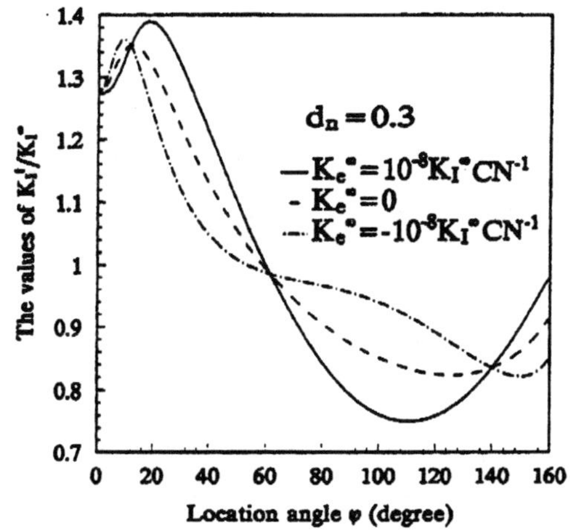

Figure 6.7. Normalized mode I *SIF* against the location angle.

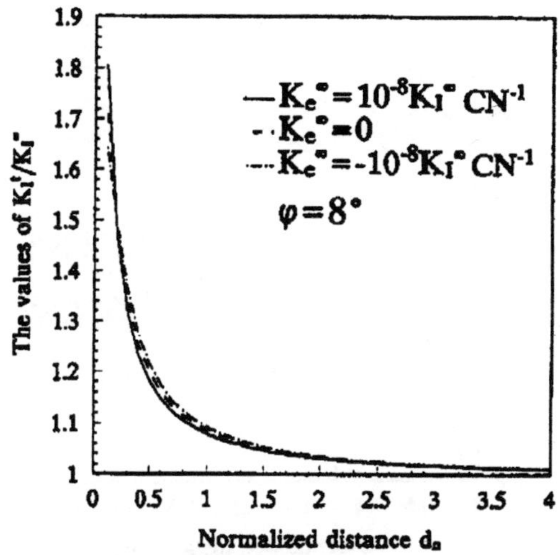

(a) $\varphi = 8^0$

electric loading are really coupled owing to the existence of the microcrack. In other words, the mode I *SIF* is no longer independent of the electric loading. For example, in the range φ of between 12 degrees and 141 degrees, the positive electric loading leads to adding the interaction effect, whilst the negative electric loading leads to impeding the effect. However, it is not always the case. In the range of φ between 0 and 12 degrees and between 141 degrees and 160 degrees, an opposite influence of the electric loading could be seen, *i.e.,* the positive electric loading leads to impeding the effect, whilst the negative electric loading leads to promoting the effect. What is more, as shown in Figure 6.7,

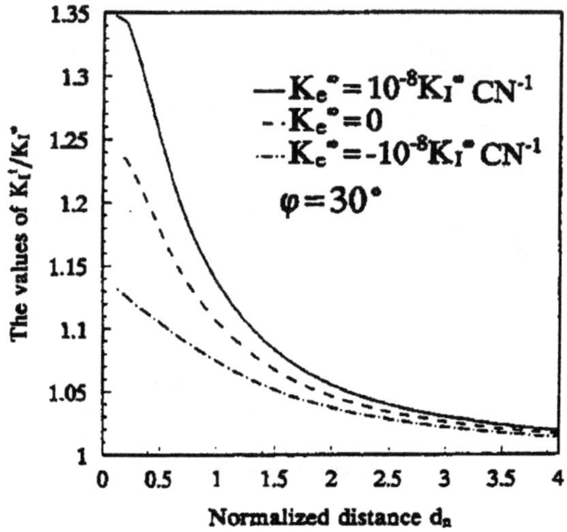

(b) $\varphi = 30^0$

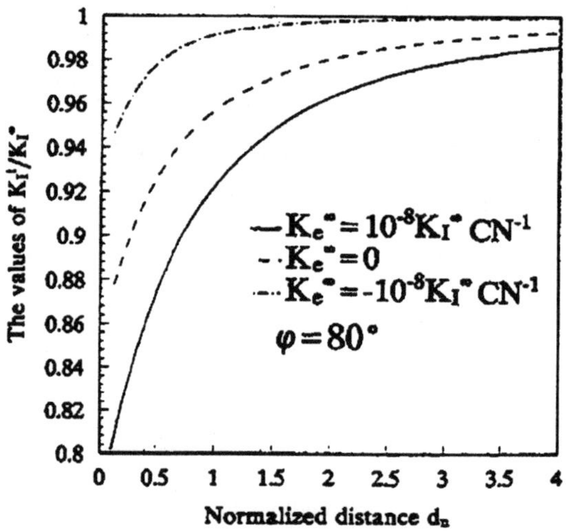

(c) $\varphi = 80^0$

there exist three special locations for the considered parallel microcrack at which neither the positive nor the negative electric loading has inf luence on the K_I^t/K_I^∞. The location angles corresponding to three special locations could be called as neutral electric loading angle (*NELA*) denoted by φ_{NE}. In the present case (taking the PZT-4 piezoelectric ceramic and setting $d_n = 0.3$ the $NELA$ φ_{NE} equals about 12 degrees, 65 degrees, and 141 degrees, respectively.

It is found also in Figure 6.7 that the electric displacement loading at inf inity plays an important role in the macrocrack microcrack interaction problem. It not only leads to shifting the maximum anti-shielding location (*MASL*, some

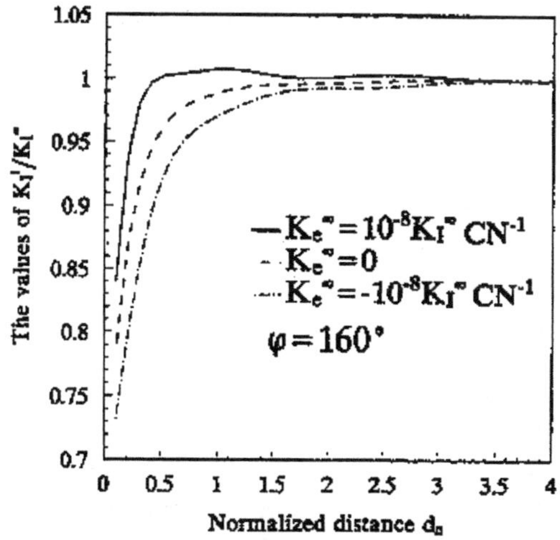

(d) $\varphi = 160^0$

Figure 6.8. Normalized mode I *SIF* against the normalized distance.

people called as the amplification effect for the same meaning of the anti-shielding effect) and amplitude when $K_I^t/K_I^\infty > 1$, but also leads to shifting the maximum shielding location (*MSL*) and amplitude when $K_I^t/K_I^\infty < 1$. It is seen in Figure 6.7 that the positive electric loading leads to increasing the *MASL* angle φ_{MA} and to decreasing the *MSL* angle φ_{MS}, whilst the negative electric loading just leads to an opposite influence on φ_{MA} and φ_{MS}.

In Figures 6.8(a,b,c,d), the mode I *SIF* against d_n is plotted. Four location angles are chosen, *i.e.*, $\varphi = 8$, 30, 80, and 160 degrees, respectively. It is seen in Figure 6.8(a) that there is a transformation distance $d_{nt} = 0.175$ at which neither the positive nor the negative electric loading has influence on mode I *SIF*. When the distance d_n is less than d_{nt}, the positive electric loading leads to increasing *SIF* and the negative loading leads to decreasing *SIF*. This nature could be called the regular influence of electric loading (*RIEL*) as found by a number of previous researchers in single crack problems. On the contrary, when d_n is larger than d_{nt}, the influence of electric loading just shows an opposite nature. This nature could be called the opposite influence of electric loading (*OIEL*), which has never been found by the previous researchers in single crack problems. However, in Figures 6.8(b, c, d) there are no such transformation distances at all. It is seen that in Figures 6.8(b, d) the variable tendencies are always the *RIEL*, whilst in Figure 6.8(c) the tendencies are always the *OIEL*. It is concluded that whether the influence of electric loading is *RIEL* or *OIEL* does depend on the location angle φ as well as the distance d_n, owing to the coupling nature of the remote mechanical loading and electric loading in the present interaction problem.

4.2 Variable tendencies of the *EDIF* against the location angle

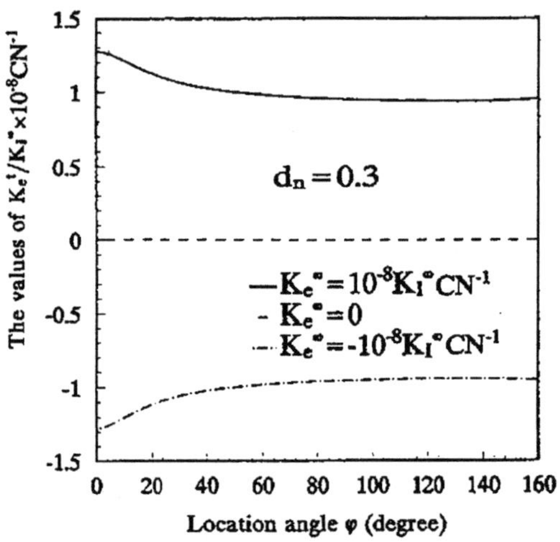

(a) The normalized distance $d_n = 0.3$.

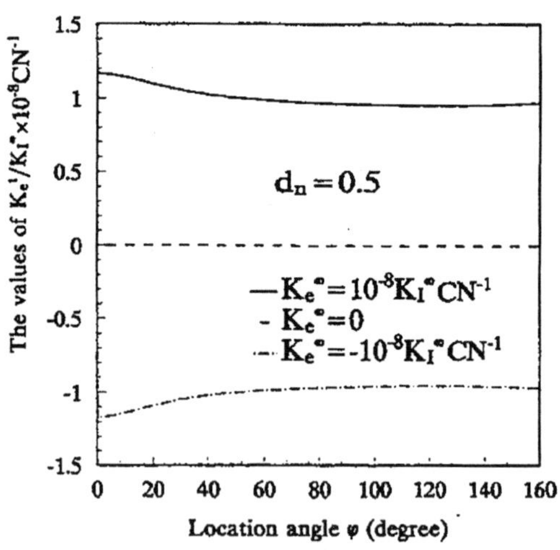

(b) The normalized distance $d_n = 0.5$.

Figure 6.9. Macrocrack-tip *EDIF* against the location angle φ.

Of the great interest are the variable tendencies of the *EIDF* against the location angle φ of the parallel microcrack. Here the *EIDF* has been normalized by $K_I^\infty \times 10^{-8} CN^{-1}$ and the distance d_n= 0.3 and 0.5, respectively. Figures 6.9(a,b) show that the mechanical loading has no influence on the *EIDF* even under strongly interacting situations with d_n= 0.3 or 0.5 (see the imaginary curves). Comparing with the conclusions derived above for the influence of electric loading on *SIF*, it is concluded that the mechanical quantities yield little influence on the electric quantities such as *EDIF*, quite contrary, the electric quantities yield large influence on the mechanical quantities such as *SIF*.

It is seen in Figures 6.9(a,b) that no matter how the microcrack location angle is, the positive electric loading always leads to increasing the *EDIF* at the macrocrack tip and the negative electric loading always leads to decreasing the *EDIF*. Moreover, it is found that the variable nature of $K_e^t / K_I^\infty \times 10^{-8} CN^{-1}$ against $K_e^\infty / K_I^\infty \times 10^{-8} CN^{-1}$ is monotonic and nearly proportional, whatever the parallel microcrack is the near-tip process zone. The *EDIF* as influenced by the microcrack is approximately equal to the remote *EDIF* for large values of the location angle, for example, $\varphi > 40$ degrees. However, for smaller values of the angle the influence of the existence of the microcrack can not be neglected and the maximum influence occurs at the collinear situation with φ= 0. This conclusion coincides well with the well known fact that the disturbance of a microcrack in the collinear case on the near-tip electric field becomes a maximum or is much larger than those in the other parallel cases corresponding to $\varphi > 0$. This reveals again that the disturbance of the near-tip electric field induced from microcracking really provides another source of shielding (Chen and Han, 1999a,b).

4.3 Variable tendencies of the mechanical strain energy release rate ($MSERR$)

Numerical results for the influence of the electric loading on the *MSERR* are plotted in Figure 6.10(a,b) under the three kinds of remote loading conditions mentioned above. Here the normalized distance d_n is taken to be 0.3 and 0.5, respectively.

The macrocrack-tip $MSERR$ G_I^{Mt} is normalized by

$$G_I^M = \frac{1}{2}(H_{22} K_I^\infty K_I^\infty + H_{23} K_I^\infty K_e^\infty), \tag{6.49}$$

where, as assumed earlier, K_I^∞ is always non-zero, K_{II}^∞ is always zero, whilst K_e^∞ is equal to $10^{-8} K_I^\infty CN^{-1}$, 0, $-10^{-8} K_I^\infty CN^{-1}$, respectively.

It is seen by making a comparison between Figure 6.7 and 6.10(a) that the variable tendencies of the macrocrack-tip *SIF* are quite different from those of

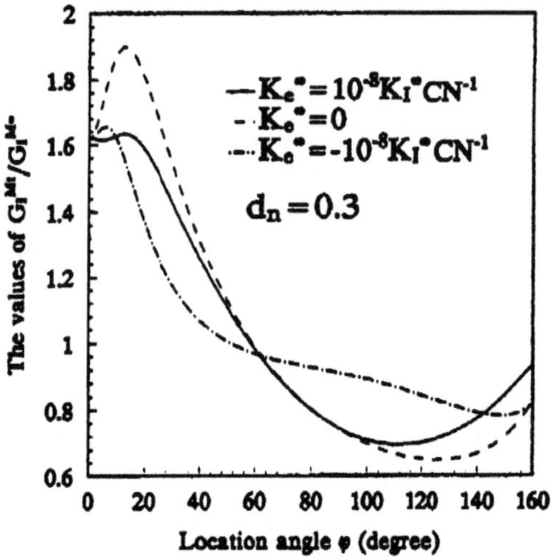

(a) The normalized distance $d_n = 0.3$.

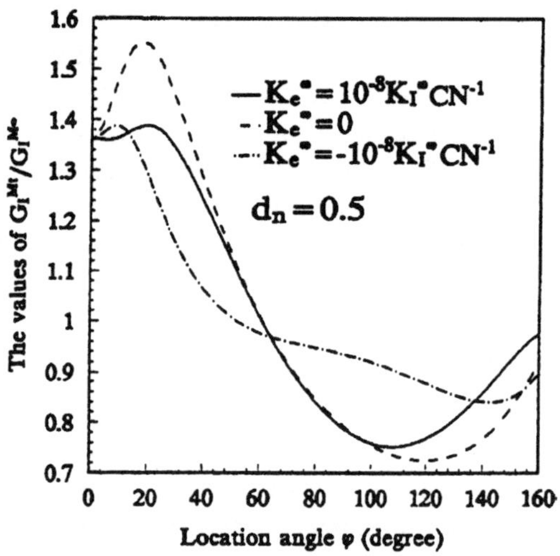

(b) The normalized distance $d_n = 0.5$.

Figure 6.10. Macrocrack-tip $MSERR$ against the location angle φ.

the *MSERR* influenced by the location angle of the parallel microcrack. For example, the maximum anti-shielding occurs at the point of the real curve in Figure 6.7 but at another point of the imaginary curve in Figure 6.10(a). This reveals that the conclusion derived by Park and Sun (1995a,b) is still held in the present situation, *i.e.*, neither the *SIF* nor the J integral are suitable to be the dominant parameter governing crack stability and growth for an impermeable crack in piezoelectric ceramics. They proposed the *MSERR* to be such parameter.

Let us discuss the major feature shown in Figures 6.10(a,b) in more detail. It is seen that as the angle φ increases from zero (corresponding to the collinear case) the *MSERR* increases until the maximum value is reached. After the maximum value the *MSERR* decreases as φ increases, until the minimum value is reached. It is seen also that the values of the *MSERR* considering the electric loading diverge far from those without considering the electric loading. This provides another evidence of the disturbance of the near-tip electric field owed to microcracking. What is more, the variable tendencies of the three curves in Figures 6.10(a,b) seem quite different in magnitude from those shown in Figure 6.7 for the mode I *SIF*. Although the electric loading still leads to shifting the maximum anti-shielding location and amplitude when $G_I^{Mt}/G_I^{M\infty} > 1$ and the maximum shielding location and amplitude when $G_I^{Mt}/G_I^{M\infty} < 1$, as contrasted with those in Figure 6.7, the electric loading, whether it is positive or negative, always leads to impeding the amplitude of the *MSERR* in the anti-shielding regions ($G_I^{Mt}/G_I^{M\infty} > 1$) and to decreasing the neutral shielding angle at which a transition from the anti-shielding effect to shielding effect occurs. Table 6.3 shows the maximum anti-shielding angles, the corresponding values of the *MSERR*, and the maximum shielding angles and the corresponding values of the *MSERR*. Table 6.4 shows the values of the *neutral shielding angle* under the three kinds of loading conditions mentioned above. Therefore, the anti-shielding region in the near-tip stress field under combined mechanical electric loading should always be smaller than that under purely mechanical loading, whatever the electric loading is positive or negative, owing to the disturbance of the microcrack in the near-tip electric field. Indeed, this conclusion reveals again that the electric loading plays an important role in microcrack shielding problems in piezoelectric solids (Chen and Han, 1999a,b).

Table 6.3. The maximum anti-shielding and maximum shielding angles and the *MSERR* values ($d_n = 0.3$).

	$K_I^\infty > 0, K_e^\infty > 0$	$K_I^\infty > 0, K_e^\infty = 0$	$K_I^\infty > 0, K_e^\infty < 0$
Maximum anti-shielding angle and the MSERR's value	12.5 degrees 1.6346	13.0 degrees 1.9024	6.0 degrees 1.6559
Maximum shielding angle and the MSERR's value	112 degrees 0.6952	126 degrees 0.6489	148.5 degrees 0.7849

Table 6.4. The neutral shielding angle.

	$K_I^\infty \succ 0, K_e^\infty \succ 0$	$K_I^\infty \succ 0, K_e^\infty = 0$	$K_I^\infty > 0, K_e^\infty < 0$
$d_n = 0.3$	58 degrees	59 degrees	51 degrees
$d_n = 0.5$	60.5 degrees	61 degrees	51.5 degrees

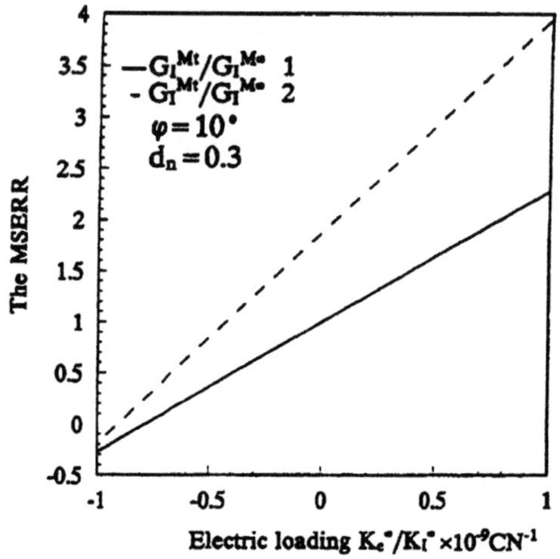

(a) $d_n = 0.3$ and $\varphi = 10^0$.

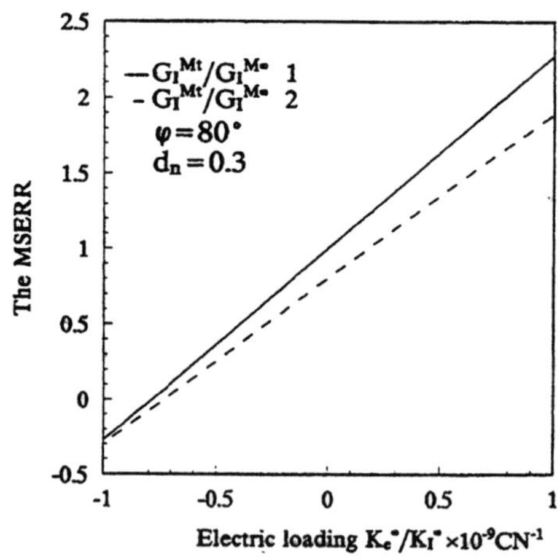

(b) $d_n = 0.3$ and $\varphi = 80^0$.

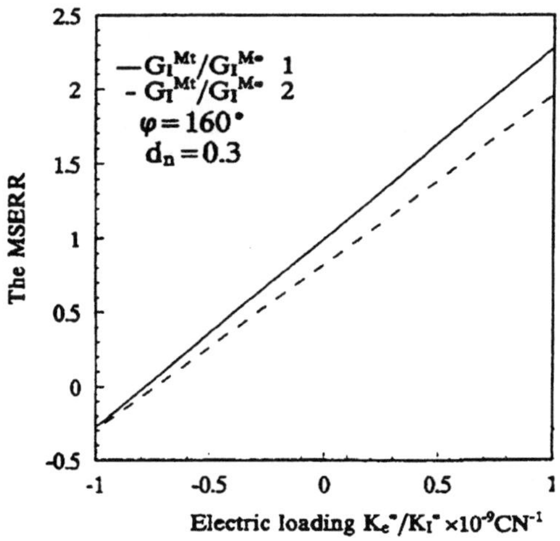

(c) $d_n = 0.3$ and $\varphi = 160^0$.

Figure 6.11. $MSERR$ against the variable electric loading.

The values of the *MSERR* against the variable electric loading are plotted in Figures 6.11(a,b,c) taking $d_n = 0.3$ and φ =10, 80, and 160 degrees, respectively. Here all the values of the *MSERR* are normalized by $G_I^{M\infty} = H_{22}(K_I^\infty)^2/4$. It is seen that the variable tendencies are also quite different from those for the traditional mode I *SIF*. In each of the figures from 6.11(a) to 6.11(c) two cases are considered for comparisons, one of which, denoted by the label 1, corresponds to the case not considering the microcrack, whilst the other denoted by label 2 corresponds to the case considering the microcrack. It is found that in both cases for every figure the variable tendencies of the *MSERR* against the electric loading are linear although the slope of the case 2 is always different from the slope of case 1. As the electric loading, *i.e.,* $K_e^\infty/K_I^\infty \times 10^{-9}CN^{-1}$ increases from -1 to $+1$, the normalized *MSERR* in both cases for every figure always increases proportionally. Detailed investigations reveal that in Figure 6.11(a) an anti-shielding effect of the microcrack could be seen since the values of the *MSERR* in the case 2 (with the microcrack) are always larger than those in the case 1 (without the microcrack), whilst in Figures 6.11(b, c) shielding effect could be seen since the values of the *MSERR* in the case 2 (with the microcrack)are always smaller than those in the case 1 (without the microcrack).

4.4 Oriented microcrack

Finally, the influence of the oriented angle of microcracking corresponding to inclined microcracks on the *MSERR* is poorly understood at present, which can be studied by using the proposed *PTED* method in a similar way.

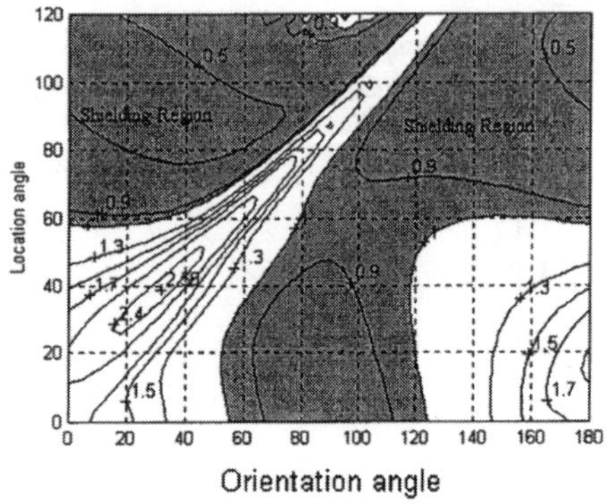

(a) Mechanical loading and positive electric loading $K_I^\infty > 0$, $K_e^\infty > 0$.

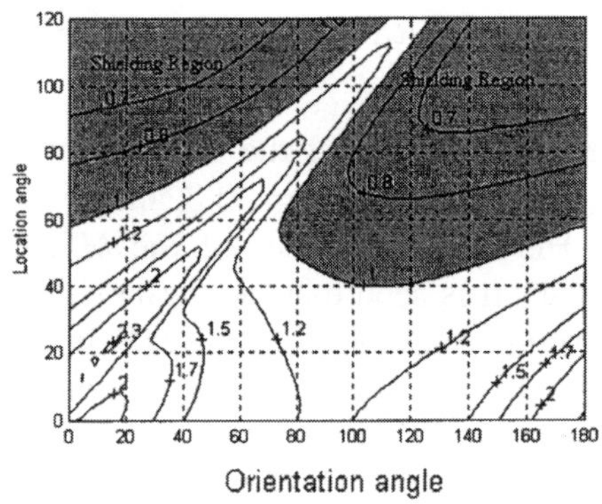

(b) Purely mechanical loading $K_I^\infty > 0$, $K_e^\infty = 0$.

The detailed manipulations are no longer repeated here for shortening the length of this chapter. Figures 6.12(a,b,c) show the anti-shielding regions (not dashed) and the shielding regions (dashed ones) corresponding to the above-mentioned three kinds of loading conditions. Indeed, it is seen that the positive electric loading always leads to increasing the shielding region and to decreasing the amplification region for inclined microcracks (Figure 6.12(a)), whilst the negative electric loading just leads to an opposite effect. What is more, we can see in Figure 6.12(c) that the shielding region is divided into three subregions rather than two subregions shown in Figures 6.12(a,b). When adopting the so called non-interacting crack assumption (*NICA*) proposed by Hutchinson (1987) for the cases of lower density of microcracking, the total shielding effect on the macrocrack tip induced from a cloud of microcracks in the near-tip

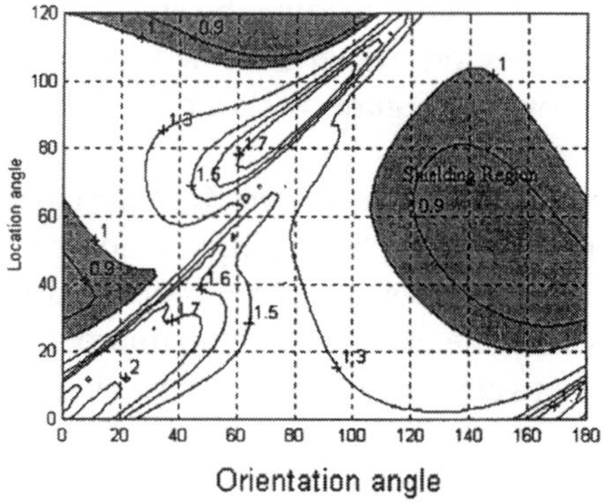

(c) Mechanical loading and negative electric loading $K_I^\infty > 0$, $K_e^\infty < 0$.

Figure 6.12. The anti-shielding and shielding regions for an oriented microcrack.

field under combined mechanical electric loading will be larger or smaller than that induced from the same cloud under purely mechanical loading. In other words, the electric loading may lead to some benefits in microcrack shielding problems in piezoelectric materials.

5. Brief Summary

(A) The conservation law of the J integral is still valid in shielding problems for piezoelectric materials. However, unlike those presented in the above chapters the J integral could not be used as a dominant parameter for an impermeable crack influenced by the near-tip microcracking. Therefore the only use of the conservation law is to provide a consistency check from which the derived results can be examined.

(B) The mechanical loading and the electric loading are coupled in the macrocrack microcrack interaction problem in piezoelectric ceramics since the microcracks in the near-tip process zone of a macrocrack not only release the near-tip residual stresses and reduce the effective elastic moduli in the zone, but also disturb the near-tip electric field. The latter provides another source of shielding.

(C) The electric loading plays an important role in microcrack shielding problems for piezoelectric ceramics. The positive electric loading always leads to increasing the shielding region and to decreasing the amplification region for inclined microcracks, whilst the negative electric loading just leads to an opposite effect. The boundaries between the shielding region and the anti-shielding region are seriously disturbed by the electric loading. For example, owing to the influence of the negative electric loading the shielding region in Figure 6.12(c) is divided into three subregions rather than two subregions under the other two kinds of loading as shown in Figures 6.12(a,b).

(D) For a fixed microcrack location the *MSERR* against the electric loading shows a linear nature as it could be in single crack problems. As the electric loading increases from a negative value to a positive value the *MSERR* increases proportionally. However, the existence of the microcrack influences the slope of the linear relation between the *MSERR* and the electric loading significantly. The slope does depend on the location of the microcrack.

(E) For an oriented microcrack, the numerical results have not been given sufficiently in this chapter owing to the limitation of length. The only conclusion can be that the electric loading may lead to some benefits in microcrack shielding problems in piezoelectric materials.

(F) It should be emphasized that only the impermeable crack (or charge-free crack) is discussed in this chapter. Other kinds of electric boundary conditions, *e.g.*, a conducting crack (or permeable crack), can yield quite different features from those discussed here. For example, Heyer *et al.* (1998) reported their important experimental data and numerical simulations for a conducting crack. They argued that the SIF (as does the J integral) at a conducting crack tip does still play a dominant parameter governing crack stability and growth. In that circumstance the redistribution of the J integral will describe the phenomenological shielding effect in piezoelectric ceramics, as concluded in the chapters above for traditional non-piezoelectric materials.

References for Chapter 6

Chen, Y.Z. (1984) General case of multiple crack problems in an infinite plate, *Engineering Fracture Mechanics*, **20**, 591 – 597.

Chen, Y. H. (1996) On the contribution of discontinuities in a near-tip stress field to J integral, *International Journal of Engineering Science*, **34**, 819–829.

Chen Y.H. and Han J.J. (1999) Macrocrack-microcrack interaction in piezoelectric materials, Part I: Basic Formulations and J-analysis, *ASME Journal of Applied Mechanics*, **66**, 514 – 521.

Chen Y.H. and Han J.J. (1999) Macrocrack-microcrack interaction in piezoelectric materials, Part II: Numerical results and Discussions, *ASME Journal of Applied Mechanics*, **66**, 522 – 527.

Chen Y.H. and Hasebe N. (1998) A consistency check for strongly interacting multiple crack problem in isotropic, anisotropic, and bimaterial solids, *International Journal of Fracture*, **89**, 333 – 353.

Heyer, V., Schneider, G.A., Balke, H., Drescher, J., and Bahr, H.A., (1998) A fracture criterion for conducting cracks in homogeneous poled piezoelectric $PZT - PIC151$ ceramic, *Acta Mater*, **46**, 6615 – 6622.

Horii, H. and Nemat Nasser, S. (1985) Elastic fields of interacting inhomogeneities, *International Journal of Solids and Structures*, **21**, 731 – 745.

Horii, M. and Nemat Nasser, S. (1987) Interacting microcracks near the tip in the process zone of a macrocrack, *Journal of the Mechanics and Physics of Solids*, **35**, 601 – 629.

Hutchinson, J. W. (1987) Crack tip shielding by microcracking in brittle solids, *Acta Metall*, **35**, 1605 – 1619.

Pak, Y. E. (1990) Crack extension force in a piezoelectric material, *ASME Journal of Applied Mechanics*, **57**, 647 – 653.

Pak, Y. E. (1992) Linear electric-elastic fracture mechanics of piezoelectric materials, *International Journal of Fracture*, **54**, 79 – 100.

Park, S. B. and Sun, C. T. (1995a) Fracture criteria for piezoelectric ceramics, *Journal of the American Ceramic Society*, **78**, 1475 – 1480.

Park, S. B. and Sun, C. T. (1995b) Effect of electric fields on fracture of piezoelectric ceramics, *International Journal of Fracture*, **70**, 203 – 216.

Park, S.B. and Carmann, G.P. (1997) Minimizing stress levels in piezoelectric media containing elliptical voids, *ASME Journal of Applied Mechanics*, **64**, 466 – 470.

Sosa, H. and Pak, Y. E. (1990) Three-dimensional eigenfunction analysis of a crack in a piezoelectric material, *International Journal of Solids and Structures*, **26**, 1 – 15.

Sosa, H. (1991) Plane problems in piezoelectric media with defects, *International Journal of Solids and Structures*, **28**, 491 – 505.

Sosa, H. (1992) On the fracture mechanics of piezoelectric solids, *International Journal of Solids and Structures*, **29**, 2613 – 2622.

Suo, Z., Kuo, C. M., Barnett, D. M. and Willis, J. R., (1992) Fracture mechanics for piezoelectric ceramics, *Journal of the Mechanics and Physics of Solids*, **40**, 739 – 765.

Chapter 7

CONSERVATION LAWS FOR MICROCRACK DAMAGE IN PIEZOELECTRIC MATERIALS

This chapter focuses on characterizing damaged anisotropic piezoelectric materials by using the principles of fracture mechanics, *i.e.*, the conservation laws and energy release rates. The interpretation and evaluation of the two components of the J_k vector along contours enclosing strongly interacting microcracks in two-dimensional piezoelectric materials are presented. The conservation laws of the J_k vector established by Budiansky and Rice (1973) in single crack problems for traditional non-piezoelectric materials and extended in the above chapters to interacting multiple cracks [Chen and Hasebe (1998) and Chen (2001a, b)] are re-examined for anisotropic piezoelectric materials containing interacting multiple cracks. The interaction problem for arrays of arbitrarily orientated and distributed microcracks subjected to combined mechanical and electrical loading is studied in detail. The contribution of the second component of the J_k vector, evaluated in the local coordinate system which is attached to each microcrack, to the J_k vector evaluated in the global coordinate system is calculated. It is found that the conservation laws of the J_k vector are still valid in damaged piezoelectric materials, although in the present problem the elastic and electric fields are coupled which add complications to the original formulations by Budiansky and Rice (1973). In other words, the total contributions from the microcracks to both components of the J_k vector evaluated in the global system vanish, provided that the selected closed contour encloses all microcracks and/or discontinuities (Chen and Lu, 2001). Detailed mathematical proof is presented and numerical results are shown in figures and tables to demonstrate the conservation laws and energy release rates.

1. General description of the present problem

As discussed in Chapter 6, crack-like defects are abundant in piezoelectric materials, and have been the subject of intensive study in the past decade (Pak,

1990, 1992; Sosa, 1991,1992; Suo *et al.*, 1992; Pak and Tobin, 1993; Dunn, 1994; Park and Sun, 1995a,b; Sosa and Khutoryyansky, 1996; Shindo *et al.*, 1997; Chung and Ting, 1996; Heyer, *et al.*, 1998; Park, *et al.*, 1998; Xu and Rajapakse, 1999). Recently, Han and Chen (1999) developed a 'pseudo-traction electric displacement' method to solve the problem of interacting multiple parallel cracks for transversely isotropic piezoelectric materials. It is shown that, as a result of crack interactions, the stress intensity factors (*SIF's*) at each crack tip are no longer independent of remote electric loading, in contrary to those for single crack problems (Sosa, 1992, Pak and Sun, 1995a,b). The formation of multiple parallel cracks next to a single crack not only relaxes the residual stresses but also disturbs the electric field in the vicinity of the crack. It is the latter source that leads to the indirect but significant influence of electric loading on the $SIF's$ as well as on the mechanical strain energy release rate $MSERR$. The work of Han and Chen (1999) is an initial attempt to solve the more general crack interaction problems, as only the first component of the J_k vector is concerned owing to the assumption that the multiple cracks are parallel with each other and perpendicular to the poling direction of the piezoelectric ceramics. For multiple cracks arbitrarily oriented and distributed in a piezoelectric ceramic such that they are not necessarily perpendicular to the poling direction of the material, the role of electric loading is unclear. Moreover, the validity of the conservation laws of the J_k vector in such situations needs to be established since Chen (2001a,b) has shown that the second component of the vector must be considered if the interacting cracks are no longer parallel (see equations (1.39) and (1.40) in Chapter 1).

This chapter presents an analytical and numerical study for conservation laws of the J_k vector (Budiansky and Rice, 1973) and their application, for strongly interacting and arbitrarily oriented microcracks in two-dimensional anisotropic piezoelectric materials. Following Suo *et al.*,(1992), the physical interpretation of the second component of the J_k vector in such materials is established (Chen and Lu, 2001). The problem is reduced to a system of integral equations, which are subsequently solved numerically by using the Chebyshev integration technique. The conservation laws of the J_k vector established for non-piezoelectric materials (Budiansky and Rice, 1973; Chen and Hasebe, 1998) are re-examined for the present interacting problem in piezoelectric materials; numerical results for a PTZ-4 ceramic are given to complement the analytical work. It is found that the total contributions from a cloud of microcracks in two-dimensional piezoelectric materials to each component of the J_k vector evaluated in a global system vanish, if the closed contour encloses all microcracks.

2. J_k vector in piezoelectric media: physical interpretation and conservation laws

2.1 Path independence of the two components of the J_k vector

For a plane crack in a two-dimensional anisotropic piezoelectric material, the J integral is given by (Suo *et.al.*, 1992):

$$J = \oint_C [(\frac{1}{2}\sigma_{ij}\varepsilon_{ij}dx_2 - n_i\sigma_{ip}\frac{\partial u_p}{\partial x_1}ds) - (\frac{1}{2}D_iE_idx_2 + n_iD_i\frac{\partial \phi}{\partial x_1}ds)], \quad (7.1)$$

where C refers to a closed contour starting from one point on the lower surface of the crack and ending at another point on the upper surface of the crack, and σ_{ij},ε_{ij}, u_p, D_i, E_i, and ϕ are the stresses, strains, displacements, electric displacements, electric field components, and electric potential, respectively. Here i, j, p =1 and 2, n_i is the outer normal to the contour C, and (x_1, x_2) is the Cartesian coordinate system.

By comparing equation (7.1) with the original interpretation of the J integral by Rice (1968) for non-piezoelectric materials, it is seen that the first two terms in equation (7.1) contributed by the mechanical quantities are identical to those in Rice's formulation, and the last two terms are additional ones introduced to account for the contributions from the electric quantities. In other words, although the electric and mechanical quantities are coupled in the constitutive equations for piezoelectric materials (see, *e.g.*, Suo *et al.*, 1992), they are not directly linked together in equation (7.1).

According to Budiansky and Rice (1973) the J integral is only the first component of the J_k vector ($k = 1, 2$), which is parallel to the crack surfaces and will be denoted below by J_1. The second component denoted by J_2 is perpendicular to the crack surfaces and can be introduced in a similar way as J_1 (Chen and Lu, 2001):

$$J_2 = \oint_C [(-\frac{1}{2}\sigma_{ij}\varepsilon_{ij}dx_1 - n_I\sigma_{Ip}\frac{\partial u_p}{\partial x_2}ds) - (-\frac{1}{2}D_iE_idx_1 + n_iD_i\frac{\partial \phi}{\partial x_2}ds)]. \quad (7.2)$$

Note that the J_2 integral is in general path dependent when the integral contour C encloses only one single tip of a crack. It is therefore likely to be far from significant as it would be in brittle materials (Herrmann and Herrmann, 1981). However, recent investigations of Chen and Ma (1997) and Chen (2001a) reveal that, for multiple crack interaction problems in traditional non-piezoelectric materials, the J_2 integral plays a role similar to that of J_1. This conclusion is reached by assuming, as suggested by Herrmann and Herrmann (1981), that the closed contour selected for calculating the J_2 integral encloses either one single crack completely or all cracks present in the material. Under this assumption, both components of the J_k vector are then path independent and hence have equal physical significance.

2.2 Conservation laws: statement

Based on the conclusions derived above in Chapters 1 and 2 for brittle materials, the new conservation laws of the J_k vector for a 2D piezoelectric material containing one or more cracks/discontinuities can be stated as: the total contributions to both J_1 and J_2 integrals evaluated along a closed path C_∞ (Figure 7.1) enclosing either a single crack completely or all cracks vanish. Mathematically, this can be written as (Chen and Lu, 2001):

$$\sum_{l=1}^{N} J_{1(l)} = 0, \tag{7.3}$$

$$\sum_{l=1}^{N} J_{2(l)} = 0, \tag{7.4}$$

where N refers to the total number of cracks/discontinuities in the material, and the subscript (l) refers to the lth crack such that $J_{1(l)}$and $J_{2(l)}$ denote the contributions of the lth crack to the first and second components of the J_k vector evaluated in a global system (x_1, x_2). The contribution of the lth crack evaluated in a local system (x_1^l, x_2^l) to the J_k vector, denoted separately by $J_{1(l)}^l$ and $J_{2(l)}^l$ (Figure 7.1), are related to $J_{1(l)}$ and $J_{2(l)}$ by:

$$J_{1(l)} = J_{1(l)}^l \cos\theta_l - J_{2(l)}^l \sin\theta_l, \tag{7.5}$$

$$J_{2(l)} = J_{1(l)}^l \sin\theta_l + J_{2(l)}^l \cos\theta_l, \tag{7.6}$$

where the superscript l is used to denote quantities that are evaluated in the local system (x_1^l, x_2^l), and θ_l is the orientation angle of the crack measured from the x_1-axis to the x_1^l-axis (Figure 7.1).

Equations (7.3)–(7.6) provide the new conservation laws of the J_k vector in strongly interacting multiple crack problems for piezoelectric materials, which

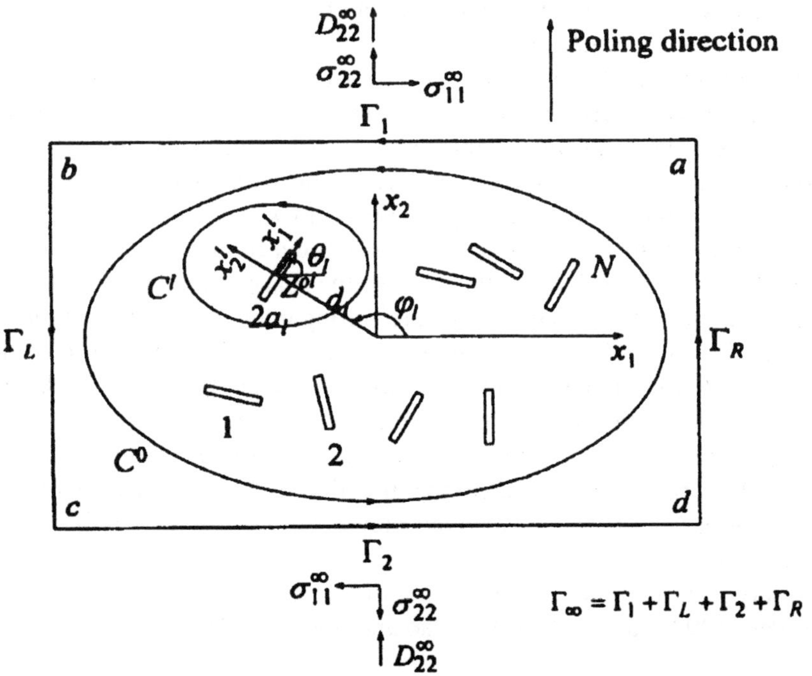

Figure 7.1. Geometry and conventions for a cloud of interacting cracks in an infinite piezoelectric ceramic.

show quite different aspects from those customarily used in single crack problems for traditional non-piezoelectric materials (see, *e.g.*, Rice, 1968, Budiansky and Rice, 1973, Herrmann and Herrmann, 1981). These have already been proved numerically (Chen and Hasebe, 1998) and analytically (Chen, 2001a) for non-piezoelectric brittle materials. Because equations (7.5) and (7.6) are concerned with the projected relations of the J_k vector between its values in the local system and those in the global system (Figure 7.1), they may also be termed as **the projected conservation laws of the J_k vector** for strongly interacting cracks in a 2D solid. For anisotropic piezoelectric materials, the conservation laws (7.3)–(7.6) should be re-examined with caution since the elastic and electric fields are coupled and the displacements and electric displacements add complications to the original formulations of the vector by Budiansky and Rice (1973).

2.3 Mathematical proof of conservation laws

Before performing lengthy mathematical manipulations to prove the conservation laws (7.3)–(7.6), it is worth defining the selection of electric boundary conditions as argued by a number of researchers in the recent years (Park and Sun, 1995a, Heyer, *et al.,* 1998).

The multiple crack interacting problem to be solved is shown in Figure 7.1, where N arbitrarily oriented microcracks are distributed in an infinitely large, two-dimensional anisotropic piezoelectric solid. Assume that all the microcracks are fully open and not intersecting with each other, and that the distance between the center of a typical microcrack, say the lth microcrack, and that of its immediate neighbor is of the same order of magnitude as the average length of the microcracks. In other words, the space among the cracks may be smaller than the crack length. Therefore, strong interaction amongst the cracks is expected. With reference to Figure 7.1, let a_l be the half length of the lth microcrack, θ_l its orientation angle, ϕ_l its location angle, and let d_l denote the distance of the center of the lth microcrack to the origin of the global system (x_1, x_2). The remote loading conditions are σ_{22}^{∞} and D_2^{∞} along the x_2-axis and σ_{12}^{∞} along the x_1-axis. The poling direction of the piezoelectric material is taken to be parallel to the x_2-axis but may not always be perpendicular to the microcracks.

For a typical microcrack, three types of electric boundary condition may be used (Pak, 1992, Suo *et al.,* 1992, Hao and Shen, 1994):

(i) Insulation boundary condition (charge free crack or impermeable crack):

$$D_n^+ = D_n^- = 0, \tag{7.7}$$

(ii) Conduction boundary condition (permeable or conducting crack):

$$D_n^+ = D_n^-, \quad \phi^+ = \phi^-, \tag{7.8}$$

(iii) Mixed boundary condition (Hao and Shen, 1994):

$$D_n^+ = D_n^-, \quad D_n^+(u_n^+ - u_n^-) = \varepsilon_v(\phi^- - \phi^+), \tag{7.9}$$

where D_n^+ and D_n^- are the upper and lower boundary values of the electric displacement on the crack, ϕ^+ and ϕ^- are the upper and lower boundary values of the electric potential on the crack, ε_v is the dielectric permittivity of the medium inside the crack (as a void), and u_n^+ and u_n^- are the upper and lower boundary values of the displacement on the crack, respectively.

Since the mechanical and electric quantities are coupled and the three different types of crack boundary condition would lead to significantly different behaviors of the energy release rate *(ERR),* there has been considerable effort to find dominant parameters that govern crack stability and growth in piezoelectric ceramics. For example, as mentioned in Chapter 6, Park and Sun (1995a,b) concluded that neither the stress intensity factor nor the J integral ($= J_1$) are suitable for describing the stability of an insulated crack; the mechanical strain energy release rate *(MSERR)* was proposed as a new fracture parameter. In contrary, Heyer *et al.*, (1998) performed a series of experimental measurements based on the conduction crack condition, and found that the stress intensity factor is still a suitable parameter to describe the stability of a conducting crack. On the other hand, Suo *et al.*, (1992) argued that the permeable condition is not realistic as there is always an electric potential drop across a crack. More recently, Xu and Rajapakse (1999) developed a unified formulation for a crack containing either air or vacuum in piezoelectric ceramics. However, despite all these efforts, the true electric boundary condition on the crack surfaces remains an issue to be resolved, and is a topic beyond the scope of this chapter. The purpose of this chapter is not to clarify the confusion regarding the three types of electric boundary condition but rather to explore if there exist conservation laws in piezoelectric materials containing strongly interacting cracks. If the conservation laws indeed exist, these should be valid irregardless of the type of electric boundary condition selected. Consequently, unless otherwise stated, the insulation boundary condition (7.7) will be used throughout the remainder of this chapter.

As there has been some doubt about the validity of equations (7.3) and (7.4), a detailed proof is given below, the key being the correct use of the remote uniform stress electric field. Introduce a closed but sufficiently large contour $\Gamma_\infty = \Gamma_1 + \Gamma_L + \Gamma_2 + \Gamma_R$ surrounding all the cracks, and introduce a smaller closed contour C_l which only encloses the lth crack completely (Figure 7.1). According to the path independent nature of the J_k integral vector, one has

$$J_1^\infty = \sum_{l=1}^{N} J_{1(l)},$$
$$J_2^\infty = \sum_{l=1}^{N} J_{2(l)}, \tag{7.10}$$

where the terms on the left hand side are calculated over Γ_∞, whilst the terms on the right hand side are calculated over C_l with l=*1, 2, . . . , N*. Since every term in the summation of equation (7.10) is in general not equal to zero owing to strong interaction amongst the cracks, it is not clear whether the summation should vanish, or not, as a number of people have suspected.

In order to take full advantage of the remote uniform stress electric field, the closed rectangular contour Γ_∞ has been chosen as large as possible. As mentioned in Chapter 1, the reason to introduce the rectangular closed contour rather than an arbitrary one is to significantly simplify the mathematical manipulations involved. However, owing to the path independent nature of the vector, the results evaluated along an arbitrary smooth closed contour C^0 (be it large or small, see Figure 7.1) should be identical to those calculated over Γ_∞. It is worth mentioning here that the second component of the vector is also path independent, because the closed contour is always chosen either to enclose all the cracks or to enclose a single crack completely (Herrmann and Herrmann, 1981).

Since the traction-free surfaces of each crack lead to nonzero contributions to the second component (Herrmann and Herrmann, 1981), the three different types of electric boundary condition will also lead to nonzero contributions to the last two terms of equation (7.2). Of course, different electric conditions are expected to yield different magnitudes of these contribution, but the path independent nature of the integral should not be altered if the closed contour is chosen as C_l (Figure 7.1). A universal formulation for a typical crack l in its local coordinate system (x_1^l,x_2^l) with θ_l as the orientation angle can always be written as

$$F_{2al}^* = -\int_{-a_l}^{a_l} [(w_m^+ - w_m^-) + (w_e^+ - w_e^-)]dx_{1*}^{(l)}, \tag{7.11}$$

where F_{2al}^* represents the contribution from the surfaces of the lth crack, the superscripts $+$ and $-$ denote the upper and lower boundary values on the crack surfaces, and the subscripts m and e refer to the mechanical energy and electric energy, respectively, given by:

$$\begin{aligned} W_m &= \frac{1}{2}\sigma_{ij}\varepsilon_{ij}, \\ W_e &= -\frac{1}{2}D_i E_i. \end{aligned} \tag{7.12}$$

Equations (7.11) and (7.12) will be calculated below by using a technique developed specially for the current situation.

Since $dy = dx_2 = 0$ on Γ_1 and Γ_2, and $dx = dx_1 = 0$ on Γ_L and Γ_R, the left hand side of equation (7.10) becomes

$$\begin{aligned} J_1^\infty = &\int_{\Gamma_1+\Gamma_2} [(-\sigma_{i2}^\infty n_2 \partial u_i/\partial x) + (-D_2 n_2 \partial\phi/\partial x)]ds \\ &+ \int_{\Gamma_L+\Gamma_R} (W_m + W_e)dy + \int_{\Gamma_L+\Gamma_R} [(-\sigma_{i1} n_1 \partial u_i/\partial x) + (-D_1 n_1 \partial\phi/\partial x)]ds, \end{aligned} \tag{7.13}$$

with that $n_2 = 1$ on Γ_1, $n_2 = -1$ on Γ_2 , $n_1 = 1$ on Γ_R and $n_1 = -1$ on Γ_L, (7.13) can be rewritten as

$$\begin{aligned} J_1^\infty &= \sigma_{i2}^\infty \left[\int_a^b u_{i,1} dx - \int_d^c u_{i,1} dx \right] + D_2^\infty \left[\int_a^b \phi_{,1} dx - \int_d^c \phi_{,1} dx \right] \\ &+ \left[\frac{1}{2} \sigma_{ij}^\infty \varepsilon_{ij}^\infty \left(\int_b^c dy - \int_a^d dy \right) \right] - \left[\frac{1}{2} D_i^\infty E_i^\infty \left(\int_b^c dy - \int_a^d dy \right) \right] \\ &+ \sigma_{i1}^\infty \left[\int_b^c u_{i,1} dy - \int_a^d u_{i,1} dy \right] + D_1^\infty \left[\int_b^c \phi_{,1} dy - \int_a^d \phi_{,1} dy \right]. \end{aligned} \tag{7.14}$$

Obviously the third and forth terms in equation (7.14) vanish because they have a term $\left[\int_b^c dy - \int_a^d dy \right] = 0$, whereas the first term depends on the value of $\left[\int_a^b u_{i,1} dx - \int_d^c u_{i,1} dx \right]$, the second term on the value of $\left[\int_a^b \phi_{,1} dx - \int_d^c \phi_{,1} dx \right]$, the fifth term on the value of $\left[\int_b^c u_{i,1} dy - \int_a^d u_{i,1} dy \right]$, and the sixth term on the value of $\left[\int_b^c \phi_{,1} dy - \int_a^d \phi_{,1} dy \right]$. It should be mentioned that $u_{i,1}$ and $\phi_{,1}$ in the integrals are defined at infinity since the rectangular closed contour has been chosen as large as possible.

Now, the remote displacement field must be linear with respect to both x- and y-coordinates so that the uniform strain field at infinity would be maintained. Similarly, the electric potential should also be linear in order that the uniform electric displacement field at infinity is sustained. In fact, the remote uniform mechanical electric fields do not depend on the configuration of the N interacting cracks, even if the array of the cracks is not symmetrical with respect to either $x = 0$ or $y = 0$. Therefore, the asymptotic values of $u_{i,1}$at infinity should be constant, *i.e.*, the integral $\left[\int_a^b u_{i,1} dx - \int_d^c u_{i,1} dx \right]$ should vanish. Similarly, the asymptotic values of $\phi_{,1}$ at infinity should also be constant and hence the integral $\left[\int_a^b \phi_{,1} dx - \int_d^c \phi_{,1} dx \right]$ vanishes.

The situation becomes much more clearer when the original problem shown in Figure 7.1 is divided into two sub-problems. The first involves no cracks, with linear remote displacement and electric potential fields with respect to both

x- and y-coordinates; the second involves N cracks with self-balanced tractions as well as self-balanced electric displacements acting on both surfaces of each individual crack, and no remote loading is applied. The detailed configuration of such subdivision is well known, as we discussed in Chapter 1, and hence is not repeated here. It suffices to mention that the second sub-problem yields zero displacement field at infinity because the stresses and strains caused by the self-balanced surface tractions have an asymptotic nature with the order of r^{-2} (for large values of $r = \sqrt{x^2 + y^2}$). Thus the remote displacement field is only dominated by the first sub-problem, resulting in $u_{i,1}$ (on Γ_1) = $u_{i,1}$ (on Γ_2). A similar result holds for the electric field at infinity. These results lead directly to the conclusion that the first, second, fifth and sixth terms in (7.14) should all vanish, no matter how many cracks are present in the finite region of an infinite 2D piezoelectric solid and whether the array of the cracks is symmetrical with respect to *x=0 (or y=0)* or not. Therefore, equation (7.3) as a new conservation law of the $J_1 (= J)$ integral has been proved. Its validity will be confirmed further in the next section using numerical examples. Similarly, a straightforward manipulation can be given for equation (7.4) representing a new conservation law of the J_2 integral. For brevity, details are not repeated here.

Obviously such a vanishing nature of the vector does not depend on the shape of the cracks enclosed by the closed contour C^0 (Figure 7.1). Therefore, other types of discontinuity such as curved cracks, bifurcated cracks, voids with arbitrary shapes, and inclusions, all enclosed by the contour C^0, will also satisfy equations (7.3) and (7.4). The detailed expressions of the contributions from different types of discontinuity to the vector may of course be different, but the total summation of the contributions should vanish. It is important to note that the manipulations performed hitherto are independent upon the type of the electric boundary condition imposed on the crack surfaces. In other words, the conservation laws are always valid, irregardless of the type of electric boundary condition selected.

3. Numerical techniques and examples

To prove the conservation laws with numerical examples, an initial attempt has been made by Han and Chen (1999) who considered parallel microcracks oriented in a direction perpendicular to the poling direction of a transversely isotropic piezoelectric material. Therefore, only the J_1 integral is calculated since the contribution of $J^l_{2(l)}$ to $J_{1(l)}$ in equation (7.5) vanishes owing to the fact that the orientation angle $\theta_l \equiv 0$ for this simple array of cracks.

For the problem shown in Figure 7.1, the situation is much more complicated than that considered by Han and Chen (1999). The randomly oriented and distributed microcracks are no longer parallel with each other; they also are not necessarily perpendicular to the poling direction of the material. It is then expected that the $J^l_{2(l)}$ integral in the local system will play an important role as it would be in non-piezoelectric materials (Chen, 2001a). Before this can be verified, it is important to first find an accurate method to calculate numerically $J^l_{2(l)}$ in the local coordinate system (x^l_1, x^l_2), since Herrmann and Herrmann (1981) have shown that, besides the two crack tips of the lth microcrack, the traction-free surfaces of the lth microcrack also contribute to $J^l_{2(l)}$. Such a method is introduced below (Chen and Lu, 2001).

To calculate $J^l_{2(l)}$, a typical crack of length $2a_l$ is shown in Figure 7.2 in a local coordinate system, where a special closed contour is introduced:

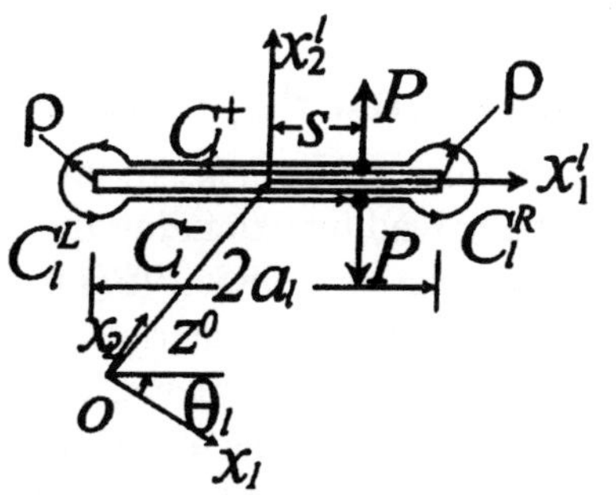

Figure 7.2. Special closed contour for a single crack in local coordinates.

$$C_l = C_l^- + C_l^R - C_l^+ - C_l^L, \tag{7.15}$$

where the superscripts $+$ and $-$ refer to the upper and lower surfaces of the crack, R and L refer to the right and left tips of the crack, and C_l^R and C_l^L denote two infinitely small circles with the radius $\rho \to 0$ (Figure 7.2), respectively.

The calculation of $J^l_{2(l)}$ along C_l can be divided into two parts, one contributed by both tips of the crack along path C_l^R and C_l^L, and the other induced from the crack surfaces along path C_l^+ and C_l^-. The special technique proposed by the present authors and co-workers (Chen and Ma, 1997) for anisotropic elastic materials is adopted here. The integration of J^t_2 along an infinitely small circle surrounding a crack tip may be obtained by using the following energy release rate relationship (Chen and Ma, 1997):

$$J_2^t = -g = \lim_{\Delta \to 0} \frac{-1}{2\Delta} \int_0^\Delta \sigma_{i1}(\Delta - \rho) u_i(\rho) d\rho, \quad (i = 1, 2), \tag{7.16}$$

where J_2^t refers to the contribution from a single crack tip to the J_2 integral and g denotes the special energy release rate as defined by Chen and Ma (1997) in anisotropic non-piezoelectric materials and by Chen and Lu (2001) in piezoelectric materials. Here readers should note that, unlike the conventional concept of ERR in fracture mechanics, the stress used in (7.16) is σ_{i1} rather than σ_{i2}.

The validity of (7.16) will be checked against numerical results presented below. As regards the contribution of both surfaces of the crack to $J_{2(l)}^l$, there are no special difficulties to be overcome as long as all mechanical and electric quantities along C^+ and C^- are known *a priori*. It will be shown in the next section that these quantities can be determined by employing the complex function method and performing the Chebyshev numerical integration.

3.1 Pseudo-traction electric displacement method

The interaction problem among N cracks in an anisotropic piezoelectric material (Figure 7.1) can be solved by extending the pseudo-traction method proposed by Horii and Nemat Nasser (1985, 1987) for solving multiple crack interaction problems in non-piezoelectric materials. The key here is to decompose the original problem into N sub-problems, each containing one single crack with pseudo-tractions and pseudo electric displacement assigned to both of its surfaces. This topic has been well addressed in Chapter 6. Once the fundamental solutions for a single crack loaded by concentrated tractions and concentrated electric displacement on both crack surfaces are found, the following set of Fredholm's integral equations can be derived from simple superposition (Han and Chen, 1999, Chen and Lu, 2001):

$$\mathbf{P}^l(s) + \sum_{j=0}^{N} \sigma^{lj}(s) = 0, \quad (l = 1, 2, 3, ..., N, j \neq l). \tag{7.17}$$

Here $\mathbf{P}^l(s)$ denotes a function vector with the unknown pseudo-tractions and electric-displacement on both surfaces of the Lth crack as its elements, and $\sigma^{l0}(s)$ refers to the stresses and electric displacement arising from the remote loading σ^∞ and D_2^∞ , given by

$$\sigma^{l0}(s) = \mathbf{R}_1 \sigma^\infty, \tag{7.18}$$

where

$$\sigma^\infty = [\sigma_{12}^\infty, \quad \sigma_{22}^\infty, \quad 0, \quad D_2^\infty]^T, \tag{7.19}$$

$$\mathbf{R}_1 = \begin{bmatrix} \frac{1}{2}\cos 2\theta, & \frac{1}{2}\sin 2\theta, & 0, & 0 \\ -\frac{1}{2}\sin 2\theta, & \cos^2\theta, & 0, & 0 \\ 0, & 0, & \cos\theta, & 0 \\ 0, & 0, & 0, & \cos\theta \end{bmatrix}. \tag{7.20}$$

Furthermore, in (7.17) $\sigma^{lj}(s)$ denotes the contribution of the rest of the microcracks to the lth crack :

$$\sigma^{li}(s) = \int_{-a_l}^{a_l} \mathbf{F}^{lj}(s,x)\mathbf{P}^{j}(X)dx, \quad (j \neq l), \tag{7.21}$$

where

$$\mathbf{P}^j(x) = [P_1^j(x), P_2^j(x), P_3^j(x), D_2^j(x)]^T, \tag{7.22}$$

and $\mathbf{F}^{li}(s,x)$ is the influence function on the lth crack induced from the jth crack (Han and Chen,1999, Chen and Lu, 2001).

The integral equation (7.17) can be solved numerically by expanding $\mathbf{P}^l(s)$ into the second Chebyshev polynomial:

$$\mathbf{P}^l_{(s)} = \sum_{k=0}^{M} \mathbf{G}_k^l \mathbf{U}_k(s/a_l), \tag{7.23}$$

where

$$\mathbf{G}_k^l = \int_{-1}^{1} 2\sqrt{1-s^2}\mathbf{P}^l(a_l x)\mathbf{U}_k(x)/\pi dx, \tag{7.24}$$

and $\mathbf{U}_k(s/a_l)$ is the second Chebyshev polynomial with order k, and M is an integer number controlling the numerical accuracy.

Once equation (7.17) is solved, the stress intensity factors *(SIF's)* and the electric displacement intensity factor *(EDIF)* at the right and left tips of the lth crack can be straightforwardly calculated as:

$$\mathbf{K}^{lR} = -\int_{-a_l}^{a_l} \mathbf{P}^l(s)(a_l+s)^{1/2}(a_l-s)^{-1/2}(\pi a_l)^{-1/2}ds, \tag{7.25}$$

$$\mathbf{K}^{lL} = -\int_{-a_l}^{a_l} \mathbf{P}^l(s)(a_l-s)^{1/2}(a_l+s)^{-1/2}(\pi a_l)^{-1/2}ds, \tag{7.26}$$

where

$$\mathbf{K}^{lR} = [K_{II}^{lR}, K_{I}^{lR}, K_{III}^{lR}, K_{e}^{lR}], \tag{7.27}$$

$$\mathbf{K}^{lL} = [K_{II}^{lL}, K_{I}^{lL}, K_{III}^{lL}, K_{e}^{lL}], \tag{7.28}$$

Here the subscripts I, II, and III refer to the Mode I, Mode II, and Mode III *SIF's*, respectively, the subscript e denotes to the electric displacement intensity factor, and the superscripts l, R, and L represent separately the lth crack, its right and left tips.

The first component of the J_k vector evaluated in the local coordinate system (x_1^l,x_2^l) originated at the center of the lth crack can be calculated as (see, for example, Suo, *et al.*, 1992)

$$J_1^{lR} = \frac{1}{4}(\mathbf{K}^{lR})^T \mathbf{H} \mathbf{K}^{lR}, \tag{7.29}$$

$$J_1^{lL} = \frac{1}{4}(\mathbf{K}^{lL})^T \mathbf{H} \mathbf{K}^{lL}, \tag{7.30}$$

On the other hand, from equation (7.16) it can be shown that the second component is given by:

$$J_2^{lR} = \frac{1}{4}(\mathbf{K}^{lR})^T \mathbf{X} \mathbf{H} \mathbf{K}^{lR}, \tag{7.31}$$

$$J_2^{lL} = -\frac{1}{4}(\mathbf{K}^{lL})^T \mathbf{X} \mathbf{H} \mathbf{K}^{lL}, \tag{7.32}$$

where the matrix **H** is

$$\mathbf{H} = 2Re(i\mathbf{A}\mathbf{B}^{-1}), \tag{7.33}$$

for which the matrices **X**, **A** , and **B** can be found in Suo *et al.*, (1992).

Note that J_2^{lR} and J_2^{lL} only represent contributions of the right and left tips of the lth crack to $J_{2(l)}^l$ in the local coordinate system (x_1^l, x_2^l); they are calculated along C_l^R and C_l^L as $\rho \to 0$ (Figure 7.2). To calculate the full value of $J_{2(l)}^l$, the contribution from both surfaces C_l^+ and C_l^- must included. However, since the integral equation (7.17) has already been solved numerically so that all the mechanical and electric quantities on the surfaces of the lth crack are now determined, the contribution of crack surfaces can be calculated numerically by using equation (7.11). Finally, the total contribution of the lth crack to the second component of the J_k vector in the local coordinate system (x_1^l, x_2^l) is obtained as

$$\begin{aligned} J_{2(l)}^l &= J_2^{lR} + J_2^{lL} + J_2^{lf}, \\ &= \frac{1}{4}(\mathbf{K}^{lR})^T \mathbf{XHK}^{lR} - \frac{1}{4}(\mathbf{K}^{lL})^T \mathbf{XHK}^{lL} + \int_{-a_l}^{a_l} (W^+ - W^-) ds, \end{aligned} \tag{7.34}$$

The contribution of the lth crack to the vector in the global system can then be evaluated by using the projected relations (7.5) and (7.6).

3.2 Numerical results

Although the results presented above are derived for general anisotropic piezoelectric materials, for simplicity, the numerical results to be given below will be limited to a transversely isotropic piezoelectric material, *i.e.*, the PTZ-4 ceramic, with its material constants given in Table 7.1. It is assumed that the poling direction of the ceramic is parallel to the x_2-axis (Figure 7.1).

Table 7.1. Material constants of PTZ-4 ceramics.

$\mathbf{C}_{11}$	$\mathbf{C}_{12}$	$\mathbf{C}_{13}$	$\mathbf{C}_{33}$	$\mathbf{C}_{44}$	**Unit**
1.4020	0.7892	0.7565	1.1577	0.2525	$(N/m^2) \times 10^{11}$
$\mathbf{e}_{31}$	$\mathbf{e}_{33}$	$\mathbf{e}_{15}$			
−5.2677	15.4455	12.0000			(C/m^2)
$\mathbf{w}_{11}$	$\mathbf{w}_{33}$				
0.6359	0.5523				$(C/Vm) \times 10^{-8}$

First, the formulation for calculating the J_2 integral induced only from one tip of a single crack embedded in the PTZ-4 ceramic is examined. Table 7.2 shows that the values derived by using equation (7.2) coincide well with those derived by using the energy release rate relation (7.16), no matter how the Mode I and Mode II stress intensity factors and the electric displacement intensity factor at infinity are combined. Judging from the results of Table 7.2 one may conclude that the derivative procedure, *i.e.*, equation (7.16), is not only valid for anisotropic elastic materials (Chen and Ma, 1997), but also valid for transversely isotropic piezoelectric ceramics.

The conservation laws of the J_k vector, *i.e.*, equations (7.3) and (7.4), are examined next by introducing an array of four microcracks (Figure 7.3), each of

Table 7.2. Numerical examination for the J_2 integral (7.2) and the energy release rate g in (7.16) under combined mechanical and electric loading conditions

$$(K_e^N = K_e/(10^{-8}K_I C N^{-1}))$$

$\frac{K_{II}}{K_I}$	$\frac{K_I}{K_I}$	$\frac{K_{III}}{K_I}$	K_e^N	$r = 0.1$		$r = 1.0$	
				J_2^t/J_1	$-g/J_1$	J_2^t/J_1	$-g/J_1$
0.5	0.5	0.0	0.0	1.31667	1.31667	1.31667	1.31667
0.5	1.0	0.0	0.0	1.06462	1.06462	1.06462	1.06462
1.0	0.5	0.0	0.0	1.04227	1.04227	1.04227	1.04227
0.5	0.5	0.0	1.0	−0.18119	−0.18119	−0.18119	−0.18119
1.0	1.0	0.0	1.0	−5.37522	−5.37522	−5.37522	−5.37522

length $2a$, with orientations and locations specified in Table 7.3. Here d_l, φ_l, θ_l , and $x_{1(l)}$, and $x_{2(l)}$ (l =1, 2, 3, and 4) denote separately the location distance from the origin, location angle, orientation angle, and position coordinates of the center of the lth microcrack in the global coordinate system (x_1, x_2). Note that the angle θ_1 is a variable, whilst other angles have certain values.

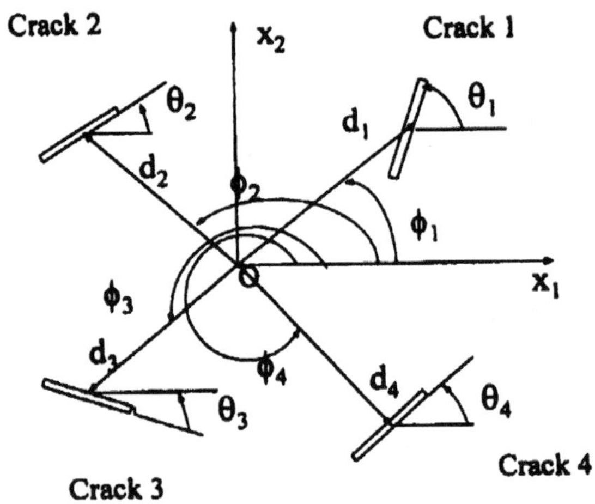

Figure 7.3. Configuration of four interacting cracks in a $PZT - 4$ ceramic.

Without loss of generality three types of remote loading condition are considered: (i) purely Mode I mechanical loading, with $\sigma_{12}^{\infty} = \sigma_{32}^{\infty} = D_2^{\infty} = 0$ and $\sigma_{22}^{\infty} > 0$; (ii) Mode I mechanical loading combining with positive electric loading, with $\sigma_{12}^{\infty} = \sigma_{32}^{\infty} = 0, \sigma_{22}^{\infty} > 0$, and $D_2^{\infty} = 10^{-8}\sigma_{22}^{\infty}CN^{-1}$; (iii) Mode I mechanical loading combining with negative electric loading, with $\sigma_{12}^{\infty} = \sigma_{32}^{\infty} = 0, \sigma_{22}^{\infty} > 0$,and $D_2^{\infty} = -10^{-8}\sigma_{22}^{\infty}CN^{-1}$. Numerical experi-

Table 7.3. Orientations and relative positions of the four microcracks.

$l=$	1	2	3	4
θ_l degrees	variable	−25	55	30
φ_l degrees	45	140	260	340
d_l	1.6a	1.4a	1.3a	1.6a
$x_{1(l)}$	1.1314a	−1.0725a	−0.2257a	1.5035a
$x_{2(l)}$	1.1314a	0.9000a	−1.2803a	−0.5472a

ments reveal that a value of 41 for the total number of terms M in the Chebyshev polynomial (7.23) yields satisfactory accuracy for all three types of remote loading condition; M=41 is therefore used in all subsequent numerical calculations.

Table 7.4. Values of $J_{1(l)}/J_O$ with (l=1,2,3,4) under combined mechanical and positive electric loading (type (ii) loading).

θ_1 degrees	0	30	60	90	120	150
$J_{1(1)}/J_O$	−0.0200	−0.2377	−0.0594	0.0032	−0.0536	−0.0020
$J_{1(2)}/J_O$	−0.0123	0.2293	0.0475	−0.0415	−0.0824	−0.0993
$J_{1(3)}/J_O$	0.0763	0.0960	0.1401	0.1278	0.0964	0.0774
$J_{1(4)}/J_O$	−0.0439	−0.0875	−0.1282	−0.0895	0.0396	0.0239
$\Sigma J_{1(l)}/J_O$	0.0000	0.0000	0.0000	0.0000	0.0000	0.0000

Table 7.5. Values of $J_{2(l)}/J_O$ with (l = 1,2,3,4) under combined mechanical and negative electric loading (type (iii) loading).

θ_1 degrees	0	30	60	90	120	150
$J_{2(1)}/J_O$	−0.1308	−0.0281	0.1020	0.0114	−0.0121	0.0144
$J_{2(2)}/J_O$	0.2673	0.1305	−0.0379	−0.0074	0.0386	0.1247
$J_{2(3)}/J_O$	−0.2342	−0.2310	−0.2538	−0.2831	−0.2639	−0.2419
$J_{2(4)}/J_O$	0.0977	0.1286	0.1897	0.2791	0.2374	0.1028
$\Sigma J_{2(l)}/J_O$	0.0000	0.0000	0.0000	0.0000	0.0000	0.0000

For type (ii) remote loading condition (Mode I mechanical loading combining with positive electric loading), Table 7.4 shows the values of $J_{1(l)}$ (l = 1,2,3,4) contributed by the four microcracks to the J integral. Here the J_k vector is normalized by

$$J_0 = \frac{1}{4}(\mathbf{K}_0)^T \mathbf{H}_0 \mathbf{K}_0, \tag{7.35}$$

where $\mathbf{K}_0$ and $\mathbf{H}_0$ are the fundamental solutions for a single crack, oriented parallel to the x_1-axis, subjected to the same type (ii) loading condition. The orientation angle θ_1 of the first microcrack in Table 7.3 is taken to be 0, 30, 60, 90, 120, and 150 degrees, respectively. It is seen from Table 7.4 that the summation of $J_{1(l)} (l = 1, 2, 3, 4)$ indeed vanishes, confirming the conservation law (7.3). Table 7.5 presents the numerical results for $J_{2(l)}$ under type (iii) loading condition (Mode I mechanical loading combining with the negative electric loading). Again, the summation of $J_{2(l)}$ (l = 1,2,3,4) is seen to vanish, as suggested by the conservation law (7.4). Thus, the conservation laws of the J_k-vector (7.3) and (7.4) for non-piezoelectric materials (Chen and Hasebe, 1998; Chen, 2001a) have been shown to be equally valid for multiple crack interaction problems in piezoelectric materials. That is, the summation of the contributions induced from all the microcracks to each component of the J_k vector always vanishes, even though the values of $J_{1(l)}$ and $J_{2(l)}$ for each individual microcrack are far from zero owing to strong crack interactions. Consequently, the conservation laws as formulated by equations (7.3) and (7.4) provide not only a consistency check of the numerical results presented in this section but also a powerful tool to examine the validity of the 'pseudo-traction electric displacement' method developed in the previous section.

4. Applications: two arbitrarily located interacting cracks

In this section the pseudo-traction electric displacement method is used to study in detail the interaction effect between two arbitrarily located microcracks (Figure 7.4) on crack tip parameters such as *SIF's* and *EDIF*.

The two cracks are embedded in an infinitely large PTZ-4 ceramic. Crack 1 is always parallel to the global coordinate axis x_1, whereas the position and orientation of crack 2 with respect to crack 1 are arbitrary. Let $2a$ denote the length of the microcracks, and let θ_2, ϕ_2 and d_{12} represent the orientation angle, location angle, and distance from tip A of crack 1 to the center O_2 of crack 2, respectively. As an illustration, the normalized distance between the two crack centers is taken as $d_{12}/a = 1.2$, the orientation angle of crack 2 is assumed to be ϕ_2= 60 degrees, whilst the location angle of crack 2, θ_2, varies from 0 to 180

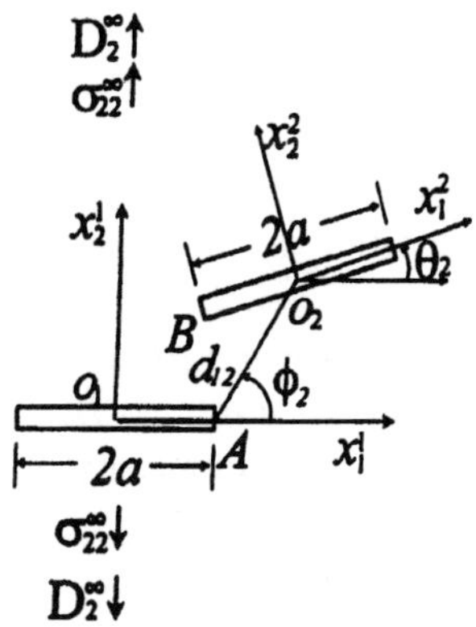

Figure 7.4. Two arbitrarily located cracks in a $PZT-4$ ceramic.

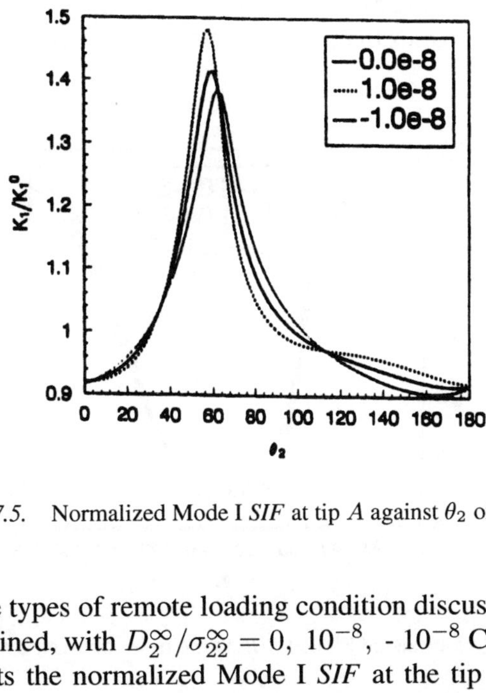

Figure 7.5. Normalized Mode I *SIF* at tip A against θ_2 of crack 2.

degrees. All three types of remote loading condition discussed in the previous sections are examined, with $D_2^\infty/\sigma_{22}^\infty = 0,\ 10^{-8}$, - 10^{-8} C/N, respectively.

Figure 7.5 plots the normalized Mode I *SIF* at the tip A of crack 1 (see Figure 7.4) as a function of the location angle θ_2 of crack 2 for type (i), (ii), and (iii) loading conditions. Although the trends are in general similar, there is considerable deviation of the dotted lines corresponding to type (ii) and (iii) loading conditions from the solid line corresponding to the pure mechanical loading (type (i)). Moreover, the positive electric loading (type (ii)) and negative electric loading (type (iii)) have opposite effects on the *SIF*. That is, for certain values of θ_2, *e.g.*, the range between 3 and 34 degrees or the range between 63 and 112 degrees, the positive electric loading tends to decrease the *SIF* whereas the negative electric loading tends to increase the *SIF*. For other ranges of θ_2, *e.g.*, the range between 34 and 62 degrees or the range between 112 and 180 degrees, the roles of the positive and negative electric loading reverse. Four electric-neutral angles *(ENA)*, 3, 34, 63 and 112 degrees, can be identified on Figure 7.5, where neither positive nor negative electric loading influences the value of *SIF*.

Of great interest is the influence of remote electric loading on the position of the maximum interacting effect angle *(MIEA)* (Figure 7.5). The *MIEA* is defined here as the location angle of crack 2 which leads to the maximum interacting effect between the two cracks. Figure 7.5 suggests that the electric loading shifts the value of *MIEA*, which is about 60 degrees under pure mechanical loading (type (i)): a positive electric loading decreases *MIEA* but increases the crack interaction effect; a negative electric loading increases *MIEA* whereas reduces the interacting effect. These are apparently nonlinear effects of the electric loading on *SIF*. In other words, although the positive and negative electric loadings have opposite effects on *SIF*, the effects are not proportional owing to crack interaction.

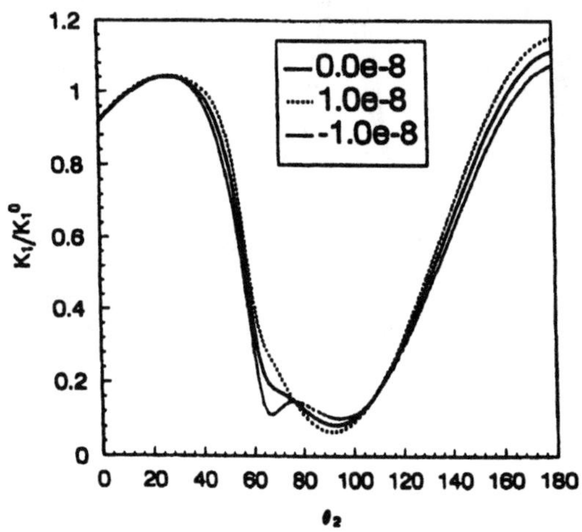

Figure 7.6. Normalized Mode I *SIF* at the tip *B* of crack 2 against θ_2.

Figure 7.6 presents the normalized Mode I *SIF* at the tip *B* of crack 2 as a function of θ_2. The trends are similar to those discussed above for Figure 7.5.

The results shown in Figures 7.5 and 7.6 for the two interacting cracks problem are quite different from those given in Sosa (1991,1992) for the single crack problem. Sosa (1991, 1992) found that the remote electric loading has no influence on the Mode I *SIF* at the tip of a single crack embedded in an infinitely large piezoelectric solid. In the present problem with two interacting cracks, the second crack centered at O_2 (Figure 7.4) not only releases stresses in the vicinity of tip A of the first crack centered at O_1, but also disturbs the near-tip electric field at A. Similarly, the stress and electric fields in the vicinity of tip B of crack 2 are disturbed by the presence of crack 1. These are believed to be the reasons behind the coupling between electric loading and Mode I *SIF* for strongly interacting cracks.

The electric displacement intensity factors *(EDIF)* at crack tips A and B are shown in Figures 7.7 and 7.8 as functions of θ_2, respectively. Again, it is seen that the positive and negative electric loadings have opposite effects on *EDIF*. However, in contrast to the case of Mode I *SIF* (Figure 7.5 and 7.6), their influence on *EDIF* is apparently proportional because the mechanical loading has little influence on *EDIF* (see the solid lines in Figures 7.7 and 7.8).

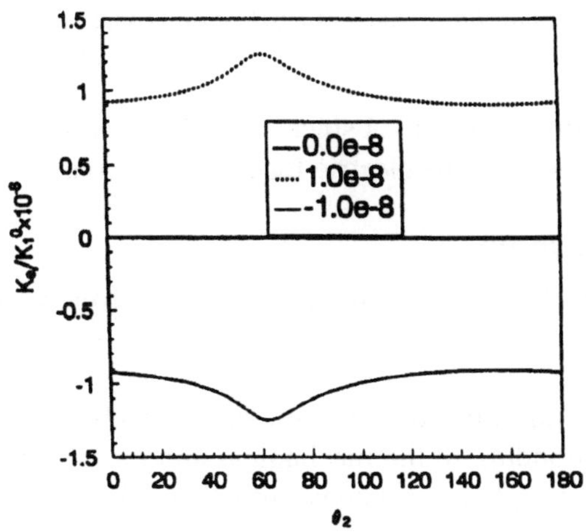

Figure 7.7. Values of *EDIF* at tip A against θ_2.

The maximum influence of remote electric loading on *EDIF* occurs at about 62 degrees for tip A (Figure 7.7) and at about 55 degrees for tip B (Figure 7.8). Moreover, the influence of electric loading on *EDIF* depends strongly upon the orientation of the second crack. An electric neutral angle of about 90 degrees, at which neither positive nor negative electric loading has influence on the *EDIF* at tip B, is seen to exist in Figure 7.8. It corresponds to the case that crack 2 is parallel both to the poling direction of the piezoelectric material and to the direction of remote electric loading. However, no such electric neutral

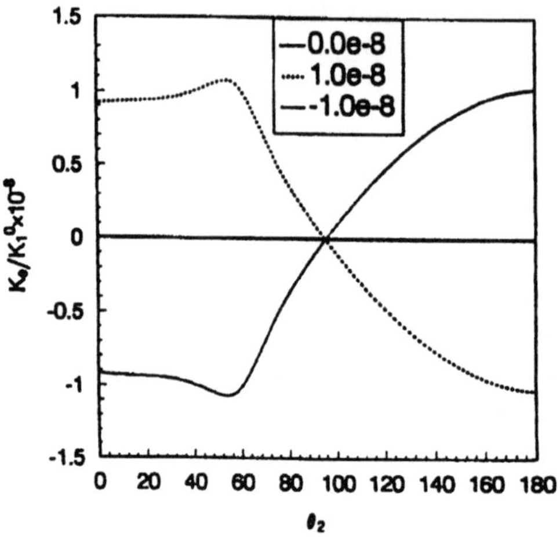

Figure 7.8. Values of *EDIF* at tip B against θ_2.

angle exists in Figure 7.7 for tip A since crack 1 is always perpendicular to the poling direction as well as the electric loading direction.

5. Brief Summary

Based on the results presented in this chapter, the following conclusions can be summarized:

(A) The J_2 integral in piezoelectric materials can be divided into two separated parts, corresponding to the mechanical and electric quantities, although both quantities are coupled in the constitutive equations of the materials.

(B) The conservation laws (1.39) and (1.40) found in non-piezoelectric materials also hold for piezoelectric materials, *i.e.,* equations (7.3) and (7.4). In other words, the total contributions to both components of the J_k vector owing to a cloud of randomly distributed, interacting microcracks in 2D piezoelectric materials vanish, providing that the closed contour along which the integration is performed encloses all the microcracks (or, equivalently, no other discontinuities exist outside of the closed contour).

(C) The conservation laws (7.3) and (7.4) can be used as a powerful tool to check the validity of numerical results and the corresponding numerical methods for strongly interacting multiple crack problems in piezoelectric materials, although they only provide two necessary conditions rather than sufficient conditions.

(D) The remote electric displacement loading has significant influence on the magnitudes of crack tip *SIF's* and *EDIF* when taking into account the interaction among many cracks. However, the trends exhibited by the *SIF's* at each crack tip subjected to combined mechanical and electric loading are similar to those under pure mechanical loading.

(E) Positive and negative electric loadings impose opposite effects on crack tip *SIF's* for interacting cracks. The influence is strongly nonlinear owing to crack interaction, depending on the relative locations and orientations of the cracks. For certain crack arrays, a positive electric loading increases the crack tip *SIF's* whereas a negative electric loading decreases the crack tip *SIF's*. The reverse is true for other crack arrays. In general, both electric and mechanical loadings can have profound effects on crack tip *SIF's*: neither can be neglected.

(F) In contrary to the crack tip *SIF's*, the effect of remote electric loadings on crack tip *EDIF* is linear, because the remote mechanical loading has little or no influence on *EDIF*. However, in the presence of remote electric loading (positive or negative), the variation trends exhibited by the *EDIF* as the positions and orientations of the cracks are varied are rather complicated, depending strongly on the orientation of each crack relative to the poling direction of the piezoelectric. Further research is needed to fully understand the mutual shielding effect in interacting cracks.

References for Chapter 7

Budiansky, B. and Rice, J. R. (1973) Conservation laws and energy-release rates. *ASME Journal of Applied Mechanics*, **40**, 201 – 203.

Chen, Y.H., (2001a) M-integral analysis for two-dimensional solids with strongly interacting microcracks, Part I. In an infinite brittle solid, *International Journal of Solids and Structures,* **38**, 3193 – 3212.

Chen, Y.H., (2001b) M-integral analysis for two-dimensional solids with strongly interacting microcracks, Part II. In the brittle phase of an infinite metal/ceramic bimaterial, *International Journal of Solids and Structures,* **38**, 3213 – 3232.

Chen, Y. H. and Hasebe, N. (1998) A consistency check for strongly interacting multiple crack problem in isotropic, anisotropic, and bimaterial solids, *International Journal of Fracture,* **89**, 333 – 353.

Chen, Y. H. and Ma, H. (1997) Explicit formulation of the J_2-integral in anisotropic materials and its application in microcrack shielding problem, *Science in China,* (E), **40**, 588 – 596.

Chen, Y.H. and Lu, T.J. (2001) Conservation laws of the J_k-vector for microcrack damage in piezoelectric materials, *International Journal of Solids and Structures,* **38**, 3233 – 3245.

Chung, M. Y. and Ting, T. C. T. (1996) Piezoelectric sold with an elliptic inclusion or hole, *International Journal of Solids and Structures,* **33**, 3343 – 3361.

Dunn, M. (1994) The effects of crack face boundary conditions on the fracture mechanics of piezoelectric solids, *Engineering Fracture Mechanics,* **48**, 25–39.

Han, J. J. and Chen, Y. H. (1999) Multiple parallel cracks interaction problem in piezoelectric ceramics, *International Journal of Solids and Structures,* **36**, 3375 – 3390.

Hao, T-H. and Shen, Z-Y. (1994) A new electric boundary condition of electric fracture mechanics and its application, *Engineering Fracture Mechanics,* **47**, 793 – 802.

Herrmann, A. G. and Herrmann, G. (1981) On energy-release rates for a plane crack. *ASME Journal of Applied Mechanics,* **48**, 525 – 528.

Heyer, V., Schneider, G.A., Balke, H., Drescher, J., and Bahr, H. A., (1998) A fracture criterion for conducting cracks in homogeneously poled piezoelectric *PZT-PIC151* ceramics, *Acta Mater.,* **46**, 6615 – 6622.

Horii, H. and Nemat Nasser, S. (1985) Elastic f ield of interacting inhomogeneities, *International Joural of Solids and Structures,* **21**, 731 – 745.

Horii, M. and Nemat Nasser, S. (1987) Interacting microcracks near the tip in the process zone of a macrocrack, *Journal of the Mechanics and Physics of Solids,* **35**, 601 – 629.

Pak, Y. E. (1990) Crack extension force in a piezoelectric material, *ASME Journal of Applied Mechanics,* **57**, 647 – 653.

Pak, Y.E. (1992) Linear electro-elastic fracture mechanics of piezoelectric materials, *International Journal of Fracture,* **54**, 79 – 100.

Pak, Y.E. and Tobin, A. (1993) On electric f ield effects in fracture of piezoelectric materials, *ASME Mechanics of Electromagnetic Materials and Structures* eds, Lee, J.S., Maugin G.A., and Shino Y, AMD-Vol. 161, MD-Vol. **42**, 51 – 62.

Park, S. B. and Sun, C. T. (1995a) Fracture criteria for piezoelectric ceramics, *Journal of the American Ceramic Society,* **78**, 1475 – 1480.

Park, S. B. and Sun, C. T. (1995b) Effect of electric f ields on fracture of piezoelectric ceramics, *International Journal of Fracture,* **70**, 203 – 216.

Park, S.B, Park, S.S, Carman, G.P, and Hahn, H. T, (1998) Measuring strain distribution during mesoscopic domain reorientation in ferrorelectric materials. *ASME Journal of Engineering Materials and Technology,* **120**, 1 – 6.

Rice, J. R. (1968) A path independent integral and the approximate analysis of strain concentration by notches and cracks, *ASME Journal of Applied Mechanics,* **35**, 379 – 386.

Shindo, Y., and Tanaka, K., and Narita, F. (1997) Singular stress and electric f ields of a piezoelectric ceramic strip under longitudinal shear, *Acta Mechanica,* **120**, 31 – 45.

Sosa, H. (1991) Plane problems in piezoelectric media with defects, *International Journal of Solids and Structures* , **28**, 491 – 505.

Sosa, H. (1992) On the fracture mechanics of piezoelectric solids, *International Journal of Solids and Structures*, **29**, 2613 – 2622.

Sosa, H. and Khutoryansky, N. (1996) New development concerning piezoelectric materials with defects, *International Journal of Solids and Structures,* **33**, 3399 – 3414.

Suo, Z., Kuo, C. M., Barnett, D. M. and Willis, J. R. (1992) Fracture mechanics for piezoelectric ceramics, *Journal of the Mechanics and Physics of Solids,* **40**, 739 – 765.

Xu, X-L. and Rajapakse, R.K.N.D. (1999) Analytical solution for an arbitrarily oriented void/crack and fracture of piezoceramics, *Acta Mater*, **47**, 1735 – 1747.

Chapter 8

SOME OTHER DEVELOPMENTS OF THE CONSERVATION LAWS AND ENERGY RELEASE RATES

This chapter deals with some recent developments of the conservation laws and energy release rates in various cases. Needless to say, in writing this book we have been able to draw upon a large number of works on the conservation laws published since 1968, and it is impossible for us to cover the whole subject given the limited space. For example, Cherepanov (1974,1979) might be the first to develop a conservation integral, an invariant Γ integral, for the combined action of mechanical stress and an electric field, Atluri (1982) studied path independent integrals in finite elasticity and inelasticity, Bakker (1984) studied the three-dimensional J integral, Park and Earmme (1986) studied the application of conservation integrals to interface cracks, Chen and Hasebe (1994a,b) studied Bueckner's integrals (1973) for interface cracks, and Fretcher (1976) and Kirshmoto *et al.,* (1980, 1982) applied the path independent integrals in dynamic and viscoelastic solids. Instead, besides the contents discussed in the above chapters, in this chapter we aim to provide recent progresses in this field, with focus placed on other new and novel applications. Because there are two apparent tendencies in recent years for the development of the laws, we attempt to discuss the major features that are worth noting and further studying. The first is the extension and application of the invariant integrals in treating multiple interacting microcracks (Chen, 2001a,b; Han and Chen, 1997, 1999, 2000a,b,c; Tian and Chen, 2000a,b,c), opening a new and rich research area in Damage Mechanics as we have discussed in Chapter 2, 3, 4, and 5. The second is associated with the direct extension of energy release rate concepts to treat crack-like defects in functional materials such as piezoelectric ceramics as we have discussed in Chapters 6, and 7, or in nanostructures (Chiu and Gao, 1993; Gao, 1994). For example, both the Energy Momentum Tensor and the Pseudo-Orthogonal Property (Chen, 1985) have been proved to be valid and useful for cracks in homogeneous piezoelectric materials or in dissimilar piezoelectric materials (Pak, 1990a,b; Ma and Chen, 2001a,b), respectively.

Besides those in the above chapters, the aim of this chapter is therefore to summarize some other developments in the above two new directions. We focus not only on the new and attractive features of the laws associated with the new applications, but also on the major difficulties that further advances in this area may be facing. In section 8.1 an application of the M integral to the Zener crack is discussed (Suo, 2000). In section 8.2 the topic as how to establish conservation laws or invariant integrals in more complex materials, *e.g.*, functional materials with defects, is discussed. At least two methods that are developed in the past decade will be discussed. The extension of the Energy Momentum Tensor (Eshelby 1956, 1970, 1975) in traditional structural materials to functional materials is presented. The illuminating results of Pak (1990a, b), Wang and Shen (1996) and others will be discussed in detail, and the main shortcoming of using the Energy Momentum Tensor to construct such integrals for functional materials is pointed out. In section 8.3 the Bueckner integral (1973) and the corresponding Pseudo-Orthogonal Property (Chen, 1985) for the eigenfunction expansion form (*EEF*) of cracks in functional materials are examined. More specifically, the Pseudo-Orthogonal Property of the *EEF* for cracks in piezoelectric materials (Ma and Chen, 2001a,b) is discussed. Also the main shortcoming of the Bueckner integral is pointed out. The inherent relations between the Bueckner (1973) Work Conjugate Integrals and the J and M integrals are presented. These are valid not only in traditional structural materials (Chen Y.Z, 1985; Chen and Shi, 1998; Chen and Tian, 1999; Chen and Ma, 2000) but also in piezoelectric ceramics as well as dissimilar materials (Ma and Chen, 2001a,b). This reveals that the J integral and the M integral are merely two special cases of the Bueckner integral associated with two different complementary states. In section 8.4 the application of the J_k integrals or energy release rates in nano-structures is reviewed (Chiu and Gao, 1993; Gao, 1994).

1. Application of the M integral to the Zener crack

Zener (1948) proposed that a dislocation pileup concentrates stress, which may cause a crack to nucleate. Any crystallographic discontinuity (*e.g.,* a grain boundary or a phase boundary) may act as an obstacle to block dislocation gliding. In a pair of bonded solids, when the interface blocks dislocations, the crack can nucleate either on the interface, or in one of the solids. The model has been analyzed by many authors in the last 50 years, as reviewed by Cottrell (1975) and more recently by Cherepanov (1993) and Fan (1994). A main result is the energy release rate of the crack, which has been obtained by solving elasticity boundary value problems for various special cases. When the two solids have dissimilar elastic properties, only the case where the crack lies on the interface has been solved. Suo (2000) calculates the energy release rate by using a path independent integral, *i.e.*, the M integral, following the procedure of Freund (1978). He gave several examples of cracks in an isotropic and homogeneous solid. Here the two solids are anisotropic and dissimilar. The crack can be either on the interface, or in one of the solids. Suo's (2000) work presents a new example of the M integral analysis for Zener's crack, which starts from the original formulation given by Knowles and Sternberg (1972), *i.e.*, equation (1.2). In fact, it should be mentioned that Kubo (1982) had already used the M integral to study cracks in dissimilar elastic materials. But his study was brief and had not received much attention in the literature.

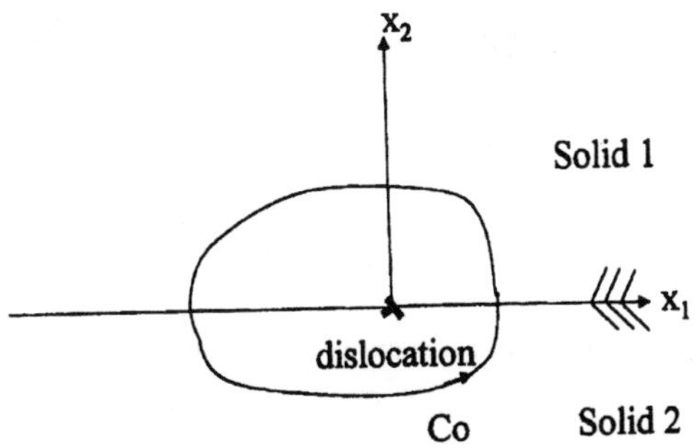

Figure 8.1. A dislocation with Burgers vector ***b*** lying on the interface.

If the solid is homogeneous along rays from the coordinate origin and if the closed contour C_0 encloses no singularity, $M = 0$; equation (1.19), that is, M is path independent. For a pair of dissimilar solids bonded together with a flat interface, the homogeneity requirement is satisfied by placing the coordinate origin at any point on the interface (see equations (2.33–2.35)). On the contrary, if C_0 encloses a singularity, M may not vanish. For example, Rice (1985) showed that the M integral along a closed curve surrounding a dislocation equals the pre-logarithmic factor of the dislocation energy. Consider a dislocation with Burgers vector **b** lying on the interface between two solids (Figure 8.1). Let the coordinate origin be placed at the dislocation and select a curve C_0 around the dislocation. The M integral is (Suo, 2000)

$$M = \frac{1}{2\pi}\mathbf{b}^{\mathrm{T}}\mathbf{H}^{-1}\mathbf{b}, \tag{8.1}$$

Here **H** is a positive-definite Hermitian matrix, and can be calculated once the elastic constants of the two solids are given (Suo, 1990). The location and shape of the curve C_0 do not affect the value of M. Furthermore, M is unaffected by the presence of an external boundary or other singularities in the solids, so long as C_0 encloses no other singularities than the dislocation such that the singular stress field owing to the dislocation prevails over the stress field owing to other sources.

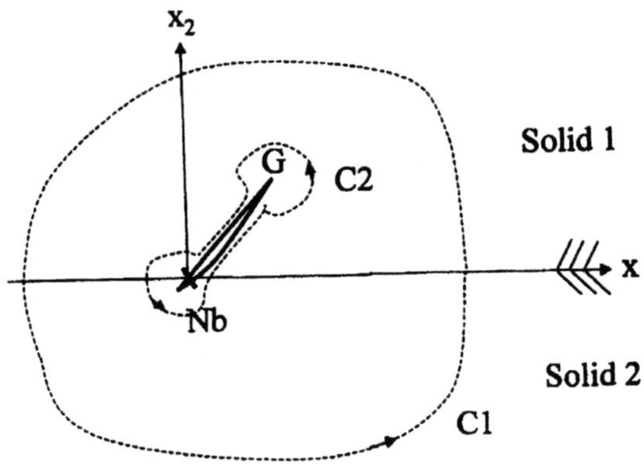

Figure 8.2. A crack of length *l* lies either on the interface or in one of the solids.

It has been shown that for a pair of anisotropic solids with a fixed relative orientation, under an in-plane coordinate rotation, **H** transforms like a second-order tensor (Tucker, 1969; Barnett and Lothe, 1974; and Ting, 1982). Consequently if the interface rotates with the coordinate, but the Burgers vector **b** remains fixed relative to the solids, the scalar number $\mathbf{b}^T\mathbf{H}^{-1}\mathbf{b}$ is invariant, and so is the value of M. Figure 8.2 illustrates a flat interface between a pair of semi-infinite solids. The interface blocks N dislocations, each having the Burgers vector ***b***. A crack of length l lies either on the interface or in one of the solids. All the dislocations glide into a crack and blunt one crack tip. The other crack tip advances either on the interface or in one of the solids. To calculate the energy release rate G at the advancing crack tip, let the origin of the coordinates be at the point where the dislocations are blocked. Because of the path independence, the M integral calculated along any curve enclosing the dislocation crack complex has the same value. First consider at the curve C_1 far away from the complex. At a distance far from the dislocations, $r \to \infty$, all the dislocations behave collectively like a single super-dislocation having a Burgers vector ***Nb***; the stress field owed to the super-dislocation decays as $1/r$, but the modification owed to the presence of the crack decays as $1/r^2$. Consequently the M integral over C_1 is identical to that over an isolated interface dislocation of the Burgers vector ***Nb***. Equation (8.1) is applicable once **b** is replaced by ***Nb***. Next consider at the curve C_2 tightly surrounding the dislocation crack complex. The stress field near the origin is now less singular than $1/r$, so that the small circle around the origin will contribute to the M integral, but the traction-free crack faces will not. It can be shown that the M integral over the small circle around the crack tip away from the origin equals lG (Freund, 1978). Equating the M integral evaluated over C_1 and C_2 we obtain that

$$G = \frac{N^2}{2\pi l}\mathbf{b}^T\mathbf{H}^{-1}\mathbf{b}. \tag{8.2}$$

For the case where the crack lies on the interface, Fan (1994) solved the elasticity boundary value problem using a complex variable method, and calculated the energy release rate from the stress field. Equation (8.2) agrees with Fan's result (1994). Furthermore, Suo (2000) demonstrated that the same equation is applicable to a crack located inside the solids. Let γ_1, γ_2, and γ_i be the surface energy per unit area of solid 1, solid 2, and the interface, respectively. When a crack advances a unit area, the surface energy increases by Γ, where $\Gamma = \gamma_1$ if the crack is in solid 1, or $\Gamma = \gamma_1 + \gamma_2 - \gamma_i$ if the crack is on the interface. Crack can nucleate if the energy release rate compensates the surface energy increase, namely, $G = \Gamma$, such that

$$\frac{N^2}{2\pi l}\mathbf{b}^T\mathbf{H}^{-1}\mathbf{b} = \Gamma. \tag{8.3}$$

The crack length l scales with N^2, everything else fixed. It is also interesting to predict crack orientation from this model. The value of Γ depends on the crystalline orientation. As previously mentioned, once the relative orientations of the two solids and the slip plane are fixed the factor $\mathbf{b}^T\mathbf{H}^{-1}\mathbf{b}$ is invariant with respect to the rotation of the interface or the crack. Consequently when the relative orientations of the two solids and the slip plane are fixed, the crack orientation is entirely determined by the anisotropy in Γ. According to this model, anisotropy of elastic constants plays no role in selecting the crack's orientation.

Equations (8.1), (8.2), and (8.3) reveal that the M integral concept can be combined with the well known dislocation concept and in this way crack nucleation and propagation on or out of an interface can be described. Obviously Suo's work (2000) can be easily extended to treat the similar Zener crack in two-phase piezoelectric materials. This research direction is worth noting and further studying.

2. Conservation laws in functional materials

Crack-like defects are abundant in functional materials such as piezoelectric ceramics and magnetostrictives, and have been the subject of intensive study in the past decade. Moreover, as discussed in the above chapters, conservation laws or invariant integral analyses have already been extended to treat cracks in functional materials. These invariant integrals can be derived in various ways, but only two commonly used ones are mentioned in this section: the Energy Momentum Tensor (Eshelby 1970, 1975) and the Bueckner Work Conjugate Integral (1973), although both were established for traditional structural materials.

2.1 Energy Momentum Tensor in piezoelectric materials

Pak (1990a,b) proposed the Energy Momentum Tensor and the J_k integral formulation in piezoelectric ceramics with a crack, built upon earlier results by Pak and Herrmann (1986). The investigation starts from Hamilton's principle (Tieresten, 1969):

$$\delta \int_V (-H)dV + \int_S (T_i \delta u_i - q\delta\phi)dS = 0, \tag{8.4}$$

where H is the electric enthalpy density, T_i is the applied surface traction, u_i is the displacement, q is the applied surface charge, and ϕ is the electric potential from which the electric field is obtained as:

$$E_i = -\phi_{,i}. \tag{8.5}$$

The variational formulation (8.4) is applied to a static linear piezoelectric material occupying a region V bounded by surface S. The electric enthalpy density is given by

$$H(\varepsilon_{ij}, E_j) = \frac{1}{2}C_{ijkl}\varepsilon_{ij}\varepsilon_{kl} - \frac{1}{2}\bar{\varepsilon}_{ij}E_iE_j - e_{ikl}\varepsilon_{kl}E_i, \tag{8.6}$$

where C_{ijkl} are the elastic moduli measured in a constant electric field, $\bar{\varepsilon}_{ij}$ are the dielectric constants measured at constant strain, e_{ikl} are the piezoelectric constants, and ε_{ij} are the strain:

$$\varepsilon_{ij} = \frac{1}{2}(u_{i,j} + u_{j,i}). \tag{8.7}$$

Pak (1990a,b) noted that the first term in (8.6) is the strain energy, the second term is the energy stored in the electric field, and the last term is the interaction energy. The variational formulation (8.4) provides the following fundamental equations for a linear piezoelectric material:

(i) Mechanical governing field equation:

$$\sigma_{ij,j} = 0, \tag{8.8}$$

(ii) Electric governing field equation:

$$D_{i,j} = 0, \tag{8.9}$$

(iii) Mechanical boundary condition:

$$\sigma_{ij} n_j = \bar{T}_i, \tag{8.10}$$

(iv) Electric boundary condition:

$$D_i n_i = -q, \tag{8.11}$$

(v) Constitutive equations:

$$\begin{aligned} \sigma_{ij} &= \frac{\partial H}{\partial \varepsilon_{ij}} = C_{ijkl}\varepsilon_{kl} - e_{kij}E_k \\ D_i &= -\frac{\partial H}{\partial E_i} = e_{ikl}\varepsilon_{kl} + \bar{\varepsilon}_{ik}E_k, \end{aligned} \tag{8.12}$$

where σ_{ij} are the stress, n_i is the unit normal vector, and D_i is the electric displacement vector.

Following the procedure proposed by Eshelby (1970) for a static elastic continuum, Pak (1990a,b) took the electric enthalpy density to be the Lagrangian density and simply differentiated it with respect to the spatial coordinate x_k:

$$\frac{\partial}{\partial x_k} H(x_i; u_{i,j}, E_i) = \left(\frac{\partial H}{\partial x_k}\right)^{\exp} + \frac{\partial H}{\partial u_{i,j}} u_{i,jk} + \frac{\partial H}{\partial E_i} E_{i,k}, \tag{8.13}$$

where $(\frac{\partial H}{\partial x_k})^{\exp}$ denotes explicit dependence of H on x_k with the other dependent and independent variables held constant. With the use of the field equations (8.8) and (8.9) and the constitutive equation (8.12), equation (8.13) becomes

$$\left(\frac{\partial H}{\partial x_k}\right)^{\exp} = [H\delta_{jk} - \sigma_{ij} u_{i,k} + D_j E_k], \tag{8.14}$$

where δ_{ik} is the Kronecke delta. When there are no discontinuities in the volume V and the material is homogeneous, then

$$(\frac{\partial H}{\partial x_k})^{\exp} = 0, \tag{8.15}$$

or, equivalently,

$$[H\delta_{jk} - \sigma_{ij}u_{i,k} + D_j E_k] = 0. \tag{8.16}$$

The quantity in the bracket is the well known Energy Momentum Tensor (Eshelby, 1970) but with the electric quantities added. By integrating the divergence of the Energy Momentum Tensor over the volume V and applying the well known divergence theorem, Pak (1990a, b) obtains the following invariant integrals (the J_k vector):

$$J_k = \int_C [H\delta_{jk} - \sigma_{ij}u_{i,k} + D_j E_k] n_j dS, \quad (k = 1, 2, 3). \tag{8.17}$$

These integrals are identical to those derived by Pak and Herrmann (1986) for elastic dielectric materials.

Consequently for a semi-inf inite crack in a two-dimensional piezoelectric material the J_k integral vector can be expressed as

$$\begin{aligned} J_1 &= J = \int_C [\frac{1}{2}(\sigma_{ij}\varepsilon_{ij})dx_1 - n_i\sigma_{ip}\frac{\partial u_p}{\partial x_1}ds] \\ &- \int_C [\frac{1}{2}D_i E_i dx_1 + n_i D_i \frac{\partial \varphi}{\partial x_1}ds], \quad (i, j, p = 1, 2), \end{aligned} \tag{8.18}$$

and

$$\begin{aligned} J_2 &= \int_C [-\frac{1}{2}(\sigma_{ij}\varepsilon_{ij})dx_1 - n_i\sigma_{ip}\frac{\partial u_p}{\partial x_2}ds] \\ &- \int_C [-\frac{1}{2}D_i E_i dx_1 + n_i D_i \frac{\partial \varphi}{\partial x_2}ds], \quad (i, j, p = 1, 2), \end{aligned} \tag{8.19}$$

where C refers to a closed contour starting from one point on the lower surface of the crack and ending at another point on the upper surface of the crack. Comparing equations (8.18) and (8.19) with the original J_k integral vector by Knowles and Sternberg (1972) for traditional non-piezoelectric materials, *i.e.,* equation (1.1) of Chapter 1, we find that the first two terms in equation (8.18) or (8.19) contributed by mechanical quantities are identical to those in equation (1.1), and the last two additional terms are introduced to account for the contributions from electric quantities. In other words, although the electric and mechanical quantities are coupled in the constitutive equations for piezoelectric materials (equation (8.12)), they are not directly linked together in equations (8.18) and (8.19) .

On the other hand, besides the piezoelectric materials studied by Pak (1990a, b) there remains the question of how to establish invariant integrals in more general cases, where all electric, magneto, thermal, and mechanical quantities are coupled in the near-tip field of a crack, or where the distinct values of crack tip parameters need to be calculated separately. Based on the Energy Momentum Tensor in traditional materials (Eshelby, 1956, 1970, 1975), Wang and Shen (1996) defined the Energy Momentum Tensor in linear electro magneto elastic media, from which invariant integrals like the J_k integral vector and the M integral are established. They also provided a method for establishing conservation laws or invariant integrals in more complicated scenarios. Ma and Chen (2001a,b) on the other hand found that there exists another way to establish invariant integrals in functional materials, which is based on Betti's reciprocal theorem (Sokolnikoff, 1956; Bueckner, 1973). It is interesting to note that some results derived from Betti's theorem are not attainable from the Energy Momentum Tensor. For example, Ma and Chen (2001a,b) proposed some weight functions for piezoelectric ceramics, which can be used to calculate the distinct values of mechanical and electric crack tip parameters without any special treatment of the near-tip singular field. Note also that Betti's reciprocal theorem was originally proved for elastic solids (Sokolnikoff, 1956) people may question whether Betti's theorem holds in the mechanical electric coupling fields. Ma and Chen (2001a,b) found that Betti's theorem can still be extended to treat crack-like defects in such coupling fields. The works by Wang and Shen (1996) and Ma and Chen (2001a,b) will be discussed in more detail below.

2.2 Energy momentum tensor in functional materials

Based on the concept of the Energy Momentum Tensor (Eshelby, 1970, 1975), Wang and Shen (1996) presented the conservation laws or invariant integrals in linear electro magneto elastic media, and proposed formulations of the J_k vector and the M integral for such coupling problems. Their manipulations started from the classical electromagnetic field and elasticity theories, with the following fundamental equations for an electro magneto elastic medium (Parton and Kudryavtsev , 1988; Mikhailov and Parton, 1990):

$$\begin{aligned}
E_i &= -\varphi_{,i}, \\
B_i &= e_{ijk}A_{k,j}, \\
\varepsilon_{ij} &= (u_{i,j} + u_{j,i})/2, \\
D_{i,i} &= 0, \\
e_{ijk}H_{k,j} &= 0, \\
\sigma_{ij,j} &= 0,
\end{aligned} \tag{8.20}$$

where φ, E_i and D_i have the same meanings as in (8.5), (8.9) and (8.12); A_i, B_i and H_i denote the magnetic potential vector, magnetic induction and field, respectively; and e_{ijk} is a permutation tensor. An alternative combination of equations (8.20) is

$$\begin{aligned}
E_i &= -\varphi_{,i}, \\
H_i &= -\psi_{,i}, \\
\varepsilon_{ij} &= (u_{i,j} + u_{j,i})/2, \\
D_{i,i} &= 0, \\
B_{i,i} &= 0, \\
\sigma_{ij,j} &= 0,
\end{aligned} \tag{8.21}$$

where ψ is the magnetic potential.

The state function, free energy ρf, ρF and complementary free energy ρg, ρG for linear functional materials can be written as:

$$\rho f(\varepsilon_{ij}, E_i, B_i) = \tfrac{1}{2}(\alpha^{ss}_{ijkl}\varepsilon_{ij}\varepsilon_{kl} - \alpha^{ee}_{ij}E_iE_j - \alpha^{mm}_{ij}B_iB_j)$$

$$-\alpha^{se}_{ijkl}\varepsilon_{ij}E_k - \alpha^{sm}_{ijk}\varepsilon_{ij}B_k - \alpha^{em}_{ij}E_iB_j,$$

$$\rho g(\sigma_{ij}, D_i, H_i) = \tfrac{1}{2}(-\xi^{ss}_{ijkl}\sigma_{ij}\sigma_{kl} - \xi^{ee}_{ij}D_iD_j - \xi^{mm}_{ij}H_iH_j)$$

$$-\xi^{se}_{ijk}\sigma_{ij}D_k - \xi^{sm}_{ijk}\sigma_{ij}H_k - \xi^{em}_{ij}D_iH_j,$$

$$\rho F(\varepsilon_{ij}, E_i, H_i) = \tfrac{1}{2}(\beta^{ss}_{ijkl}\varepsilon_{ij}\varepsilon_{kl} - \beta^{ee}_{ij}E_iE_j + \beta^{mm}_{ij}H_iH_j) \tag{8.22}$$

$$-\beta^{se}_{ijk}\varepsilon_{ij}E_k + \beta^{sm}_{ijk}\varepsilon_{ij}B_k - \beta^{em}_{ij}E_iH_j,$$

$$\rho G(\sigma_{ij}, D_i, B_i) = \tfrac{1}{2}(-\eta^{ss}_{ijkl}\varepsilon_{ij}\varepsilon_{kl} + \eta^{ee}_{ij}E_iE_j - \eta^{mm}_{ij}B_iB_j)$$

$$-\eta^{se}_{ijk}\varepsilon_{ij}E_k - \eta^{sm}_{ijk}\varepsilon_{ij}B_k - \eta^{em}_{ij}E_iB_j,$$

where the coeff icients α^{ss}_{ijkl}, β^{ss}_{ijkl}, ξ^{mm}_{ij}, η^{mm}_{ij} denote the characteristic features of the materials (Parton and Kudryavtsev , 1988; Mikhailov and Parton, 1990). Then

$$\sigma_{ij} = \frac{\partial \rho f}{\partial \varepsilon_{ij}}, \quad D_i = -\frac{\partial \rho f}{\partial E_i}, \quad H_i = -\frac{\partial \rho f}{\partial B_i}, \tag{8.23}$$

or

$$\varepsilon_{ij} = -\frac{\partial \rho f}{\partial \sigma_{ij}}, \quad E_i = \frac{\partial \rho f}{\partial D_i}, \quad B_i = \frac{\partial \rho f}{\partial H_i}, \tag{8.24}$$

or

$$\sigma_{ij} = \frac{\partial \rho F}{\partial \varepsilon_{ij}}, \quad D_i = -\frac{\partial \rho F}{\partial E_i}, \quad B_i = \frac{\partial \rho F}{\partial H_i}, \tag{8.25}$$

or

$$\varepsilon_{ij} = -\frac{\partial \rho G}{\partial \sigma_{ij}}, \quad E_i = \frac{\partial \rho G}{\partial D_i}, \quad B_i = -\frac{\partial \rho G}{\partial H_i}, \tag{8.26}$$

and

$$(\rho f)_{,k} = \frac{\partial \rho f}{\partial \varepsilon_{ij}}\mathbf{u}_{i,jk} - \frac{\partial \rho f}{\partial E_i}\varphi_{,ik} + \frac{\partial \rho f}{\partial B_i}\mathbf{e}_{ilm}\mathbf{A}_{m,lk}. \tag{8.27}$$

Using the constitutive relation (8.23) and the fundamental equations (8.8, 8.9, 8.10, 8.11), Wang and Shen (1996) found the following conservation equation:

$$(\rho f)_{,k} - (\sigma_{ij}\mathbf{u}_{i,k} + D_j\varphi_{,k} + \mathbf{e}_{ijm}H_m\mathbf{A}_{i,k})_{,j} = 0 \qquad (8.28)$$

or in the integral form,

$$\oint_C (\rho f\delta_{kj} - \sigma_{ij}\mathbf{u}_{i,k} - D_j\varphi_{,k} - \mathbf{e}_{ijm}H_m\mathbf{A}_{i,k})n_j ds = 0, \qquad (8.29)$$

It is worth noting that equations (8.28) and (8.29) hold only if the constitutive relation (8.23) is available. This implies that it is not necessary for ρf to satisfy equation (8.20), so either (8.28) or (8.29) is suitable for problems with non-linear constitutive relations, provided that the state function exists.

Wang and Shen (1996) emphasized that, in order to establish invariant integrals in more general cases, the key step is to define a new Energy Momentum Tensor as Eshelby (1970, 1975) did in purely elastic problems:

$$\mathbf{P}_{1kj} = \rho f\delta_{kj} - \sigma_{ij}\mathbf{u}_{i,k} - D_j\varphi_{,k} - \mathbf{e}_{ijm}H_m\mathbf{A}_{i,k}, \qquad (8.30)$$

When only considering electric field, (8.30) reduces to

$$\mathbf{P}_{1kj} = D_jE_k - \frac{1}{2}E_iD_i\delta_{kj}, \qquad (8.31)$$

This Energy Momentum Tensor is just the so called Maxwell stress tensor (Parton and Kudryavtsev , 1988; Mikhailov and Parton, 1990).

From equation (8.22) and the fundamental equation (8.8), equation (8.31) can be rewritten as

$$(x_j\mathbf{P}_{1jk})_{,k} = 3\rho f - (\sigma_{ij}\mathbf{u}_{i,j} + D_j\varphi_{,j} + \mathbf{e}_{ijm}H_m\mathbf{A}_{i,j}), \qquad (8.32)$$

where the free energy of a linear material is

$$\rho f = \frac{1}{2}(\sigma_{ij}\mathbf{u}_{i,j} + D_j\varphi_{,j} + \mathbf{e}_{ijm}H_m\mathbf{A}_{i,j}). \qquad (8.33)$$

Combining (8.32) with (8.33) leads to another conservation equation:

$$[x_j\mathbf{P}_{1jk} - \frac{1}{2}(\sigma_{ik}\mathbf{u}_i + D_k\varphi + \mathbf{e}_{ikm}H_m\mathbf{A}_i)]_{,k} = 0. \qquad (8.34)$$

Its integration form is expressed as

$$\oint_C [x_j \mathbf{P}_{l,jk} - \frac{1}{2}(\sigma_{ik}\mathbf{u}_i + D_k\varphi + \mathbf{e}_{ikm}H_m\mathbf{A}_i)]n_k ds = 0, \tag{8.35}$$

or

$$\begin{aligned}&\oint_C [x_k\rho f - x_j(\sigma_{ik}\mathbf{u}_{i,j} + D_k\varphi_{,j} + \mathbf{e}_{ikm}H_m\mathbf{A}_{i,j})]n_k ds \\ &-\tfrac{1}{2}\oint_C [\sigma_{ik}\mathbf{u}_i + D_k\varphi + \mathbf{e}_{ikm}H_m\mathbf{A}_i)]n_k ds = 0,\end{aligned} \tag{8.36}$$

which directly lead to the following invariant integrals:

$$J_k = \oint_C [\rho f\delta_{kj} - \sigma_{ij}\mathbf{u}_{i,k} - D_j\varphi_{,k} - \mathbf{e}_{ijm}H_m\mathbf{A}_{i,k})]n_j ds, \tag{8.37}$$

$$\begin{aligned}M &= \oint_C [x_k\rho f - x_j(\sigma_{ik}\mathbf{u}_{i,j} + D_k\varphi_{,j} + \mathbf{e}_{ikm}H_m\mathbf{A}_{i,j})]n_k ds \\ &- \frac{1}{2}\oint_C [\sigma_{ik}\mathbf{u}_i + D_k\varphi + \mathbf{e}_{ikm}H_m\mathbf{A}_i)]n_k ds = 0,\end{aligned} \tag{8.38}$$

For a piezoelectric material with no magneto-effects, equations (8.37) and (8.38) are reduced to

$$J_k = \oint_C [\rho f\delta_{kj} - \sigma_{ij}\mathbf{u}_{i,k} - D_j\varphi_{,k}]n_j ds, \tag{8.39}$$

$$\begin{aligned}M &= \oint_C [x_k\rho f - x_j(\sigma_{ik}\mathbf{u}_{i,j} + D_k\varphi_{,j})]n_k ds \\ &- \frac{1}{2}\oint_C [\sigma_{ik}\mathbf{u}_i + D_k\varphi]n_k ds = 0,\end{aligned} \tag{8.40}$$

Equation (8.39) represents the J_k vector in piezoelectric materials, which coincides well with equations (8.18) and (8.19) derived by Pak (1990a,b) in single crack problems and has been applied by Han and Chen (1999), Chen and Han (1999a,b), Chen and Lu (2001) in interacting crack problems as we have shown in Chapters 6 and 7. Moreover, when disregarding the electromagnetic field, equations (8.39) and (8.40) are in agreement with those derived by Knowles and

Sternberg (1972), Budiansky and Rice (1973), and Fletcher (1976) in elastic solids.

McMeeking (1990, 1999, 2001) did analytical treatments and Heyer *et al.*, (1998) conducted experiments on a conducting crack in dielectric or piezoelectric ceramics. They found that the *SIF* (and hence the J integral) could still be used as a suitable fracture criterion for conducting cracks. However, it is not the case for insulated cracks (Park and Sun, 1995a,b). The reasons are cited below.

Since the mechanical and electric quantities are coupled and the three different types of crack boundary condition (see, equations (7.7), (7.8), and (7.9) would lead to significantly different behaviors of the energy release rate *(ERR)* per unit crack tip advance, there has been considerable effort to find dominant parameters that govern crack stability and growth in piezoelectric ceramics. A typical crack in an infinite plane piezoelectric ceramic subjected to remote loading conditions $\sigma_{12}^{\infty}, \sigma_{22}^{\infty}$, and D_2^{∞} is shown in Figure 8.3. Liu and Sheng (2001) divided the J integral (*i.e.*, the *ERR*) of (8.18) into two distinct parts: the mechanical part and the electric part. The first part is only associated with the mechanical quantities in (8.18), whilst the second is only associated with the electric quantities in (8.18) although, as mentioned earlier, these quantities are all coupled in the constitutive equations. Numerical results of the two parts are plotted in Figure 8.4 as functions of crack oriented angle α for both impermeable and permeable cracks.

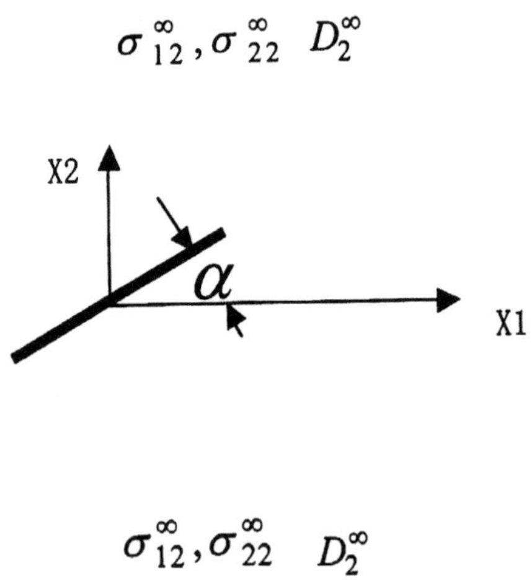

Figure 8.3. A crack rotation under remote mechanical electric loading.

For the impermeable crack the electric *ERR* (the dotted curve) against the crack oriented angle α is seen to be always negative, which coincides well with those obtained by Pak (1990a) for anti-plane fracture. Note that a negative electric energy implies that the energy is absorbed rather than released per unit crack tip advance. This conclusion is important since it is far from the classical concept of *ERR* for brittle non-piezoelectric materials. Furthermore, it explains why the *ERR* or the J integral could not be used as a fracture criterion for impermeable cracks (Park and Sun, 1995a, b). This is owing to the different roles that the mechanical *ERR* and the electric *ERR* play for the impermeable crack: the former releases whereas the latter absorbs when the crack grows. Consequently, the J integral can no longer be considered as a crack extension force because the electric *ERR* plays a role opposite to that of the mechanical *ERR*.

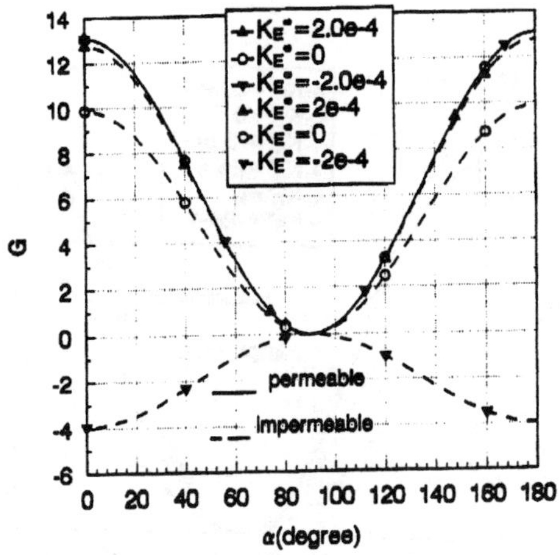

(a) Total energy release rates.

On the contrary, it is seen from Figure 8.4 that for the permeable crack the electric *ERR* (the real curve in Figure 8.4) is always much smaller than the mechanical *ERR*, and hence can be neglected. This result coincides well with Heyers' (1998) experimental observations and McMeeking's analytical treatments (1990, 1999, 2001). Consequently, the *ERR* or the J integral can still be used as a fracture criterion for conducting cracks in piezoelectric materials.

After clarifying the role that the J integral plays in piezoelectric materials we should point out another shortcoming of using the Energy Momentum Tensor. As mentioned earlier, this is mainly concerned with the inconvenience of calculating invariant integrals (J_k or M) in (8.37–8.40) since they can not distinguish the portion contributed by the *SIF's* ,K_I, and K_{II} from that caused by

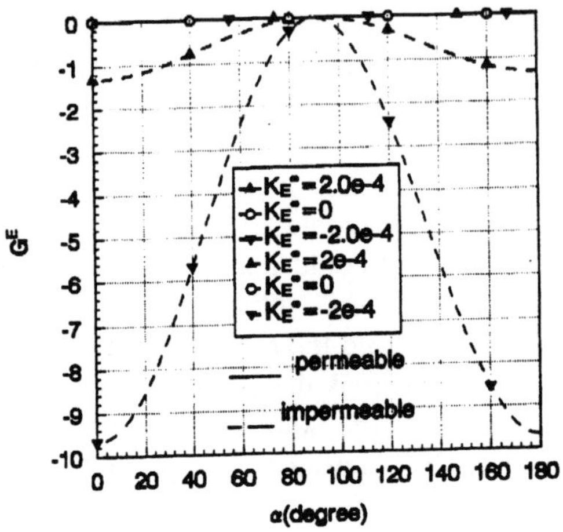

(b) Electrical *ERR*.

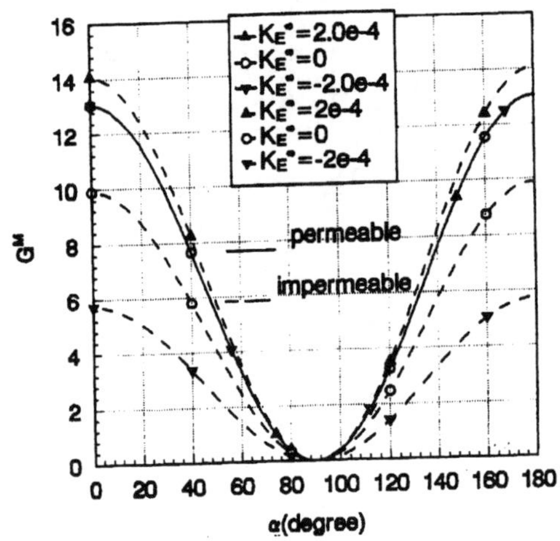

(c) Mechanical *ERR*.

Figure 8.4. Different aspects of *ERR's* for an impermeable crack and a permeable crack (α is the oriented angle of the crack with respect to the x_1-axis perpendicular to the poling direction and applied loading).

the crack electric displacement intensity factor (*EDIF*) K_e (Sosa, 1991, 1992). It is therefore disadvantageous to use them to study mixed mode fracture, *e.g.*, crack kinking and deflection in piezoelectric materials. Obviously, an alternative method should be proposed just as Bergez (1974), Freund (1978), and Stern *et al.*(1976) did for traditional materials, which can not only be used to compute path independent contour integrals, but also the separate values of K_I, K_{II}, and the K_e.

2.3 Bueckner's work conjugate integral in piezoelectric materials

An alternative method of establishing conservation laws or invariant integrals in functional materials is based on Betti's reciprocal theorem (Sokolnikoff, 1956; Bueckner, 1973; Rice, 1989). In this section this topic is reviewed by using piezoelectric ceramics as an example (Ma and Chen, 2001a,b). Some useful results and advantages of this method are discussed in detail, together with its shortcomings.

With reference to a fixed Cartesian coordinates system x_i (i=1,2,3), the linear constitutive relations and equilibrium equations for piezoelectric solids have been given by equations (8.8−8.12), respectively. For two-dimensional problems where u_k and ϕ depend on x_1 and x_2 only, the Stroh complex potentials (1958) can be used to find a general solution (Ting, 1982, 1986, 1990; Suo, 1990, 1992):

$$\mathbf{u} = \left[\ u_1,\ \ u_2,\ \ u_3,\ \ \phi\ \right]^{\mathrm{T}} = \mathbf{a} f\left(z\right),\ z = x_1 + p x_2, \tag{8.41}$$

with

$$\left[\mathbf{Q} + p\left(\mathbf{R} + \mathbf{R}^{\mathrm{T}}\right) + p^2 \mathbf{T}\right] \mathbf{a} = 0. \tag{8.42}$$

The 4×4 matrices **Q, R** and **T** are related to the elastic moduli, piezoelectric and dielectric coefficients of a piezoelectric material under consideration. $f(z)$ is a complex function. It is noted that D_i plays a role similar to σ_{ij} in (8.18) and (8.19) and u_i plays a role similar to ϕ in (8.18) and (8.19). Therefore it is convenient to introduce the generalized stress concept for D_i and generalized displacement concept for ϕ so that the formulations can be put in a compact form:

$$\begin{bmatrix} \sigma_{12} \\ \sigma_{22} \\ \sigma_{32} \\ D_2 \end{bmatrix} = \left(\mathbf{R}^{\mathrm{T}} + p\mathbf{T}\right) \mathbf{a} f'\left(z\right), \quad \begin{bmatrix} \sigma_{11} \\ \sigma_{21} \\ \sigma_{31} \\ D_1 \end{bmatrix} = \left(\mathbf{Q} + p\mathbf{R}\right) \mathbf{a} f'\left(z\right), \tag{8.43}$$

Solving the eigenvalue problem (8.42) for a piezoelectric material to obtain four pairs of complex roots denoted by $p_\alpha \quad (\alpha = 1, 2, 3, 4)$, Suo (1990), Suo *et al.*, (1992) and Sosa (1991, 1992) formulated all physical quantities as follows:

$$\mathbf{u} = \left[\ u_1, u_2, u_3, \phi\ \right]^{\mathrm{T}} = 2Re[\mathbf{A}\mathbf{f}(z)], \tag{8.44}$$

$$\varphi = [\ \varphi_1, \varphi_2, \varphi_3, \varphi_4\ ]^{\mathrm{T}} = 2Re[\mathbf{Bf}(z)], \tag{8.45}$$

$$[\sigma_{i2}] = [\varphi_{i,1}] = 2Re[\mathbf{Bf}'(z)], \tag{8.46}$$

$$[\sigma_{i1}] = -[\varphi_{i,2}] = 2Re[\mathbf{Lf}'(z)], \tag{8.47}$$

$$\mathbf{T} = [\ T_1, T_2, T_3, T_4\ ]^{\mathrm{T}} = -2Re[\mathbf{Bf}(z)], \tag{8.48}$$

where the matrices **A, B,** and **L** defined by Suo (1990) and Suo *et al.* (1992) have the following orthogonal property (Ting, 1986, 1990; Deng and Meguid, 1998):

$$\begin{array}{l} \mathbf{A}^{\mathrm{T}}\mathbf{B} + \mathbf{B}^{\mathrm{T}}\mathbf{A} = \overline{\mathbf{A}}^{\mathrm{T}}\overline{\mathbf{B}} + \overline{\mathbf{B}}^{\mathrm{T}}\overline{\mathbf{A}} = \mathbf{I} \\ \mathbf{B}\mathbf{A}^{\mathrm{T}} + \overline{\mathbf{B}\mathbf{A}}^{\mathrm{T}} = \mathbf{A}\mathbf{B}^{\mathrm{T}} + \overline{\mathbf{A}\mathbf{B}}^{\mathrm{T}} = \mathbf{I} \\ \mathbf{A}\mathbf{A}^{\mathrm{T}} + \overline{\mathbf{A}\mathbf{A}}^{\mathrm{T}} = \mathbf{B}\mathbf{B}^{\mathrm{T}} + \overline{\mathbf{B}\mathbf{B}}^{\mathrm{T}} = \mathbf{0} \\ \mathbf{A}^{\mathrm{T}}\overline{\mathbf{B}} + \mathbf{B}^{\mathrm{T}}\overline{\mathbf{A}} = \overline{\mathbf{A}}^{\mathrm{T}}\mathbf{B} + \overline{\mathbf{B}}^{\mathrm{T}}\mathbf{A} = \mathbf{0}. \end{array} \tag{8.49}$$

Consider a semi-infinite crack in an infinite piezoelectric material (see Figure 8.5) and assume the crack be both traction-free and charge-free (insulating crack or impermeable crack). Continuity across the crack:

$$[\sigma_{i2}]^{\pm} = 0 \tag{8.50}$$

leads to the following eigenfunction expansion form (*EEF*):

$$\begin{aligned} \mathbf{f}(z) &= \sum_{n=-\infty}^{\infty} \mathbf{f}_n(z) = \sum_{n=-\infty}^{\infty} [\ f_{n1}(z_1),\ \ f_{n2}(z_2),\ \ f_{n3}(z_3),\ \ f_{n4}(z_4)\ ]^{\mathrm{T}} \\ &= \sum_{n=-\infty}^{\infty} \left[\ A_n z_1^{\frac{n}{2}},\ \ B_n z_2^{\frac{n}{2}},\ \ C_n z_3^{\frac{n}{2}},\ \ D_n z_4^{\frac{n}{2}},\ \right]^{\mathrm{T}}, \end{aligned} \tag{8.51}$$

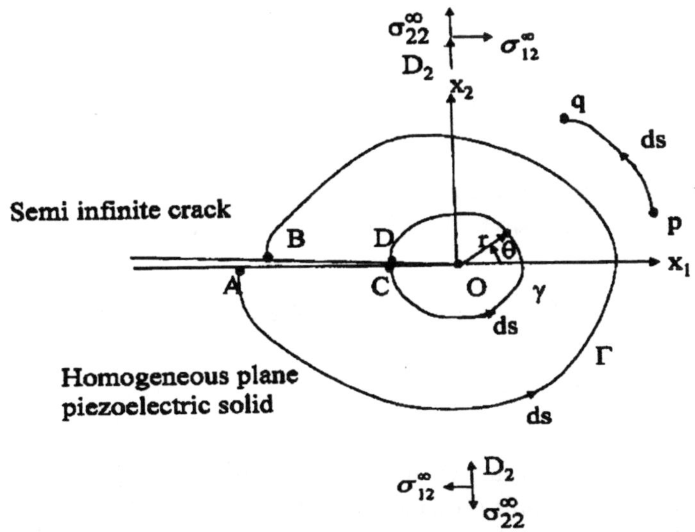

Figure 8.5. A semi-infinite crack in a plane piezoelectric solid.

where n is an integer and A_n, B_n, C_n, D_n are complex coefficients satisfying certain dependent relations. Here, the negative eigenvalues with $n < 0$ are considered for establishing invariant integrals (Chen, 1985; Chen and Hasebe, 1994a,b, 1995; Hui and Riana, 1995; Chen and Hasebe, 1997). Without going into details, the Bueckner Work Conjugate Integral (1973) is adopted below along the closed contour Γ, provided that the Betti's reciprocal theorem (Sokolnikoff, 1956) holds in the present mechanical electric coupling problem (see Figure 8.5):

$$B = \int_{\Gamma} \left(u_i^{\alpha} \sigma_{ij}^{\beta} - u_i^{\beta} \sigma_{ij}^{\alpha} \right) n_j ds + \int_{\Gamma} \left(\phi^{\alpha} D_k^{\beta} - \phi^{\beta} D_k^{\alpha} \right) n_k ds, \quad (i = 1, 2, 3; j, k = 1, 2). \tag{8.52}$$

With the generalized stress $\sigma_{4j} = D_j$ and generalized displacement $u_4 = \phi$, (8.52) can be rewritten in a new compact form, as:

$$B = \int_{\Gamma} \left(u_i^{\alpha} \sigma_{ij}^{\beta} - u_i^{\beta} \sigma_{ij}^{\alpha} \right) n_j ds, \quad (i = 1, 2, 3, 4; \quad j = 1, 2; \quad u_4 = \phi, \quad \sigma_{41} = D_1, \quad \sigma_{42} = D_2) \tag{8.53}$$

or

$$B = \int_{\Gamma} \left(u_i^{\alpha} dT_i^{\beta} - u_i^{\beta} dT_i^{\alpha} \right), \qquad (i = 1, 2, 3, 4), \qquad (8.54)$$

where the superscripts α and β refer to two different mechanical electric fields satisfying (8.50). The two fields yield:

$$\begin{aligned} B = & \int_{\Gamma} \left[\mathbf{Af}(z) + \overline{\mathbf{Af}(z)} \right]^{\beta \mathrm{T}} d \left[\mathbf{Bf}(z) + \overline{\mathbf{Bf}(z)} \right]^{\alpha} \\ & - \int_{\Gamma} \left[\mathbf{Af}(z) + \overline{\mathbf{Af}(z)} \right]^{\alpha \mathrm{T}} d \left[\mathbf{Bf}(z) + \overline{\mathbf{Bf}(z)} \right]^{\beta}, \end{aligned} \qquad (8.55)$$

where Γ refers to an arbitrary closed contour starting from point A on the lower surface and ending at point B on the upper surface of the crack. Integrating each term in (8.55) by parts and noting the orthogonal property (8.49), Ma and Chen (2001a) separated the coupled complex arguments, z_1, z_2, z_3 and z_4, defined in (8.41) for the plane piezoelectric material. This uncoupling technique was developed by Chen and Hasebe (1995) and Chen and Ma (2000) for anisotropic materials. The final result in an explicit function form with the four separated complex arguments is given as:

$$B_{\gamma} = B_{\Gamma} = \int_{\Gamma} \left[\mathbf{f}^{\beta^{\mathrm{T}}} d\mathbf{f}^{\alpha} + \overline{\mathbf{f}}^{\beta^{\mathrm{T}}} d\overline{\mathbf{f}}^{\alpha} \right] = 2Re \int_{\Gamma} \left[\mathbf{f}^{\beta^{\mathrm{T}}} d\mathbf{f}^{\alpha} \right], \qquad (8.56)$$

where γ is another closed contour (see Figure 8.5).

In fact, the above manipulations provide a special case of Betti's reciprocal theorem in linear mechanical electric coupling problems, provided that the electric displacements and electric potential can be treated as generalized tractions and generalized displacement, respectively. We note that the mechanical electric work done by the generalized tractions of field α through the generalized displacements of field β equals to the mechanical electric work done by the generalized tractions of field β through the generalized displacements of field α.

Now we are ready to review a variety of invariant integrals established by Ma and Chen (2001b) for different purposes. Of most interest is that the Pseudo-Orthogonal Property of the *EEF* found in purely elastic materials (Chen, 1985; Chen and Hasebe, 1994a, 1995; Chen and Ma, 2000) is still valid in homogeneous piezoelectric materials. Without going into the details, Ma and Chen (2001a) derived that the term with $l/2$ and the term with $k/2$ in the *EEF* (8.51) have the following property:

$$B = \begin{cases} -2\pi l Im\left[\mathbf{E}_l^{\mathrm{T}}\mathbf{E}_k\right], & l+k=0, \\ 0\,, & l+k\neq 0. \end{cases} \tag{8.57}$$

where $\mathbf{E}_l = [A_l, B_l, C_l, D_l]$ and $\mathbf{E}_k = [A_k, B_k, C_k, D_k]$ denote two constant vectors corresponding to the two mechanical electric states arising from eigenvalues $l/2$ and $k/2$ in the *EEF* (8.51), respectively. For example, upon selecting a closed contour surrounding the crack tip with starting point A and ending point B (Figure 8.5) and taking any two terms of the *EEF* (8.51) as the $(\alpha)-$ state and $(\beta)-$ states,

$$\begin{aligned} \mathbf{f}^{\alpha}(z) &= \left[f_1^{\alpha}(z_1),\ f_2^{\alpha}(z_2),\ f_3^{\alpha}(z_3),\ f_4^{\alpha}(z_4)\right]^{\mathrm{T}} \\ &= \left[A_l z_1^{\frac{l}{2}},\ B_l z_2^{\frac{l}{2}},\ C_l z_3^{\frac{l}{2}},\ D_l z_4^{\frac{l}{2}}\right]^{\mathrm{T}}, \end{aligned} \tag{8.58}$$

$$\begin{aligned} \mathbf{f}^{\beta}(z) &= \left[f_1^{\beta}(z_1),\ f_2^{\beta}(z_2),\ f_3^{\beta}(z_3),\ f_4^{\beta}(z_4)\right]^{\mathrm{T}} \\ &= \left[A_k z_1^{\frac{k}{2}},\ B_k z_2^{\frac{k}{2}},\ C_k z_3^{\frac{k}{2}},\ D_k z_4^{\frac{k}{2}}\right]^{\mathrm{T}}. \end{aligned} \tag{8.59}$$

Equation (8.56) is readily reduced to equation (8.57).

Ma and Chen (2001a) found that the property (8.57) provides a similarly effective tool to establish conservation laws or invariant integrals, which will cover those established from the Energy Momentum Tensor (Eshelby, 1970, 1975) and those can not be deduced from the Energy Momentum Tensor. From equation (8.57) we see that although the mechanical and electric quantities are coupled, the Pseudo-Orthogonal Property of the *EEF* derived from the Bueckner integral still holds. Dramatically, this property plays a role similar to that in purely elastic cases (Chen, 1985, Chen and Hasebe, 1995; Chen and Ma, 2000). For example, by using the property, the relation between the Bueckner integral and the J integral in piezoelectric solids can be easily obtained as shown below.

Assume that the α-state is a physical state (including both mechanical and electric quantities):

$$\mathbf{f}^{\alpha}(z) = \sum_{l=0}^{\infty} \begin{bmatrix} A_l z_1^{\frac{l}{2}} \\ B_l z_2^{\frac{l}{2}} \\ C_l z_3^{\frac{l}{2}} \\ D_l z_4^{\frac{l}{2}} \end{bmatrix}. \tag{8.60}$$

Again, assume that the β-state is constructed by differentiating $\mathbf{f}^{\alpha}(z)$ with respect to the x_1-axis:

$$\mathbf{f}^{\beta}(z) = \frac{\partial \mathbf{f}^{\alpha}(z)}{\partial x_1} = \sum_{l=0}^{\infty} \begin{bmatrix} \frac{l}{2} A_l z_1^{\frac{l}{2}-1} \\ \frac{l}{2} B_l z_2^{\frac{l}{2}-1} \\ \frac{l}{2} C_l z_3^{\frac{l}{2}-1} \\ \frac{l}{2} D_l z_4^{\frac{l}{2}-1} \end{bmatrix}. \tag{8.61}$$

which leads to the following cross-relations with the α field:

$$\sigma_{ij}^{\beta} = \frac{\partial \sigma_{ij}^{\alpha}}{\partial x_1}, \quad u_i^{\beta} = \frac{\partial u_i^{\alpha}}{\partial x_1}, \quad \phi^{\beta} = \frac{\partial \phi^{\alpha}}{\partial x_1}, \quad D_j^{\beta} = \frac{\partial D_j^{\alpha}}{\partial x_1}. \tag{8.62}$$

Upon substituting (8.60) and (8.61) into the Bueckner integral (8.56) and noting the Pseudo-Orthogonal Property (8.57) it is found that only the first term in the *EEF* contributes to the integral, yielding

$$B = -\pi Im\left[\mathbf{E}_1^{\mathrm{T}}\mathbf{E}_1\right] = -\frac{1}{2}\mathbf{K}^{\mathrm{T}} Im\left(\mathbf{B}^{-\mathrm{T}}\mathbf{B}^{-1}\right)\mathbf{K} = \frac{1}{2}\mathbf{K}^{\mathrm{T}}\mathbf{H}\mathbf{K} = 2J, \tag{8.63}$$

where $\mathbf{E}_1 = [A_1, \ B_1, \ C_1, \ D_1]^{\mathrm{T}}$ and $\mathbf{K} = [K_{II}, K_I, K_{III}, K_e]^T$ is the *SIF's* vector ,with

$$\mathbf{E}_1 = \frac{1}{\sqrt{2\pi}}\mathbf{B}^{-1}\mathbf{K}. \tag{8.64}$$

Here $\mathbf{B}$ is a matrix depending on the material constants (Suo *et al.,* 1992). Consequently we conclude that the J integral is merely a special case of the Bueckner integral in plane piezoelectric materials when using the complementary state (8.61). Equation (8.63) is similar to that derived by Chen (1985) for purely elastic materials.

Similar conclusion has been reached for the M integral along a closed contour that surrounds a finite crack completely (see, *e.g.*, the closed contour C in Figure 1.1 of Chapter 1 but in a plane piezoelectric solid). Taking the following complementary state denoted by a superscript β

$$\sigma_{ij}^{\beta} = \sigma_{ij}^{\alpha} + x_l \frac{\partial \sigma_{ij}^{\alpha}}{\partial x_l}, \quad u_i^{\beta} = x_l \frac{\partial u_i^{\alpha}}{\partial x_l}, \quad \phi^{\beta} = x_l \frac{\partial \phi^{\alpha}}{\partial x_l}, \quad D_j^{\beta} = D_j^{\alpha} + x_l \frac{\partial D_j^{\alpha}}{\partial x_l}, \tag{8.65}$$

where the α state is a physical state for the finite crack in Figure 1.1 but for a plane piezoelectric solid, Chen and Tian (1999) found that the Pseudo-Orthogonal Property (8.57) yields

$$B = 2M. \tag{8.66}$$

In other words, the M integral is merely a special case of the Bueckner integral as Chen and Shi (1998) pointed out for a plane elastic solid. Indeed, almost all invariant integrals derived from the Energy Momentum Tensor can be reduced from the Bueckner integral as we have shown in (8.63) for the J integral and (8.66) for the M integral. However, some invariant integrals derived from the Bueckner integral seem very difficult (if is not impossible) to obtain by using the Energy Momentum Tensor, *e.g.*, (8.16) or (8.33). As Bueckner (1973, 1989) and Chen (1985, 1994a,b) pointed out, the most useful advantage of the work-conjugate integrals is that weight functions for distinctly calculating the crack tip dominant parameters can be easily constructed from the Pseudo-Orthogonal Property (8.57). This task seems very difficult (if it is not impossible) to be accomplished by using the Energy Momentum Tensor, *e.g.*, (8.16) or (8.33). Moreover, Ma and Chen (2001b) have extended the Bueckner integral analysis to treat interface cracks between dissimilar piezoelectric materials. They have proved that equations (8.63) and (8.66) hold in interface crack problems when some suitable complementary mechanical electric states are used respectively in the dissimilar piezoelectric materials.

For further illustration, an example taken from Ma and Chen (2001a) is presented below. Following the work by Bueckner (1973, 1989) in purely elastic materials, the complementary mechanical electric state is selected for establishing weight functions:

$$\mathbf{f}^{\alpha}(z) = \mathbf{f}_s^{\alpha}(z) + \mathbf{f}_r^{\alpha}(z), \tag{8.67}$$

is selected, where $\mathbf{f}_s^\alpha(z)$ represents the so called higher order singular terms (abbreviated as *HOST's*, see, textite.g., Hui and Riana, 1995; Chen and Hasebe, 1997), given by

$$\mathbf{f}_s^\alpha(z) = \begin{bmatrix} f_{1s}^\alpha(z_1) \\ f_{2s}^\alpha(z_2) \\ f_{3s}^\alpha(z_3) \\ f_{4s}^\alpha(z_4) \end{bmatrix} = \begin{bmatrix} A_{-1}^* z_1^{-\frac{n}{2}} \\ B_{-1}^* z_2^{-\frac{n}{2}} \\ C_{-1}^* z_3^{-\frac{n}{2}} \\ D_{-1}^* z_4^{-\frac{n}{2}} \end{bmatrix},$$

$$\mathbf{f}_r^\alpha(z) = \begin{bmatrix} f_{1r}^\alpha(z_1) \\ f_{2r}^\alpha(z_2) \\ f_{3r}^\alpha(z_3) \\ f_{4r}^\alpha(z_4) \end{bmatrix} = \sum_{n=0}^{\infty} \begin{bmatrix} A_n^* z_1^{-\frac{n}{2}} \\ B_n^* z_2^{-\frac{n}{2}} \\ C_n^* z_3^{-\frac{n}{2}} \\ D_n^* z_4^{-\frac{n}{2}} \end{bmatrix}. \tag{8.68}$$

For example, a stronger singular stress as well as electric displacement field with order $O\left(r^{-\frac{3}{2}}\right)$ is present at the crack tip, corresponding to the case $n = 1$. It is emphasized again that the term $\mathbf{f}_s^\alpha(z)$ in (8.67) is only valid outside a small circular zone surrounding the tip (*i.e.*, outside the nonlinear zone). The purpose of introducing the second term $\mathbf{f}_r^\alpha(z)$ is to cancel the generalized tractions (*i.e.*, including the electric displacement) on the external boundary induced from the first term, so that the potential (8.67) denotes a special state having no generalized traction on the external boundary for a given geometric configuration (Bueckner, 1973).

Now let the physical state be denoted by $u_i^\beta, \sigma_{ij}^\beta, \phi^\beta, D_j^\beta$, which can be expressed by using the so called inside expansion (Rice, 1988) as:

$$\mathbf{f}^\beta(z) = \begin{bmatrix} f_1^\beta(z_1) \\ f_2^\beta(z_2) \\ f_3^\beta(z_3) \\ f_4^\beta(z_4) \end{bmatrix} = \sum_{k=0}^{\infty} \begin{bmatrix} A_k z_1^{-\frac{k}{2}} \\ B_k z_2^{-\frac{k}{2}} \\ C_k z_3^{-\frac{k}{2}} \\ D_k z_4^{-\frac{k}{2}} \end{bmatrix}. \tag{8.69}$$

where A_k, B_k, C_k and D_k are unknown coefficients yet to be determined, although they satisfy some dependent relations to ensure the traction-free and charge-free conditions on crack surfaces. Substituting (8.67) and (8.69) into the Bueckner integral (8.56) and noting the Pseudo-Orthogonal Property (8.57) yields

$$
\begin{aligned}
B\left(u_i^{\alpha}, \sigma_{ij}^{\alpha}, u_i^{\beta}, \sigma_{ij}^{\beta}\right) &= 2\pi Im\left(A_{-1}^{*}A_1 + B_{-1}^{*}B_1 + C_{-1}^{*}C_1 + D_{-1}^{*}D_1\right) \\
&= 2\pi Im\left(\mathbf{E}_{-1}^{*\mathrm{T}}\mathbf{E}_1\right) \\
&= \sqrt{2\pi} Im\left(\mathbf{E}_{-1}^{*\mathrm{T}}\mathbf{B}^{-1}\right)\mathbf{K} \\
&= \int_{\Gamma} u_i^{\alpha}\sigma_{ij}^{\beta} n_j ds,
\end{aligned}
\tag{8.70}
$$

where equation (8.64) has been used.

Owing to the arbitrary nature of the vector $\mathbf{E}_{-1}^{*} = \left[A_{-1}^{*}, B_{-1}^{*}, C_{-1}^{*}, D_{-1}^{*}\right]$ in equation (8.68), we can set

$$
Im\left(\mathbf{E}_{-1}^{*\mathrm{T}}\mathbf{B}^{-1}\right) = [\delta_1, \delta_2, \delta_3, \delta_4] \tag{8.71}
$$

and rewrite equation (8.70) in an alternative form:

$$
[\delta_1, \delta_2, \delta_3, \delta_4]\begin{bmatrix} K_{\mathrm{II}} \\ K_{\mathrm{I}} \\ K_{\mathrm{III}} \\ K_e \end{bmatrix} = \frac{1}{\sqrt{2\pi}}\int_{\Gamma} u_i^{\alpha}\sigma_{ij}^{\beta} n_j ds \tag{8.72}
$$

where Γ is the outside boundary of the cracked piezoelectric body, along which the traction and the electric displacement are prescribed.

The Mode I, Mode II, Mode III *SIF's* and the electric displacement intensity factor K_e *(EDIF)* can be distinguished by using (8.72) and setting the following four cases: (i) $\delta_2 = \delta_3 = \delta_4 = 0$ and $\delta_1 = 1$ for calculating K_{II}; (ii) $\delta_1 = \delta_3 = \delta_4 = 0$ and $\delta_2 = 1$ for calculating K_I ; (iii) $\delta_1 = \delta_2 = \delta_4 = 0$ and $\delta_3 = 1$ for calculating K_{III} ; (iv) $\delta_1 = \delta_2 = \delta_3 = 0$ and $\delta_4 = 1$ for calculating K_e .

Consequently for any given geometric configuration of a plane anisotropic piezoelectric solid with an edge crack subjected to an arbitrary combination of traction and electric charge on its external boundary, the *SIF's* as well as the *EDIF* can be calculated from (8.72) together with corresponding potentials (8.67) and (8.68) without any special treatment of the near-tip region. Invariant integrals derived from the Energy Momentum Tensor, *e.g.*, (8.16) or (8.33), do not have such advantage. However, we should note that the Bueckner integral has an obvious shortcoming. This is because the *pseudo-orthogonal property* (8.57) is always based on the complex potential theory and the assumption of linear materials, otherwise the Bueckner integral would not have such a powerful effect on the construction of invariant integrals for functional materials.

3. Application of conservation laws of invariant integrals in nano-structures

Finally, we discuss the recent developments of conservation laws or invariant integrals in nano-structures and nano-materials (as another kind of functional materials). As is well known, the micro electro mechanical systems (*MEMS*) and advanced materials in the nanometer scalar become increasingly popular owing to their widely potential applications in constructing new micro-instruments and highly effective memory materials. These motivate a new interest in nano-mechanics. A detailed review has been given by Ortiz (1999). In this section, we focus our attention only on those concerned with conservation laws or invariant integrals developed in these new techniques. For example, Chiu and Gao (1993) and Gao (1994) presented an application of the J integral in nanomechanics with defects (singularities). Chiu and Gao (1993) found the role that the J_2 integral plays in roughness thin film. Whereas the subject of Gao's work (1994) is concerned with a heteroepitaxial thin film, which refers to a very thin single-crystal layer of one material being deposited onto a single-crystal substrate of another material with the same crystalline structure but different lattice spacing. In general, the film thickness is typically ranging from a few nanometers to a few micrometers. As the film changes from a perfectly flat layer into an undulating configuration, the interface monolayer initially remains fully covered by the film material, eventually may become partially exposed after the surface touches the interface. During the non-exposure stage, there is an important property of surface evolution concerned with the conservation law about the J integral customarily adopted in macro-continuum mechanics, which is stated below (Gao, 1994):

During the evolution the projected average of strain energy density along the surface remains constant.

Mathematically this property can be written as

$$\langle w(surface)\rangle = w_0 \tag{8.73}$$

where the operator

$$\langle .....\rangle = \lim_{L\to\infty} \frac{1}{(2L)^2} \int_{-L}^{L}\int_{-L}^{L} (.....)dxdz \tag{8.74}$$

defines the 'projected average' of a quantity over the entire $x - z$ (film) plane. The constant, *i.e.*, the strain energy density w_0 is to be taken as w_0^{3D} for a biaxially stressed film, and as w_0^{2D} in a 2D plane strain problem (Chiu and Gao, 1993).

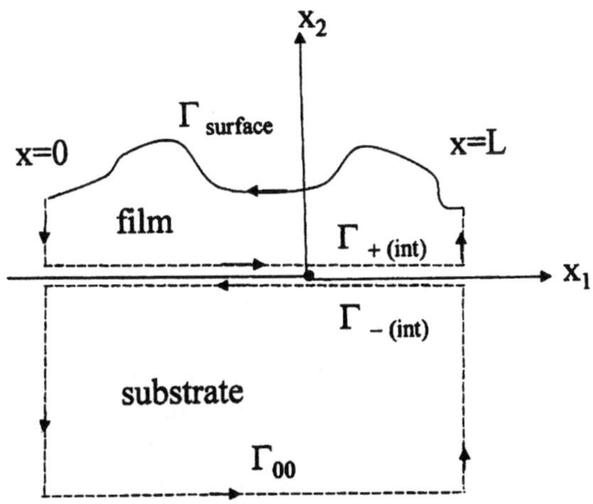

Figure 8.6. Contour selection for a roughness thin film on substrate.

Of great interest is that the property (8.73) involves application of the conservation law about the J integral (Rice, 1968a,b; Gao, 1994). Figure 9.6 shows the contour selection of the J integral application in film/substrate system. According to the coordinates chosen by Gao (1994), the integral should be rewritten in a 3D space as

$$J = \oint_C (w dx dz - n_j \sigma_{ij} \frac{\partial u_i}{\partial y} ds), \qquad (i, j = 1, 2, 3) \ . \tag{8.75}$$

Here it is should be emphasized that there exist three components of the J_k vector in a 3D space:

$$J_k = \oint_C (w n_k - u_{i,k} T_i) ds, \qquad (k = 1, 2, 3) \ , \tag{8.76}$$

whilst Γ refers to a closed curved surface and, in fact, equation (8.75) is only the second component of equation (8.76) with $k = 2$ in the 3D coordinate system ($x_1 = x, x_2 = y, x_3 = z$) (see Figure 8.6). As in a 2D space, the J integral in the 3D space vanishes along any closed surface enclosing no elastic singularities, *i.e.*, equation (1.18) in Chapter 1.

To prove the property (8.73) it is convenient to first consider a doubly periodic surface with period $2L$ and then approach an arbitrary rough surface by extending L to infinity. This is reminiscent of the usual procedure of extending a Fourier series to a Fourier integral. According to the discussion mentioned in Chapter 2 we know that the most significant step is the selection of the suitable integral contour, otherwise the conservation law may be broken. Gao (1994) applied the integral along a closed contour composed of the traction-free rough surface $\Gamma_{surface}$, two horizontal plane surfaces on the '+' and '-' sides of the interface, Γ^+_{int} and Γ^-_{int} (see the similar treatments for interface by Smelse and Gurtin (1977) and Zhao and Chen (1997), and a plane surface at infinite depth, Γ_∞, all within the period $(-L < x < L, -L < z < L)$. To form a closed contour, a number of vertical plane surfaces parallel to the y-direction can be added, but these surfaces make no net contribution to J because they cancel one another owing to their periodicity and opposite senses. Figure 8.6 illustrates a 2D-configuration of such a contour. The contour segment, Γ_∞, makes no contribution to J because the stress field vanishes at an infinite depth away from the film. The contribution to J coming from the traction-free surface is

$$-\int_{-L}^{L} w dx dz \tag{8.77}$$

and that from the interface contours is (see the same treatment in section 2.4 of Chapter 2 or Zhao and Chen, 1997):

$$J(\Gamma^+_{int}) + J(\Gamma^-_{int}) = \int_{-L}^{L}\int_{-L}^{L} [(w^+ - w^-) - n_j\sigma_{ij} + \frac{\partial(u^+ - u^-)}{\partial y}] dx dz. \tag{8.78}$$

After performing some algebraic manipulations, equation (8.78) can be reduced to

$$J(\Gamma^+_{int}) + J(\Gamma^-_{int}) = (2L)^2 w_0 + \sigma_0 \int_{-L}^{L}\int_{-L}^{L} [\partial u_x^-/\partial x + \partial u_z^-/\partial z] dx dz \tag{8.79}$$

where the second term on the right hand side vanishes owing to the periodicity of the displacement field. Thus the conservation of J requires

$$\int_{-L}^{L} w dx dz = (2L)^2 w_0. \tag{8.80}$$

Dividing both sides by $(2L)^2$ and then letting the length L in equation (8.80) approaches infinity directly yields equation (8.73).

We conclude that the J integral does still play an important role in nanomechanics and subsequent investigations for the M integral and the L integral remain adequately treated especially in applications of the so called 'quasicontinuum method' (Ortiz, 1999).

4. SUMMARY

From the above lengthy manipulations and discussions, we summarize the main conclusions derived and discussed in this book:

(1) Although the basic concepts of the J_k vector, the M integral, and the L integral, as well as the Energy Momentum Tensor or Work Conjugate Integral, were established more than 30 or 40 years ago and developed by a number of researchers in the 20th century, they will still play quite important roles at present in cracked solids or in microcracking solids, even in modern functional materials and nanometer structures and materials.

(2) The Energy Momentum Tensor (Eshelby, 1970, 1975) seems very convenient to be adopted to establish conservation laws or invariant integrals in linear functional materials such as linear electro magneto elastic materials. However, the derived invariant integrals always represent the relative energy release rates, from which many parameters at a crack tip such as *SIF's* and *EDIF* can not be distinguished. This is the major shortcoming of the Energy Momentum Tensor.

(3) The Bueckner Work Conjugate Integral (Bueckner, 1973) has been proved to be suitable for treating much more complicated problems, *e.g.*, crack kinking or deflects in homogeneous piezoelectric materials and interface cracks in dissimilar piezoelectric materials (Ma and Chen, 2001b). However, mathematically this method requires the complete Williams *EEF* in the considered materials and the reduced Pseudo-Orthogonal Property of the *EEF*. This significantly restricts its application in functional materials. Nevertheless, based on the work done by Rice (1989) and Bueckner (1989), a direct extension of weight function theory can be expected from 2D crack analysis to 3D crack analysis in piezoelectric ceramics.

(4) Of great interest is the existence of the Pseudo-Orthogonal Property in all the associated crack problems in linear piezoelectric mateials. This property has been proved to be inherent and universal, which does play a powerful tool and the simplest way to establish path independent integral or weight functions for distinctly calculating the dominant parameters at crack tips without any

special treatment of the near tip region. Moreover, this property yields such a conclusion that the J integral and the M integral are merely two special cases of the Bueckner integral. Using two different complementary states, respectively, either J or M integral can always be reduced from the Bueckner integral. Therefore, besides the Energy Momentum Tensor in Continuum Mechanics (Eshelby, 1970), Betti's reciprocal theorem (Sokolnikoff, 1956) provides another way to establish conservation laws or path independent integrals and weight functions even in modern functional materials.

(5) The new conservation laws of the J_k vector established by Chen (2001a, b) for microcracking damage solids, *i.e.,* equations (1.39) and (1.40) in Chapter 1, provide some new concepts and aspects of invariant integrals, which seem far apart from the original definitions by Rice (1968a, b), Knowles and Sternberg (1972), and Eshelby (1970, 1975). This is because, unlike the traditional concepts of conservation laws or path independent integrals, they do hold when a closed contour surrounds all discontinuities completely. Moreover, they not only hold in infinite solids, but also hold in finite solids, providing that the outside boundaries are treated as a special kind of discontinuities, *i.e.,* the interface between air and the solids or between rigid solids and the microcracking solids (see, *e.g.,* sections 2.4 and 2.5 in Chapter 2). It is proved that only in this way could the analyses of the vector, the M integral, and the L integral be performed for finite microcracking solids.

(6) It is concluded that, besides the well known *effective moduli theory*, the M integral analysis provides a new description for microcracking damage. The M integral has been proved to be twice the change of the total potential energy owing to the formation of microcracks, *i.e.,* $M = 2U$. Thus, from the phenomenological point of view, the M integral does represent the energy dissipation owing to microcracking and in turn it does describe the damage level.

(7) Of most interest is the simple relation between the M integral and the L integral, *i.e.,* $L = -\frac{1}{2}\frac{\partial M}{\partial \gamma}$, even in strongly interacting situations. Unlike the traditional concepts of these two integrals, the present discussion reveals that both integrals are not independent, quite contrary, they are closely related to each other although they are well known to represent two different kinds of energy release rates (Budiansky and Rice, 1973; Herrmann and Herrmann, 1981).

(8) The new conservation laws of the J_k vector are also established in microcracking damaged piezoelectric materials, *i.e.,* equations (7.3, 7.4). Detailed manipulations and numerical results verify their validity under the mechanical electric coupling situations. Although the mechanical quantities and electric

quantities show quite different behaviors, they could be treated in a similar way by introducing the concepts of the generalized stress (corresponding to the electric displacement) and the generalized displacement (corresponding to the electric potential) as addressed in section 8.2.

(9) Besides Rice (1968a,b), Eshelby (1956,1970,1975), Chelepanov (1974, 1979) and many others, the work done by Herrmann and Herrmann (1981) provides a basic understanding of the conservation laws and energy release rates, which seems quite useful for the further investigations. In particular, they clarified the role played by the J_2 integral to avoid its path dependence and then emphasized the more natural description of the *ERR* provided from the M and L integrals than the J_k vector. On the other hand, as mentioned in Chapter 1, King and Herrmann (1981) provide an experimental tool for evaluating the M integral. Readers can directly see that this method will be used to evaluate the damage level for a plate with many in-plane and through cracks.

(10) This book reveals that adequately substantial investigations for conservation laws and path independent integrals are absolutely necessary, especially in modern materials with defects, *e.g.*, in piezoelectrics, magnetostrictives, electrostrictives, and shape memory alloys, as well as in nanostructures. Only after doing so can the physical significance and practical applications of them in modern functional materials be recognized and understood clearly. Therefore the purpose of this book can be considered as to supply the lack of potential developments in this 'old' topic and to motivate further investigations especially for functional materials with defects or/and microcrack damage in the new century.

References for Chapter 8

Atluri, S.N. (1982) Path independent integrals in finite elasticity and inelasticity, with body forces, inertia, and arbitrary crack-face conditions, *Engineering Fracture Mechanics,* **18**, 341 – 364.

Bakker, A. (1984) *The three-dimensional J-integral,* Ph. D. Thesis, Technical University Delf.

Barnett, D.M., and Lothe, J. (1974) An image force theorem for dislocations in anisotropic bi-crystals, *Journal of Physics F: Metal Physics,* **4**, 1618 – 1635.

Bergez, D., (1974) Determination of stress intensity factors by use of path independent integrals, *Mechanics Research Communications,* **1**, 179 – 180.

Budiansky, B. and Rice, J.R. (1973) Conservation laws and energy release rates, *ASME Journal of Applied Mechanics*, **40**, 201 – 203.

Budiansky, B., and O'Connell, R.J., (1976) Elastic Moduli of a Cracked Solid,. *International Journal of Solids and Structures*, **12**, 81 – 97.

Bueckner, H.F. (1973) *Field Singularities and Related Integral Representations,* in *Mechanics of Fracture*, Edited by G.C.Sih, Vol.1. 239 – 314, Noordhoff, Leyden.

Bueckner, H.F. (1989) Observations on weight functions*, Engineering Analysis with boundary Elements*, **6**, 3 – 18.

Chaboche, J.L. (1988a) Continuum Damage Mechanics: Part I-General Concepts, *ASME Journal of Applied Mechanics*, **55**, 59 – 64.

Chaboche, J.L. (1988b) Continuum Damage Mechanics: Part II-Damage Growth, Crack Initiation, and Crack Growth, *ASME Journal of Applied Mechanics*, **55**, 65 – 72.

Chen, Y.H. and Hasebe, N. (1994a) Further investigation of Comninou's *EEF* for an interface crack with completely closed faces, *Int. J. Eng. Science,* **32**, 1037 – 1046.

Chen Y.H., and Hasebe,N. (1995) Investigation of *EEF* properties for a crack in a plane orthotropic elastic solid, *Engineering Fracture Mechanics,* **50**, 249 – 259.

Chen Y.H. , and Hasebe N. (1997) Explicit formulations of J-integral considering higher order singular terms in eigenfunction expansion form, Part I. Analytical treatment, *International Journal of Fracture,* **85**, 11 – 14.

Chen Y.H. ., and Shi J.P. (1998) On the relation between the M-integral and the Bueckner work conjugate integral, *ACTA MECHANICA SINICA*, **30**, 495 – 502, In Chinese.

Chen Y.H. ., and Tien W.Y. (1999) On the Bueckner work conjugate integral and its relations to the J-integral and M-integral in piezoelectric materials, *ACTA MECHANICA SINICA,* **31**, 625 – 632, In Chinese.

Chen Y.H. and Han J. J. (1999a) Macrocrack microcrack interaction in piezoelectric materials, Part I. Basic formulations and J-analysis, *ASME Journal of Applied Mechanics,* **66**, 514 – 521.

Chen Y.H. and Han J.J. (1999b) Macrocrack microcrack interaction in piezoelectric materials, Part II. Numerical results and Discussions, *ASME Journal of Applied Mechanics,* **66**, 522 – 527.

Chen Y.H. , (2001a) M-integral analysis for two-dimensional solids with strongly interacting cracks, Part I. In an inf inite brittle sold, *International Journal of Solids and Structures,* **38**, 3193 – 3212.

Chen Y.H. , (2001b) M-integral analysis for two-dimensional solids with strongly interacting cracks, Part II. In the brittle phase of an inf inite metal/ceramic bimaterial, *International Journal of Solids and Structures,* **38**, 3213 – 3232.

Chen, Y.H. and Ma, L.F. (2000) Bueckner work conjugate integrals and weight functions for a crack in anisotropic solids, *Acta Mechanica Sinica*, (English Edition), **16**, 240 – 253.

Chen Y.H., and Lu T.J. (2001) Conservation laws of the Jk-vector for microcrack damage in piezoelectric materials, *International Journal of Solids and Structures,* **38**, 3233 – 3249.

Chen, Y.Z. (1985) New path independent integrals in linear elastic fracture mechanics, *Engineering Fracture Mechanics,* **22**, 673 – 686.

Chen, Y.Z. and Hasebe, N. (1994b) Eigenfunction expansion and higher order weight functions of interface cracks, *ASME Journal of Applied Mechanics,* **61**, 843 – 849.

Cherepanov, G.P. (1974) *Mechanics of Brittle Fracture Mechanics,* Moscow, Publishing House 'Nauka', English translation published by McGraw-Hill, (1979).

Cherepanov, G.P. (1993) Interface microcrack nucleation, *Journal of Mechanics and Physics of Solids,* **42,** 665 – 680.

Chiu, C.H., and Gao, H. (1993) Stress singularities along a crycloid rough surface, *International Journal of Solids and Structures,* **30**, 2983 – 3012.

Cottrell, A.H. (1975) *Fracture, The Physics of Metals,* Vol. 2: *Defects,* P.B. Hirsch, ed., Camridge University Press, Cambridge, UK, 247 – 280.

Deng, W. and Meguid, S. A. (1998) Analysis of conducting rigid inclusion at the interface of two dissimilar piezoelectric materials, *ASME Journal of Applied Mechanics,* **65**, 76 – 84.

Eshelby, J.D. (1956) *The Continuum Theory of Lattice Defects, Solid State Physics,* **3**, F. Seitz and D. Turnbull (editors), Academic, New York, 79 – 141.

Eshelby, J.D. (1970) *The Energy Momentum Tensor in Continuum Mechanics, Inelastic Behavior of Solids,* edited by Kanninen, M.F., *et al.,* McGraw-Hill, New York.

Eshelby, J. D. (1975) The elastic energy-momentum tensor, *J. Elasticity* **5**, 321 – 335.

Fan, H. (1994) Interfacial Zener Stroh crack, *ASME Journal of Applied Mechanics* **61**, 829 – 834.

Fretcher, D.C., (1976) Conservation laws in linear elastodynamics, *Arch. Rational Mech. Anal.,* **60,** 329 – 353.

Freund, L.B. (1978) Stress intensity factor calculations based on a conservation integral, *International Journal of Solids and Structures,* **14**, 241 – 250.

Gao, H. (1994) Some general properties of stress-driven surface evolution in a heteroepitaxial thin film structure, *Journal of Mechanics and Physics of Solids,* **42**, 741 – 772.

Han J. J. and Chen Y.H. , (1997) On the contribution of a micro-hole in the near-tip stress field to the J-integral, *International Journal of Fracture,* **85**, 169 – 183.

Han J. J. and Chen Y.H. , (1999) Multiple parallel crack interaction problem in piezoelectric ceramics, *International Journal of Solids and Structures,* **36**, 3375 – 3390.

Han J.J. and Chen Y.H. , (2000a) Interface crack interacting with micro-voids in the near-tip process zone, *International Journal of Fracture,* **102**, 223 – 245.

Han J.J. and Chen Y.H. , (2000b) T-effect for the interaction problem of an interface macrocrack with a near-tip micro-void, *International Journal of Fracture,* **102**, 205 – 222.

Han J.J. and Chen Y.H. , (2000c) M-integral analysis for damaged anisotropic materials, *Archive of Applied Mechanics,* **70**, 625 – 634.

Herrmann, G.A. and Herrmann, G. (1981) On energy release rates for a plane cracks, *ASME Journal of Applied Mechanics,* **48**, 525 – 530.

Heyer, V., Schneider, G.A., Balke, H., Drescher, J., and Bahr, H. A. (1998) A fracture criterion for conducting cracks in homogeneously poled piezoelectric *PZT-PIC151* ceramics, *Acta Mater.,* **46**, 6615 – 6622.

Hui, C.Y., and Riana, A. (1995) Why K? Higher order singularities and small scale yielding, *International Journal of Fracture,* **72**, 97 – 120.

King, R.B., and Herrmann, G. (1981) Nondestructive evaluation of the J and M integrals, *ASME Journal of Applied Mechanics,* **48**, 83 – 87.

Kirshmoto, K., Aoki, S., and Sakata, M. (1980) On the path independent integral-$\hat{J}$., *Engineering Fracture Mechanics,* **13**, 841 – 850.

Kirshmoto, K., Aoki, S., and Sakata, M. (1982) Use of $\hat{J}$-integral in dynamic analysis of cracked linear viscoelastic solids by *FEM, ASME Journal of Applied Mechanics,* **13**, 841 – 850.

Knowles, J.K. ans Stermberg, E. (1972) On a class of conservation laws in linearized and finite elastostatics, *Arch. Rat. Mech. Anal.,* **44**, 187 – 211.

Kubo, S. (1982) An application of M integral to cracks in dissimilar elastic materials, *International Journal of Fracture,* **20**, $R27$ – $R30$.

Liu, Q. D., and Sheng, J. N., (2001) On effect of different electric boundary conditions along crack surfaces., submitted.

Ma, L.F. and Chen, Y.H. (2001a) Bueckner work conjugate integrals and weight functions for a crack in piezoelectric solids, *ACTA MECHANICA SINICA*, (English Series), **17**, in press.

Ma, L.F. and Chen, Y.H. (2001b) Weight function for interface cracks in dissimilar anisotropic piezoelectric materials, *International Journal of Fracture,* **111**, 263 – 279.

McMeeking, R.M. (1990), A J integral for the analysis of electrically induced mechanical stress at cracks in elastic dielectrics., *International Journal of Engineering Science*, **28**, 605 – 613.

McMeeking, R.M. (1999) Crack tip energy release rate for a piezoelectric compact tension specimen, *Engineering Fracture Mechanics*, **64**, 217 – 244.

McMeeking, R.M. (2001) Towards a fracture mechanics for brittle piezoelectric and dielectric materials, *International Journal of Fracture,* **108**, 25 – 41.

Mikhailov, and Parton V.Z. (1990) *Electromahnetoelasticity*, Gordon and Breach, New York.

Ortiz, M. (1999) Nanomechanics of defects in solids, *Advances in Applied Mechanics,* **36**, 2 – 79.

Pak, Y. E. (1986) Conservation laws and the material momentum tensor for the elastic dielectric, *International Journal of Engineering Science,* **24**, 1365 – 1374.

Pak, Y. E. (1990a) Crack extension force in a piezoelectric material, *ASME Journal of Applied Mechanics,* **57**, 647 – 653.

Pak, Y.E. (1990b) Force on a piezoelectric screw dislocation, *ASME Journal of Applied Mechanics,* **57**, 863 – 869.

Park, J.W, and Earmme, Y.Y. (1986) An application of conservation integrals to interface crack problems, *Mechanics of Materials,* **5**, 261 – 276.

Park, S. B, and Sun, C. T. (1995a) Fracture criteria for piezoelectric ceramics, *Journal of the American Ceramic Society,* **78**, 1475 – 1480.

Park, S. B, and Sun, C. T. (1995b) Effect of electric fields on fracture of piezoelectric ceramics, *International Journal of Fracture,* **70**, 203 – 216.

Parton, V.Z., and Kudryatvsev, B.A. (1988) *Electromahnetoelasticity*, Gordon and Breach, New York.

Rice, J.R. (1968a) A path independent integral and the approximation analysis of strain concentration by notches and cracks, *ASME Journal of Applied Mechanics,* **35**, 297 – 320.

Rice, J.R. (1968b) *Mathematical Analysis in the Mechanics of Fracture,* in *Fracture*, edited by Liebowitz, H., **2**, 191 – 311, Academic Press, New York.

Rice, J.R. (1985) *Conserved Integrals and Energetic Forces,* in *Fundamentals of Deformation and Fracture*, Bilby, B.A., Miller, K.J., and Willis, J.R., (editors), Cambridge University Press, Cambridge, UK, 33 – 56.

Rice, J.R. (1988) Elastic fracture mechanics concepts for interfacial cracks, *ASME Journal Applied Mechanics,* **55**, 98 – 103.

Rice, J.R. (1989) *Weight function theory for three-dimensional elastic crack analysis, Fracture Mechanics, Perspectives and directions* (Twentieth Symposium), ASTM-STP-1020, Edited by Wei, R.P., Gangloff, R.P., America Society for Testing and Materials, Philadelphia, 29 – 57.

Smelser R. E., and Gurtin E. (1977) On the J-integral for bi-material bodies, *International Journal of Fracture,* **13**, $R382 - R384$.

Sokolnikoff, S., *Mathematical Theory of Elasticity,* McGraw-Hill New York (1956).

Sosa, H. (1991) Plane problems in piezoelectric media with defects, *International Journal of Solids and Structures,* **28**, 491 – 505.

Sosa, H. (1992) On the fracture mechanics of piezoelectric solids, *International Journal of Solids and Structures,* **29**, 2613 – 2622.

Stern, M., Becker, E.B., and Dunham, R.S. (1976) A contour integral computation of mixed-mode stress intensity factors, *International Journal of Fracture,* **12**, 359 – 368.

Stroh, A. N. (1958) Dislocations and cracks in anisotropic elasticity, *Philosophical Magazine* **3**, 625 – 646.

Suo, Z., (1990) Singularities, interfaces and cracks in dissimilar anisotropic media. *Proceedings of Royal Society of London,* **A727**, 331 – 358.

Suo, Z., Kuo, C. M., Barnett, D. M. and Willis, J. R.(1992) Fracture mechanics for piezoelectric ceramics, *Journal of the Mechanics and Physics of Solids,* **40**, 739 – 765.

Suo, Z. (2000) Zener's crack and the M-integral, *ASME Journal of Applied Mechanics,* **67**, 417 – 418.

Tian W.Y., and Chen Y.H. , (2000a) Interaction between an interface crack and a parallel sub-interface cracks in dissimilar anisotropic materials, *International Journal of Fracture,* **102**, 305 – 322.

Tian W.Y., and Chen Y.H. , (2000b) A semi-infinite interface crack interacting with sub-interface matrix cracks in dissimilar anisotropic materials, Part I, Fundamental formulations and the J-integral analysis, *International Journal of Solids and Structures,* **37**, 7717 – 7730.

Tian W.Y., and Chen Y.H. , (2000c) Interaction between an interface crack and sub-interface microcracks in metal/piezoelectric bimaterials, *International Journal of Solids and Structures,* **37**, 7743 – 7757.

Tieresten, H.F., (1969) *Linear Piezoelectric Plate Vibrations*, Plenum Press, New York.

Ting, T.C.T. (1982) Effects of change of reference coordinates on the stress analysis of anisotropic elastic materials, *International Journal of Solids and Structures*, **18**, 139 – 152.

Ting, T. C. T. (1986) Explicit solution and invariance of the singularities at an interface crack in anisotropic composites, *International Journal of Solids and Structures,* **22**, 965 – 983.

Ting, T.C.T. (1990) Interface Crack in Anisotropic Bimaterials, *Journal of Mechanics and Physics of Solids,* **38,** 505 – 513.

Tucker, M.O. (1969) Plane boundaries and straight dislocations in elastically anisotropic materials, *Philosophical Magazine* , **19**, 1141 – 1159.

Wang, X.M., and Shen, Y.P. (1996) The conservation laws and path independent integrals with an application for linear electro-magneto-elastic media, *International Journal of Solids and Structures,* **33**, 865 – 878,

Zhao L. G, and Chen Y.H. , (1997) Further investigation of sub-interface cracks, *Archive of Applied Mechanics,* **67**, 393 – 406.

Zener, C. (1948) *The micro-mechanism of Fracture*, Fracturing of Metals, America Society of Metals, Cleveland, OH, 3 – 31.

Index